CHEMISTRY OF ADVANCED ENVIRONMENTAL PURIFICATION PROCESSES OF WATER

CHEMISTRY OF ADVANCED ENVIRONMENTAL PURIFICATION PROCESSES OF WATER

Fundamentals and Applications

Edited by

Erik G. Søgaard

Aalborg University Esbjerg

ELSEVIER

AMSTERDAM • BOSTON • HEIDELBERG • LONDON
NEW YORK • OXFORD • PARIS • SAN DIEGO
SAN FRANCISCO • SINGAPORE • SYDNEY • TOKYO

Elsevier
Radarweg 29, PO Box 211, 1000 AE Amsterdam, The Netherlands
The Boulevard, Langford Lane, Kidlington, Oxford OX5 1GB, UK

First edition 2014

British Library Cataloguing in Publication Data
A catalogue record for this book is available from the British Library

Library of Congress Cataloging-in-Publication Data
A catalog record for this book is available from the Library of Congress

For information on all Elsevier publications
visit our web site at store.elsevier.com

Printed and bound in Poland
14 15 16 17 18 10 9 8 7 6 5 4 3 2 1

ISBN: 978-0-444-53178-0

Dedication

This book is dedicated to my students and PhD fellows in chemical engineering from my years of teaching and research at Esbjerg Teknikum and Aalborg University Esbjerg since 1990.

Esbjerg, February 2014
Erik G. Søgaard

Contents

Contributors

Project leader Daniel Anobaah Ankrah Dansk GeoservEx A/S, Aarhus Denmark

Project leader Lars R. Bennedsen Rambøll A/S, Vejle, Denmark

Ph.D. fellow Krzysztof P. Kowalski Section of Chemical Engineering, Department of Biotechnology, Chemistry and Environmental Engineering, Aalborg University Esbjerg, Niels Bohrs Vej 8, DK 6700 Esbjerg, Denmark

Ph.D. fellow Henrik Tækker Madsen Section of Chemical Engineering, Department of Biotechnology, Chemistry and Environmental Engineering, Aalborg University Esbjerg, Niels Bohrs Vej 8, DK 6700 Esbjerg, Denmark

Assistant Professor Jens Muff Section of Chemical Engineering, Department of Biotechnology, Chemistry and Environmental Engineering, Aalborg University Esbjerg, Niels Bohrs Vej 8, DK 6700 Esbjerg, Denmark

Assistant Professor Rudi P. Nielsen Section of Chemical Engineering, Department of Biotechnology, Chemistry and Environmental Engineering, Aalborg University Esbjerg, Niels Bohrs Vej 8, DK 6700 Esbjerg, Denmark

Assistant Professor Morten E. Simonsen Section of Chemical Engineering, Department of Biotechnology, Chemistry and Environmental Engineering, Aalborg University Esbjerg, Niels Bohrs Vej 8, DK 6700 Esbjerg, Denmark

Professor MSO Erik G. Søgaard Section of Chemical Engineering, Department of Biotechnology, Chemistry and Environmental Engineering, Aalborg University Esbjerg, Niels Bohrs Vej 8, DK 6700 Esbjerg, Denmark.

Introduction

The book contains different aspects of water and its treatment. All chapters can be read independently.

Chapter 1 is a short introductory chapter containing information on different types of water and the water cycle. Chapter 2 has its focus on soil and groundwater treatment by help of chemical oxidation, whereas Chapter 3 covers the possibilities for electrochemical purification of soil and groundwater and different kinds of polluted waters. These two chapters also show important areas of advanced oxidation technology (AOT). Chapter 4 goes on with AOT, now by including the methodology of photochemical and photocatalytic oxidation. The chapter also contains information on nano-particle production in the form of the photocatalytic semiconductor TiO_2 technology. Chapter 5 contains information of water in its near-critical and supercritical phase where it totally changes its character — a property which can be applied for different purposes including purification of water. Chapter 6 is dedicated membranes and membrane filtration, a methodology that can be combined with any one of other purification methods from the other chapters. Chapter 7 is focused on drinking water and drinking water treatment with special emphasis on the application of iron-oxidising bacteria for iron removal. Chapter 8 covers the problems that the world has discovered in the last couple of decades with respect to the naturally present arsenic in ground water, its toxicity and its removal.

The co-authors of this book were all PhD students of the editor during the period 2005—2014. We hope that the collaboration and research in our group during these years can be of interest for other researchers in chemical engineering, scientist with interests in environmental chemistry and other researchers with a focus on water and water remediation technologies and mainly for their students that are going to be trained in chemical aspects of water, its pollution and remediation principles for the development of sustainable solutions for societies of all kinds.

The authors want to acknowledge Arunan Sritharan and Paula Epure for their help with some of the figures in the book.

CHAPTER

1

Water and Water Cycle

Professor MSO Erik G. Søgaard

Section of Chemical Engineering, Department of Biotechnology, Chemistry and Environmental Engineering, Aalborg University Esbjerg, Niels Bohrs Vej 8, DK 6700 Esbjerg, Denmark

1.1. ORIGIN OF WATER

Where water on the Earth originally came from is in fact not known with security. There exist two competing theories of which the most popular is based on the fact that comets, which contain up to 90% water, should have delivered the most important parts of the oceans. The oceans are calculated to make up about 1% of the mass of the Earth. It has been shown that water from some of the comets that during their passage of the Earth were observed by spectroscopic measurements have contents of deuterium, which is close to the content of deuterium in the oceans. A part of the theory also goes on to state that impacts of meteors in the form of carbon chondrites had a content of organic material including amino acids which could have made up the indigenous organic material present in the oceans.

The alternative theory is based on the fact that indigenous water came from the mantle of the Earth. The mantle is divided into a lower part consisting of fixed rocks, the mesosphere, and an upper part, the asthenosphere, which comprises plastic rocks. The transition zone between the two parts is not expected to separate them with regard to composition of the rocks but rather with regard to physical conditions which is crucial for the understanding of the model. Especially, the chemical constituents that make up the mineral olivine are involved in the theory. This mineral consisting of Fe and Mg silicates constitutes a major part of the occurrence in the mantle of the Earth. At the transition zone between the lower and the upper mantle from 410 km below the surface of the Earth to about 1050 km, it can exist in a more specific part the β-phase of olivine (Wadsleyit) that contains water as hydrate water and potential

Chemistry of Advanced Environmental Purification Processes of Water
http://dx.doi.org/10.1016/B978-0-444-53178-0.00001-8

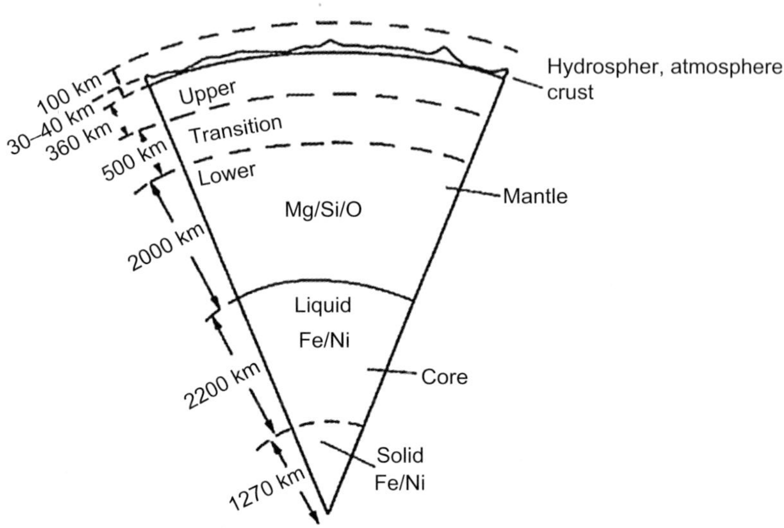

FIGURE 1.1 **Structure of the Earth. J.E. Fergusson (1985).** *Reprinted with permission from Pergamon Press.*

water in form of hydroxyl groups that is taken up as a part of the mineral. Hydrogen ions originally captured in the transition zone interact with the hydroxyl groups. Through interaction with the radioactive decays of unstable atomic nuclei the hydroxyl groups are supposed to escape and create more hydrate water. A part of this water has evaporated out through the upper part of the mantle and the crust by volcanic activities and has formed the oceans. Drillings in the crust has been performed down to between 8.000 and 9.000 m. Contrary to what was expected, water was found all the way down with a high content of minerals and gaseous components. The size of the well should have been deeper but it was stopped due to higher temperature gradient than expected combined with the natural limited robustness of the equipment (Figure 1.1).

1.1.1. Formation Water

Drilling for oil and gas also leads to contact with the formation water that is captured by layers of sediments together with oil and gas. The sediments act like a cap from which oil, gas or water cannot escape (Figure 1.2). The water can also be called fossil water similar to fossil oil and gas. Fossil oil and gas are produced from remains of organic materials from organism originally living in the sea. Together with seawater they were buried in the sediments and due to high pressure and temperature they were converted into the fossil oil and gas that we drill for today. The formation water changed its properties in the sediments due to the high

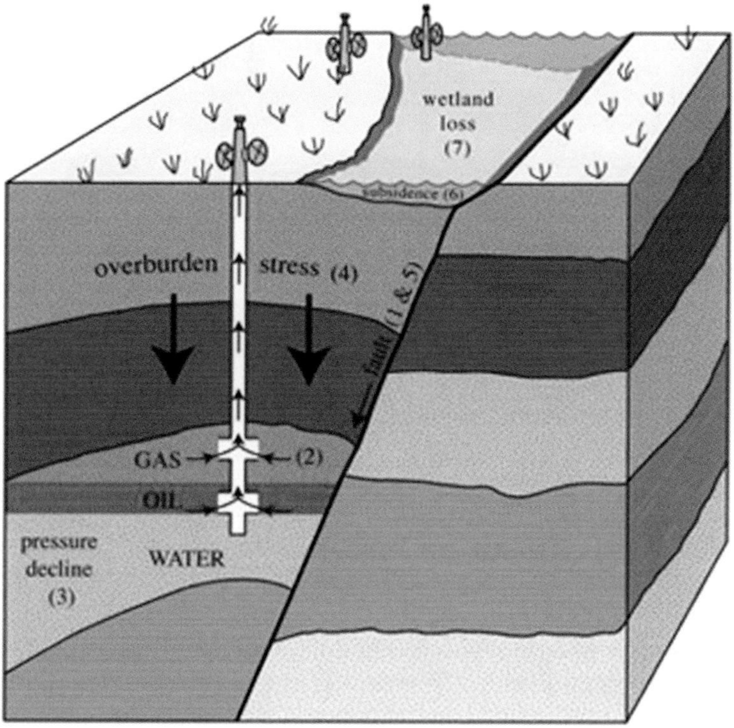

FIGURE 1.2 **Presence of fossil water together with fossil oil and gas.** *Reprinted with permission from U.S. Geological Survey.*

temperature and became much more saline compared to seawater. In the reservoirs of the crust of the Earth formation water can move in sandstone or chalk and bring itself into equilibrium with these minerals. Therefore, details of water–mineral interfaces and water–oil interfaces in the reservoirs are of great importance especially for the recovery of the oil.

1.1.2. Produced Water

Together with oil and gas the oil companies also produce water from the reservoirs. From a new oil well this water will be the formation water but after sometime it is necessary to push the oil to the production well and this is done by water flooding (Figure 1.3).

Using an injector well, water is injected into the reservoir and a water flooding scenario is started. After some time the injected water will also become part of the produced water that will shift from being only the original water from the reservoir to a mixture of the two kinds of water. The injected water can be fresh seawater, produced water from the oil production platform, or a mixture of both.

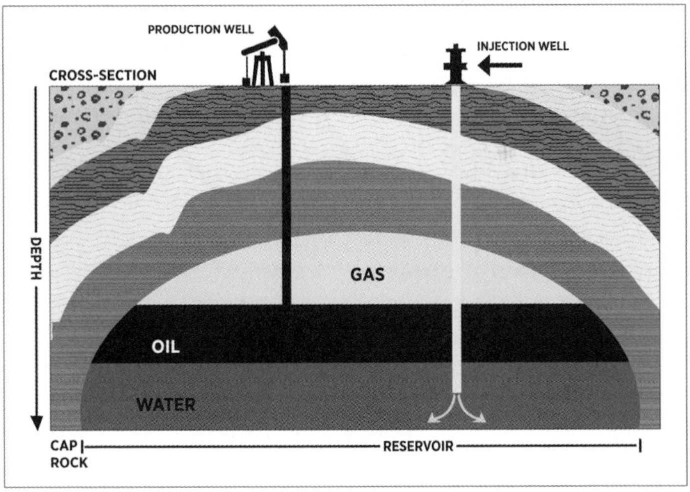

FIGURE 1.3 Water from injector well is flooding the oil reservoir by increasing pressure of the formation water and pushing up the oil. After some years, the injected water will be produced together with the oil. (For colour version of this figure, the reader is referred to the online version of this book.)

"Smart water flooding" is a methodology where the natural composition of ionic compounds in the seawater is changed by adding ions that can enhance oil production by changing the wettability and adsorbance of the oil to the reservoir minerals that normally are either chalk or sandstone. Water flooding and, therefore, also oil recovery can be enhanced by adding polymers, surfactants, or particles to the injection water. These chemicals have different targets in the reservoirs and the produced water will after some have a content of them. Besides the separation of oil and water in separators at the production platform it is also necessary to separate these chemicals from the produced water before it can be discharged into the sea or reused for water flooding by injection if they stick to the oil or participate in an emulsion of oil and water.

1.2. RAINWATER, GROUNDWATER AND DRINKING WATER

Rainwater is the most important part of the freshwater activity on Earth. When water evaporates from surface of the oceans it looses the content of ionic species and therefore changes from being seawater to freshwater. A kind of distillation process has taken place. On land, freshwater can evaporate from lakes and streams or sublimate from snow. Therefore, the atmosphere always has a content of water. This water takes part in the absorbance of the long-wave radiation from the Earth and as such plays a role in the greenhouse effect together with carbon dioxide and other gases. When the evaporated water is condensed into clouds or aerosols due to local decrease of temperature, droplets of water can be formed and due to gravity they will fall as rain (precipitation). The first rainwater that hits the ground will contain lots of small solid colloidal aerosol particles that may also contain adsorbed sulphuric and nitrous oxides. These colloidal dust particles come from the ground and are present due to the combination of winds and dry soils. Therefore, they often contain ammonium sulphate and ammonium nitrate that were used as fertilisers for agricultural purposes. Later, the rain is more pure and only contains the mentioned oxides together with hydrogen carbonate. In that way, these kinds of fertilisers in the form of nitrous and sulphuric oxides also can find their way into the soil in very small amounts. Another source for solid aerosol particle production is burning of forests and wood in general. The water cycle is shown in Figure 1.4.

Rainwater can be harvested and directly used for drinking water and for different purposes where limited amounts of ionic species in the water is wanted, e.g. for washing cars or similar purposes. The particles present are supposed to sediment in the rainwater harvesting container after sometime. Normally, it will also be saturated with dioxygen and dinitrogen. Use of rainwater is without any costs. When rainwater hits the ground, some of it will run to rivers and streams and in this way be rather quickly transported back into the sea. However, a very important part of it will penetrate the soil and become a part of the groundwater. Groundwater also flow towards lakes, rivers, and streams but with a much lower velocity due to resistance from the soil particles. The soil containing groundwater is divided into the saturated zone where all capillaries,

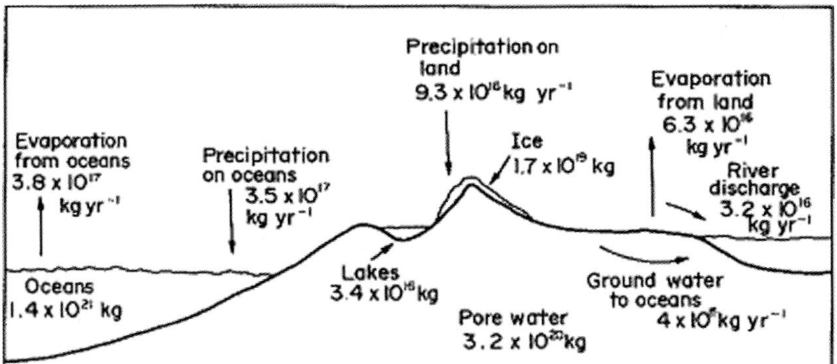

FIGURE 1.4 **The water cycle with rainwater in the form of precipitation, groundwater, and pore water from the inner part of the crust.** The numbers for flows and reservoirs are estimations with high uncertainties. J.E. Fergusson (1985). *Reprinted with permission from Pergamon Press.*

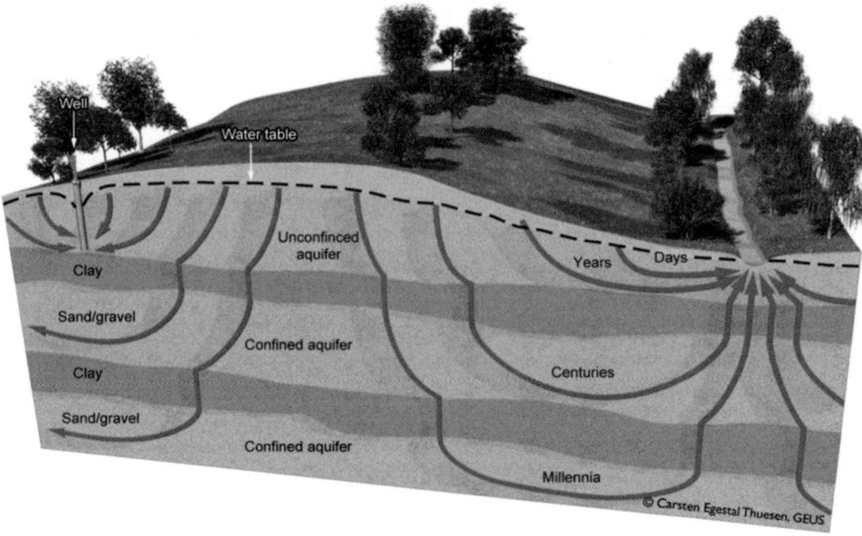

FIGURE 1.5 Groundwater penetrates soil with high permeability and some of it even penetrates the clay with low permeability but high porosity and becomes a part of confined aquifers where water can stay for millennia. The water table is the interface between saturated and unsaturated aquifers. *(Reprinted with permission from USGS)* (For colour version of this figure, the reader is referred to the online version of this book.)

pores, and cracks are filled with groundwater and the unsaturated zone on top of the saturated zone where water is only partly present depending on the amount of rain. The interface between the two zones changes its position depending on precipitation, evaporation, and transportation of the groundwater. It is called the water table. Groundwater can stay in the aquifer for days, years, or millennia depending on its possibility to penetrate the soil and confinements (Figure 1.5).

1.2.1. Species in Groundwater

The composition of components in the groundwater is a result of the interaction between the water and the solid aquifers that it penetrates. If groundwater stays long enough it can become saturated with dissolved ions and other compounds from the aquifer. The main natural components are shown in the Table 1.1. The ions, ammonium, nitrate and phosphate, in some cases, can be a result of fertilisers washed down into the aquifer. However, ammonium can also be a result of natural degradation of humic substances.

In many countries groundwater is the main source of drinking water. It can be pumped directly from the aquifer and often it is drinkable without any treatment. More often the content of iron and manganese needs to be

TABLE 1.1 Typical Components in Groundwater

Group		Constituents
Main components	Cations	Ca^{2+}, Na^{+}, NH_4^{+}, K^{+}, Mg^{2+}, Fe^{2+}, Mn^{2+}
	Anions	HCO_3^{-}, NO_3^{-}, SO_4^{2-}, PO_4^{3-}, Cl^{-}
Uncharged species		H_4SiO_4
Trace components		Al^{3+}, Ni^{2+}, Zn^{2+}, F^{-}, H_3AsO_3, and others
Gases		CO_2, H_2S, CH_4, O_2
Organic compounds		Humus

lowered due to health problems if the intake of these metals is too high. Normally, groundwater used as raw water for drinking has no content of oxygen because the water is pumped from 20 to >100 m-deep wells. By aeration of the water, oxygen will oxidize Fe(II) and Mn(II) to Fe(III) hydroxides and Mn(III,IV) oxides. These oxides will precipitate in the sand filters of waterworks built for the purpose. If humic substances are present in the raw water they normally also will be adsorbed to the sand grains in the filters together with iron.

Left are the macroions: Ca^{2+}, Na^{+}, K^{+}, Mg^{2+}, HCO_3^{-}, SO_4^{2-}, and Cl^{-}. In their most often range of concentration they are harmless and important for the humans. H_2S, CH_4, and CO_2 will become stripped off during the aeration. If not special treatments are necessary because hydrogen sulphide is toxic, methane can cause heavy problems with bacteria producing biopolymers in the filtration systems and too much carbon dioxide can decrease pH and result in corrosion of the water distribution systems often made of steel. Of the trace components from Table 1.1, only Ni^{2+}, F^{-}, and H_3AsO_3 sometimes are in such a high content in the raw water after sometime of production of drinking water that it is necessary to close down the well. Only minor amounts of these and the other trace components are allowed. However, treatment methodologies for them exist. A typical groundwater and its threshold limit for drinkable water can be seen in Table 1.2.

1.3. WASTEWATER

In the cities drinking water from the water distribution system is also often used for the sewage system. In this way drinking water ends up as sewage containing a lot of different compounds whose levels are measured with the help of chemical oxygen demand (COD) and biological oxygen demand (BOD). If storm water is not separated from sewage

TABLE 1.2 Raw Water for Drinking Water Content

Parameter	Unit	Groundwater	Threshold Limit
Ca^{2+}	mg/l	10–200	<200
Mg^{2+}	mg/l	2–30	50
Hardness	°dH		5–30
Na^{+}	mg/l	10–100	175
NH_4^+	mg/l	0.08–6	0.05
Fe	mg/l	0.02–40	0.05
Mn	mg/l	0.001–3	0.02
HCO_3^-	mg/l	10–400	>100
Cl^-	mg/l	30–70	250
NO_3^-	mg/l	0.5–110	50
NO_2^-	mg/l	–	0.01
SO_4^{2-}	mg/l	20–100	250
H_2S	mg/l	–	0.05
Agg. CO_2	mg/l	–	2
CH_4	mg/l	–	0.01
O_2	mg/l	0	10>5

water then sewage will also contain elements from the surroundings of the households. From kitchens, bathrooms and toilets the sewage in the future will be transported to the wastewater treatment plant in its own sewer transportation system and storm water in a separate system. This way the two systems containing very different wastewater can be treated in a much more sustainable way. Principally, the sewage from household contains organics and therefore gives rise to a high COD and BOD. Contrary to this, storm water has higher contents of inorganic compounds in the form of insoluble particles as sand, clay, and dust together with the macroions also found in groundwater.

1.3.1. Storm Water

When solid waste is separated from the soluble compound in the storm water only heavy metals should be a problem before discharging the water into the nearby surroundings, which could be the sea, a river, a fjord, or similar surroundings. The heavy metals in question comprise mercury, lead, and cadmium.

Mercury is produced not only by the use of coal for electricity power plants but also from other industries such as those producing chlorine, electrical devices, paints, etc. However, many of these applications have been phased out or are on their way to be so in the future and the main part of this mercury can be recycled. The principal part of mercury comes from degassing of the Earth's crust. The numbers are uncertain but may up to 150 metric tonnes per year. Compared to this anthropogenic quantities are much smaller but more concentrated. In storm water mercury will be present either as Hg^{2+} or is converted into CH_3Hg^+ (methylmercury). In the latter case it is more toxic for human beings but the aerobic conversion process from inorganic to organic mercury is slow.

Lead will adsorb to clayish compounds or be a part of insoluble particles. Only a minor part of lead will be present in real solution pairing with chloride, hydroxide or carbonate. Therefore, the main part goes with the solid waste at the wastewater treatment plant and the rest will absorb to a sludge created at the plant. Even if mercury is less adsorbable to particles than lead it will similarly end up adsorbed to sediment sludge.

Cadmium also pairs with chloride and in this way participates in soluble compounds that will be transported to the sewage system with the storm water. Addition of carbonate will reduce Cd^{2+} to a 1000-fold lower level.

In storm water several contaminants with an organic origin can also be present. These comprise compounds as herbicides, insecticides, polycyclic aromatic hydrocarbons (PAHs), polychlorinated biphenyls (PCBs), phenols, aliphatic and aromatic compounds from gasoline and oil, chlorinated ethenes from dry cleaners and many others depending on the place and activities in the local area of the wastewater treatment plant. However, if chlorinated ethenes should be present in storm water they should come from some inexpedient accident. This argument does not go for the other mentioned compounds whose presence is due to exterior use for combating weeds and insects, burning gasoline, barbecue, smoking, PCBs in materials from houses, etc.

The typical way to degrade these compounds at the wastewater treatment plant is aerobic biodegradation. However, if the compounds go with the storm water and not the sewage they will probably not become treated at the wastewater treatment plant but will be stored in the sediment of a storm water reservoir together with the heavy metals mentioned above. After sedimentation the water can be discharged into a river or a fjord or used as secondary water. The sediment containing heavy metals and slowly degradable organics from storm water will after sometime become dewatered and stored at landfills. If the sediment has high contents of heavy metals it also could be added together with coal into a coal-fired electricity power plant so that the metals can end up in the fly ash. The sustainability of this method in the long run depends on

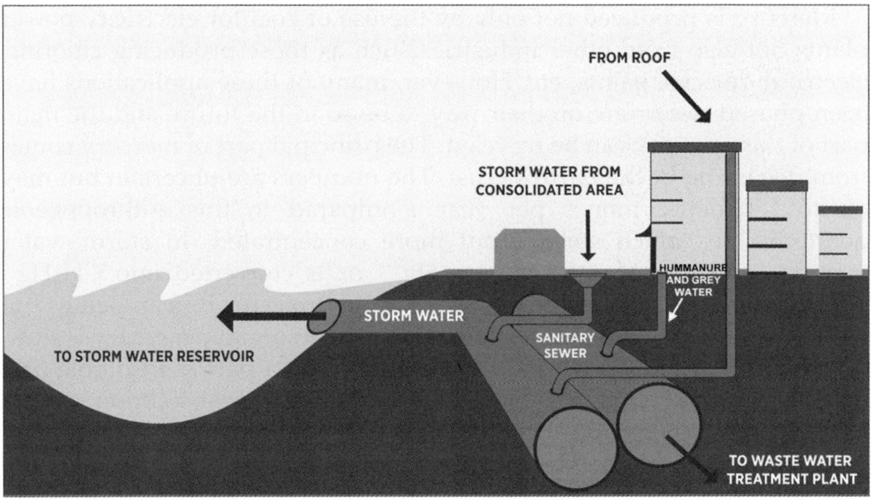

FIGURE 1.6 Separation of sanitary sewage water in form of humanure and graywater from storm water. (For colour version of this figure, the reader is referred to the online version of this book.)

whether coal-fired power plants will continue using coal. Alternatively, the sludge can be a part of fertilisers for growing of plants that are not meant for feed (Figure 1.6).

1.3.2. Sewage from Households

Contrary to the storm water sewage from households separated from storm water will be treated at the wastewater treatment plant by aerobic biodegradation. The sewage can be divided into humanure and gray-water where the first part comes from toilets and the other part from water from kitchens sinks, bathing facilities, and washing machines. It consists of solid compounds as inorganic or organic particles and paper, colloidal organic particles, surfactants, as well as fats and organic compounds in real aqueous solution. These sewage sources have a high BOD. After screening and pretreatment the sewage will be piped to an activated sludge reactor where a biological floc composed of bacteria will take care of the biodegradation on addition of air or oxygen to the reactor. Microbes living there in aerobic conditions will oxidise organic waste to carbon dioxide and water and at the same time oxidise nitrogen-containing compounds to nitrate (Figure 1.7).

Other microbes living at anoxic and anaerobic conditions can take care of the denitrification process and the degradation of more persistent organic pollutants. If anaerobic degradation also takes place as a pre-treatment before the aerobic degradation then the burden on aerobic

FIGURE 1.7 **Wastewater treatment-activated sludge reactor (Shutterstock image id 153095855. Courtesy of Shutterstock).** (For colour version of this figure, the reader is referred to the online version of this book.)

biodegradation can become relieved. Normally, it is a problem to degrade some pharmaceuticals, xenobiotics, e.g. hormones, and some PAHs in water treatment systems. Therefore, it can be necessary to add other methods like advanced oxidation processes in the form of, e.g. photochemical degradation, to get rid of the problem. Also, lignin can be a problem. A portion of the sludge will be returned and maintained in the reactors. Other portions will become dewatered and often used as fertiliser for agricultural purposes depending of the contents of heavy metals (Figure 1.8).

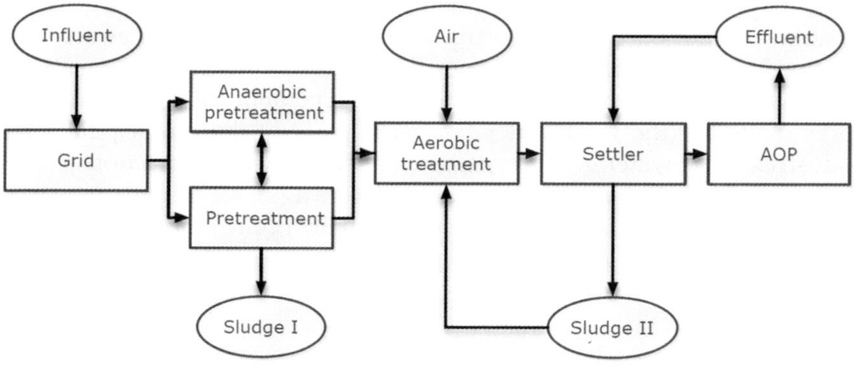

FIGURE 1.8 **Sketch of anaerobic/aerobic treatment unit at a wastewater treatment plant where pollutants can get degraded either by anaerobic or aerobic means.**

1.3.3. Industrial Wastewater

Like households use water for different purposes, industries use water for their production and in most cases it is much more than for the households in the cities or neighbourhoods. Normally, treated drinking water is used but for agricultural activities normally its own water wells are used for irrigation of the fields.

The waste from industrial wastewater comprises a long row of inorganic and organic pollutants depending on the industry. Some of the waste is not real waste but more an overproduction of nonhazardous chemicals, e.g. Na^+, Ca^{2+}, Cl^-, and CO_3^{2-}, from water produced in oil and gas industry or from the mining of chalk for different applications. It could be for the desulphurization plants at electricity power plants where another stream of wastewater is produced containing gypsum, fly ash, heavy metals from the coals, etc. F^- can be present from glass etching and CN^- from metal processing activities. Other inorganics are metals that come from the manufacturing of metal-based goods as cars and plating industry in general. Some of it is solids. However, other parts of it will end up in the industrial wastewater stream and some of it in storm water. The main components are Fe, Al, Cu, and Zn and in minor amounts Hg, Pb, Co, Cd, Ni, As, and Se.

The organic part of wastewater contains some volatile organic compounds not very much soluble in water as benzene, toluene and the xylenes; organic solids as fats and grease can also be present together with colloids and organics in real solution. Depending on the industry in question it can contain sugars, starch, dyes, mercaptans, and many others. The important part of the treatment is supposed to take place at the site of manufacture so only smaller amounts of waste will be transported to the wastewater treatment plant.

Development of legislation, rules and regulations for the use of chemicals in industrial production and their faiths in wastewater treatment is a steady ongoing process to avoid inappropriate pollution of neither the local industrial area nor the wastewater treatment system or its discharge area. The producers are encouraged to arrange their own local wastewater treatment systems and are paying for both amounts and contents in their wastewater to be treated at the wastewater treatment plant of the municipality. The industrial wastewater goes together with sewage water from households. The separation of water in storm water and sewage concentrate the sewage so it is easier to treat and at the same time almost pure storm water can be used as secondary water, e.g. cooling towers, heating, etc.

Reference

Fergusson, J.E., 1985. Inorganic Chemistry and the Earth. Pergamon Press.

CHAPTER

2

In situ Chemical Oxidation

The Mechanisms and Applications of Chemical Oxidants for Remediation Purposes

Project leader Lars R. Bennedsen

Rambøll A/S, Vejle, Denmark

Contamination of the subsurface by persistent organic contaminants remains a significant problem, even after decades of research on remediation technologies (Watts et al., 1999b, Watts and Teel, 2005). First, approaches focused on excavation, pump and treat via activated carbon, bioremediation, and natural attenuation. In the 1990s the first reports on in situ chemical oxidation (ISCO) were published, which is a technique involving the introduction of chemical oxidants into the subsurface in order to transform contaminants into less harmful substances. Hydrogen peroxide was the first chemical oxidant investigated and used in full scale. Shortly thereafter ozone and permanganate came into use. In the past few years persulphate has provided yet another option.

In this chapter, the chemical reactions of the most common chemical oxidants used in ISCO are reviewed and the applicability of the two most relevant, modified Fenton's reagent (MFR) and activated sodium persulphate (ASP), are demonstrated using the Kærgaard Plantation megasite in Denmark as case study. This site represents one of the most difficult remediation challenges in Scandinavia and, therefore, regulatory agencies have been evaluating remediation techniques for source area remediation.

Chemistry of Advanced Environmental Purification Processes of Water
http://dx.doi.org/10.1016/B978-0-444-53178-0.00002-X

2.1. INTRODUCTION

2.1.1. Soil and Ground Water Contamination

Freshwater comprises only 3% of all water on the Earth and around 20% of this small fraction occurs as ground water, which is a critical resource throughout the world because of its use as drinking water, for agricultural applications, for irrigation of crops and for industrial activities. Ground water serves as a significant source of drinking water ranging from 15% in Australia to 75% in Europe (Morris et al., 2003). In Denmark close to 100% of the drinking water originates from ground water.

Today soil and ground water contamination is a widespread and challenging problem threatening ground water resources throughout the world. The contamination originates from introduction of xenobiotic chemicals to the environment or from naturally occurring sources and the most common chemical contaminants found in the environment are petroleum hydrocarbons, solvents, coal tar, heavy metals and pesticides. An overview of contaminants affecting soil and ground water in Europe is given in Figure 2.1. The contamination originates from a broad range of sources with landfills, above/underground storage tanks, septic systems, dry cleaners and industrial facilities being among the frequent sources; see more details in Figure 2.2.

In North America and Western Europe most countries have a legal framework to identify and deal with these issues. In the United States alone there are estimated to be in excess of 200,000 sites requiring some form of

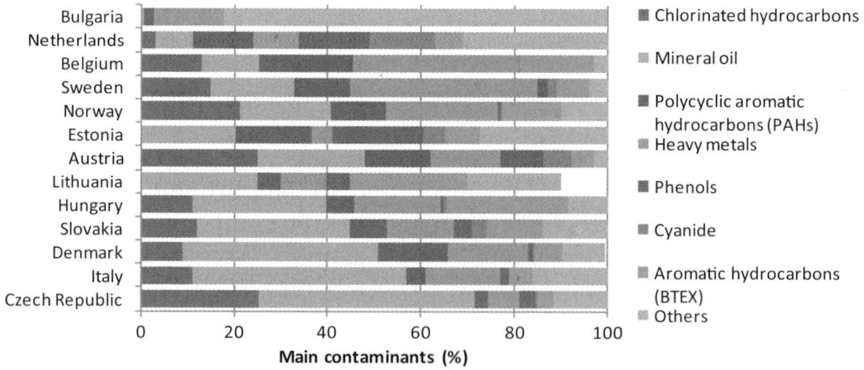

FIGURE 2.1 Overview of the main contaminants affecting soil and ground water in Europe shown as percentage of contaminated sites. *Data from European Environmental Agency (2007).* (For colour version of this figure, the reader is referred to the online version of this book.)

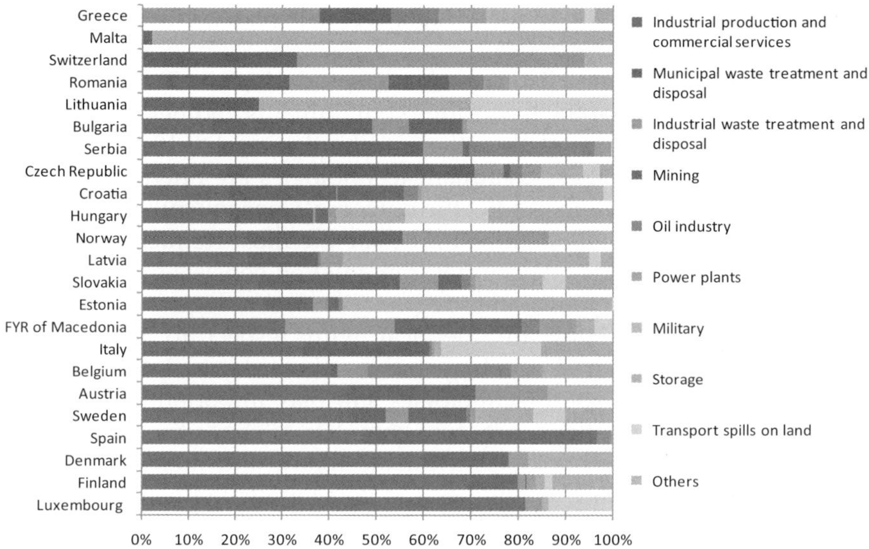

FIGURE 2.2 **Overview of activities causing soil contamination in Europe shown as percentage of contaminated sites.** *Data from European Environmental Agency (2007).* (For colour version of this figure, the reader is referred to the online version of this book.)

remediation (ITRC, 2005). In Europe this number is approximately 250,000 sites in the European Environmental Agency member countries and the number is expected to grow. Potentially polluting activities are estimated to have occurred at nearly 3 million sites in Europe (including the 250,000 sites already mentioned) and investigation is needed to establish whether remediation is required. If current investigation trends continue, the number of sites needing remediation will increase by 50% in 2025. By contrast, more than 80,000 sites have been cleaned up in the past 30 years in the countries where data on remediation is available (European Environmental Agency, 2007).

2.1.2. Remediation Technologies

Contamination of the subsurface by persistent organic contaminants remains a significant problem, even after decades of research on remediation technologies (Watts et al., 1999; Watts and Teel, 2005). New types of remediation technologies and modifications of existing technologies are continuously developed by scientists and engineers. First approaches for remediation focused on excavation, pump and treat via activated carbon, bioremediation, and natural attenuation. In the 1990s the first reports on ISCO were published. An overview of the different types of remediation technologies divided into three major categories is presented in Figure 2.3.

Only the technologies listed under 'Degradation' in Figure 2.3 will result in degradation of the hazardous contaminants and hence eliminate the toxicity and because of this, these technologies should be the ones preferred. In practice these treatment technologies are often complex to control and it is difficult to predict the outcome. Consequently, excavation and other low-tech methods are still by far the most frequently applied remediation technologies. Very often remediation consists of a combination of containment, removal and degradation technologies.

No contaminated sites are identical with respect to the type of contaminant, geology, hydrogeology and geochemistry and, therefore, all the mentioned remediation technologies still play important roles. However, some contaminations are too complex to be remediated cost-effectively with the technologies available today, and therefore, further improvements of existing techniques and development of new techniques are still needed. An important task in the present Ph.D. work was to gain a better understanding of the processes occurring in situ in the contaminated soil and ground water. One of the newer technologies attracting a lot of interest and possessing a great potential is ISCO.

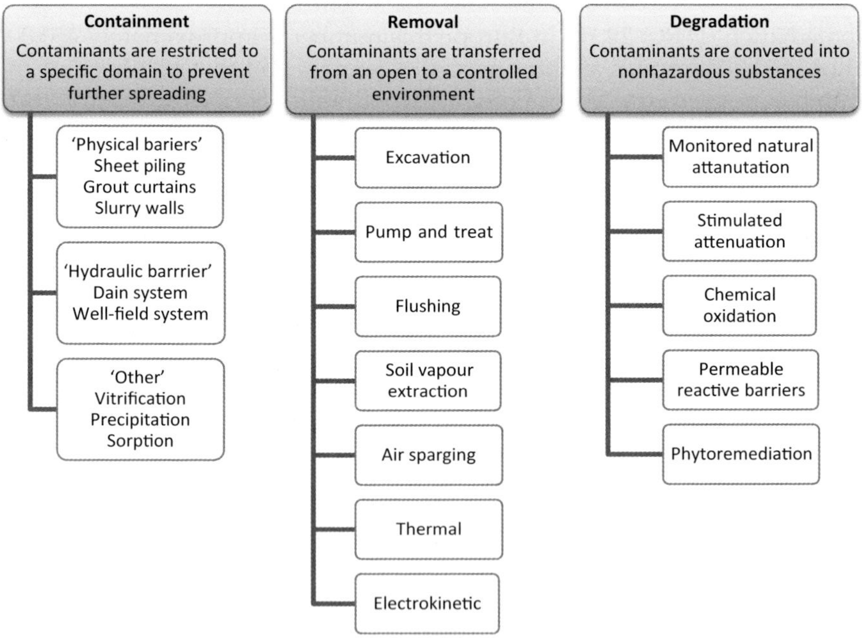

FIGURE 2.3 **Overview of remediation technologies divided into three major categories.** (For colour version of this figure, the reader is referred to the online version of this book.)

2.1.3. In situ Chemical Oxidation (ISCO)

ISCO is a technology involving the introduction of chemical oxidants into the subsurface in order to transform contaminants into less harmful compounds. Hydrogen peroxide was the first chemical oxidant investigated and used for full-scale treatment. Shortly after, ozone and permanganate came into use. In the past few years persulphate, percarbonate and other peroxygens have provided more options.

The use of permanganate for ISCO is considered an almost fully developed technique, is relatively easy to apply compared to other oxidants and has been used at numerous sites with well-documented results. Hydrogen peroxide has also been deployed at a large number of sites. However, the use of this oxidant is much more complex because of the numerous reactive intermediates and mechanisms occurring in the subsurface, which makes the process difficult to control and predict in the field. Ozone has only been used in a much more limited number of applications. Persulphate has, during the past 5–10 years, been used in an increasing number of full-scale remediation projects (Krembs et al., 2010).

Since no sites are identical, site-specific conditions regarding contaminant, geology, hydrogeology and geochemistry always have to be taken into consideration and evaluated in order to decide the remediation approach and technique to be used. This also means that there is no universal ISCO technique that will be useful at every site. ISCO for site remediation provides some advantages, but as most technologies, ISCO has limitations that should be recognised. When facing a remediation, it is important to screen if ISCO is an option for the site. Development of a comprehensive technology practices manual, including protocols and tools, to enable site-specific engineering of ISCO is almost finished (Crimi, 2010).

The use of permanganate for ISCO is an almost fully developed technique and is relatively easy to apply compared with other oxidants and it has been used at numerous sites with well-documented results. Hydrogen peroxide has also been deployed at a large number of sites. However, the use of this oxidant is much more complex because of the numerous reactive intermediates and mechanisms occurring in the subsurface, which makes the process difficult to control in the field. Ozone has only been used in much more limited applications. Persulphate has during the past 5–10 years been used for an increasing number of full-scale remediation projects.

Since no sites are identical, site-specific conditions always have to be taken into consideration and evaluated in order to decide the remediation approach and technique to be used. This also means that there is no universal ISCO technique, which will be useful at every site. ISCO for site

TABLE 2.1 Advantages and Disadvantages for ISCO (Huling and Pivetz, 2006)

Advantages:	Disadvantages:
• Applicable to a wide range of contaminants • Contaminants are destroyed in situ • In situ treatment may reduce costs incurred by other technologies such as pump and treat, monitored natural attenuation, etc. • Aqueous, sorbed, and nonaqueous phases of contaminants are transformed • Enhanced mass transfer (enhanced desorption and nonaqueous phase liquid dissolution) • Heat from H_2O_2 reactions enhances mass transfer, reaction rates and microbial activity • Potentially enhances postoxidation microbial activity and natural attenuation • Cost competitive with other candidate technologies • Relatively fast treatment	• Oxidant delivery problems due to reactive transport and aquifer heterogeneities • Natural oxidant demand may be high in some soil/aquifers • Short persistence of some oxidants due to fast reaction rates in the subsurface. • Health and safety issues regarding the handling of strong oxidants • Potential contaminant mobilisation • Potential permeability reduction • Limitations for application at heavily contaminated sites • Contaminant mixtures may require treatment trains • May have less oxidant/hydraulic control relative to other remedial technologies

remediation provides some advantages, but as most technologies, ISCO has limitations that should be recognised, see Table 2.1.

2.2. FUNDAMENTALS OF ISCO

In oxidative treatment systems, numerous reactions could potentially occur, including acid/base reactions, adsorption/desorption, dissolution, hydrolysis, ion exchange, oxidation/reduction, precipitation, etc. (Huling and Pivetz, 2006). All these reactions can make ISCO a challenging technique to apply in field scale. In the following sections chemical oxidation focussing on hydrogen peroxide and persulphate applied for soil and ground water remediation are reviewed.

2.2.1. Chemical Oxidation

All the ISCO technologies are based on the oxidative power of specific chemicals (oxidants) and the strength of the oxidants can be compared based on their standard reduction potential $E_0(V)$ as shown in Table 2.2. This potential is also known as the oxidation/reduction potential (ORP) and is a measure of the tendency of a chemical species to acquire electrons and thereby be reduced. The standard hydrogen electrode is the reference

TABLE 2.2 Standard Reduction Potentials E_0 of Reactive Species Involved in ISCO

	Standard Reduction Potential (E_0)		
PERMANGANATE			
$MnO_4^- + 4H^+ + 3e^- \rightarrow MnO_2 + 2H_2O$	1.68 V	Permanganate ion	(2.1)
HYDROGEN PEROXIDE			
$H_2O_2 + 2H^+ + 2e^- \rightarrow 2H_2O$	1.78 V	Hydrogen peroxide	(2.2)
$2OH^{\bullet} + 2H^+ + 2e^- \rightarrow 2H_2O$	2.76 V	Hydroxyl radical	(2.3)
$HO_2^{\bullet} + 2H^+ + 2e^- \rightarrow 2H_2O$	1.7 V	Perhydroxyl radical	(2.4)
$O_2^{\bullet -} + 4H^+ + 3e^- \rightarrow 2H_2O$	−2.4 V	Superoxide radical	(2.5)
$HO_2^- + H_2O + 2e^- \rightarrow 3OH^-$	−0.88 V	Hydroperoxide anion	(2.6)
OZONE			
$O_3 + 2H^+ + 2e^- \rightarrow O_2 + H_2O$	2.08 V	Ozone	(2.7)
PERSULPHATE			
$S_2O_8^{2-} + 2e^- \rightarrow 2SO_4^{2-}$	2.01 V	Peroxodisulphate	(2.8)
$HSO_5^- + 2e^- + 2H^+ \rightarrow HSO_4^- + H_2O$	1.82 V	Peroxomonosulphate	(2.9)
$SO_4^{\bullet -} + e^- \rightarrow SO_4^{2-}$	2.6 V	Sulphate radical	(2.10)
OTHER OXIDANTS			
$O_2 + 4H^+ + 4e^- \rightarrow 2H_2O$	1.23 V	Oxygen	(2.11)
$Cl_2 + 2e^- \rightarrow 2Cl^-$	1.36 V	Chlorine	(2.12)

from which all reduction potentials are measured and the potential of this electrode has therefore been defined as 0 V.

Even though the standard reduction potentials of the oxidants presented in Table 2.2 are high enough to oxidise most organic contaminants, these potentials do not indicate how the oxidants will perform under field conditions. Four major factors determine whether success will be obtained in the field: kinetics, thermodynamics, stoichiometry, and contact between oxidant and contamination. The reactions might be thermodynamically favourable based on standard reduction potentials, but under field conditions, the reaction rates can be very slow since they depend on several variables such as temperature, pH, concentrations of reactants, catalyst, reaction by-products, natural organic matter (NOM), oxidant scavengers, etc. (ITRC, 2005).

The present research focused on hydrogen peroxide and persulphate. Both oxidants have an oxygen–oxygen single bond (peroxy group) and therefore belong to the group of peroxygens. The oxygen atoms in the peroxy group are in oxidation state −1 and the oxygen–oxygen bond is unstable and splits easily into other reactive species. Several different forms of peroxygens exist, e.g. peroxides, persulphate, percarbonate, perborate and perphosphate. It should be noted that permanganate and perchlorate are not peroxygens. In Table 2.3 an overview of some of the peroxygens used for ISCO and examples of suppliers with product names is given. When the term persulphate is used in this chapter it refers to peroxodisulphate.

2.2.2. Hydrogen Peroxide

The use of hydrogen peroxide for soil and ground water remediation is an extension of the advanced oxidation processes (AOP), which have been applied for years to treat different waste streams ex situ. Although hydrogen peroxide is a strong oxidant, with a standard reduction potential of 1.78 V (Reaction (2.2)), the rate of direct oxidation of most hazardous organics is too slow to be useful for remediation (Watts and Teel, 2005).

2.2.2.1. *Fenton's Reagent*

In order to produce more reactive oxidants, hydrogen peroxide needs to be activated. In 1894 H. J. H. Fenton discovered that addition of Fe(II) dramatically increased the oxidation rate of tartaric acid with hydrogen peroxide (Fenton, 1894) and in 1934 it was proposed that production of hydroxyl radicals ($OH^•$) caused this increased reactivity (Haber and Weiss, 1934), according to the initiating Reaction (2.13), known as the classical Fenton's reaction. The Fenton's reagent is defined as a procedure where dilute hydrogen peroxide is slowly added to a rapidly stirred, degassed Fe(II) substrate (Walling, 1975). This results in a near 1:1 conversion of hydrogen peroxide to hydroxyl radicals:

$$H_2O_2 + Fe^{2+} \rightarrow Fe^{3+} + OH^\bullet + OH^- \quad (2.13)$$

The hydroxyl radical is a short-lived extremely reactive oxygen species with a high oxidation potential (2.76 V, Reaction (2.3)) that can rapidly degrade most organic contaminants. Hydroxyl radicals react with >95% of contaminants of concern (COC) at near diffusion-controlled rates, i.e. $k > 10^9$/M s (Buxton et al., 1988; Haag and Yao, 1992).

2.2.2.2. *Propagation and Termination Reactions*

A classical Fenton's system cannot be readily created in soil and ground water since it is too difficult to maintain a well-mixed, low-concentration hydrogen peroxide system in the subsurface. In practice more concentrated solutions of hydrogen peroxide (often 2–12% (Watts and Teel, 2005)) and varying types of catalyst are injected. This modified

TABLE 2.3 Overview of Peroxygens Used for Remediation

Compound and Supplier	Structure	Comments
PEROXIDE (INORGANIC)		
Hydrogen peroxide H_2O_2	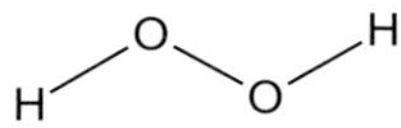	
Magnesium peroxide MgO_2 [Cool-Ox™, DeepEarth Tec.] [ORC®, Regenesis] [IXPER® M, Solvay]	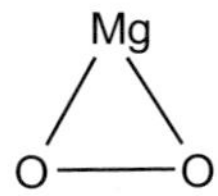	Solid peroxide. Insoluble in water but will slowly release oxygen at its natural pH>10: $MgO_2 + H_2O \rightarrow Mg(OH)_2 + ½O_2$ Lower pH (below natural pH 10) ⇒ more H_2O_2 is produced. The active oxygen can be liberated within minutes in acid: $MgO_2 + 2H^+ \rightarrow Mg^{2+} + H_2O_2$
Calcium peroxide CaO_2 [PermeOx® Plus, FMC] [IXPER® C, Solvay]	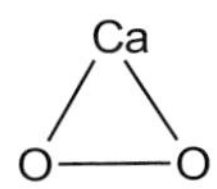	Same as above
PERSULPHATE		
Sodium peroxodisulphate $Na_2S_2O_8$ [Klozur™, FMC]	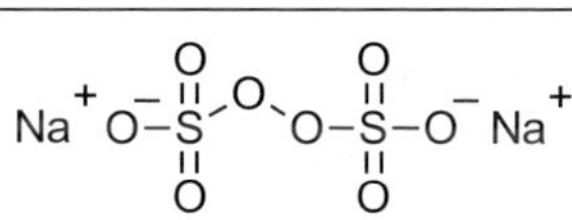	
Potassium peroxymonosulphate $KHSO_5$ [Oxone®,DuPont]	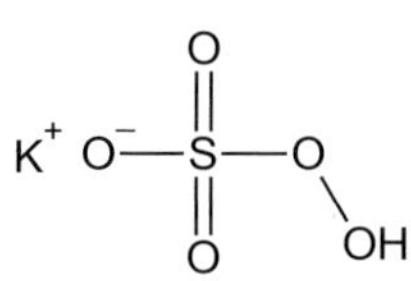	
PERCARBONATE		
Sodium percarbonate $Na_2CO_3 \cdot 1.5H_2O_2$ [RegenOx™, Regenesis] [ENVIRO First™, Solvay]	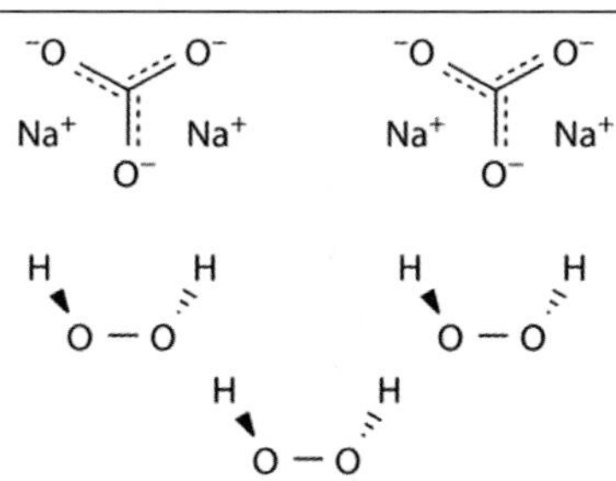	Despite the name, it is, in fact, a crystalline water-soluble chemical compound of sodium carbonate and hydrogen peroxide: $2(Na_2CO_3 \cdot 1.5H_2O_2) \rightarrow 2Na_2CO_3 + 3H_2O_2$ Part of its reactivity might be attributed to the formation of 'true' percarbonate (Mckillop and Sanderson, 1995): $HCO_3^- + H_2O_2 \rightleftarrows HCO_4^- + H_2O$

Fenton's system is often referred to as catalysed hydrogen peroxide propagation or MFR. When these high concentrations are used, a series of propagation reactions start, Reactions (2.14–2.19):

$$H_2O_2 + OH^{\bullet} \rightarrow HO_2^{\bullet} + H_2O \tag{2.14}$$

$$HO_2^{\bullet} \rightleftarrows O_2^{\bullet -} + H^+ \quad pK_a = 4.8 \tag{2.15}$$

$$HO_2^{\bullet} + Fe^{2+} \rightarrow HO_2^- + Fe^{3+} \tag{2.16}$$

$$RH + OH^{\bullet} \rightarrow H_2O + R^{\bullet} \rightarrow \text{Further oxidation} \tag{2.17}$$

$$R^{\bullet} + H_2O_2 \rightarrow ROH + OH^{\bullet} \tag{2.18}$$

$$R^{\bullet} + O_2 \rightarrow ROO^{\bullet} \tag{2.19}$$

$$H_2O_2 \rightleftarrows H^+ + HO_2^{\bullet -} \quad pK_a = 11.62 \tag{2.20}$$

Chain propagating sequences generate perhydroxyl radicals ($HO_2^{\bullet}$), superoxide radical anions ($O_2^{\bullet -}$), hydroperoxide anions (HO_2^-), and organic radicals ($R^{\bullet}$). These propagation reactions become important when hydrogen peroxide concentrations are relatively high (>0.3 M) (Watts and Teel, 2005). Processes using these high concentrations are very different from the classical Fenton's reagent because reactive species other than hydroxyl radicals are formed resulting in a wide range of reactivity. It should be noticed that although some of these reactive species (e.g. superoxide) are not very reactive in deionised water, their reactivity is significantly increased in systems that have solutes such as hydrogen peroxide, which provides a solvent effect (probably by changing its solvation shell), increasing the reactivity of superoxide (Smith et al., 2004). Similar to the addition of solvents, it has been shown that the presence of surfaces in water–solid matrices also enhances the reactivity of superoxide in water, possibly by altering the superoxide solvation shell (Furman et al., 2009).

For years all contaminant destructions obtained with MFR was ascribed to hydroxyl radicals. It has now been demonstrated that other reactive species are involved in the processes and MFR can treat compounds such as carbon tetrachloride and chloroform that were previously considered untreatable by MFR since these are unreactive with hydroxyl radicals. It was shown that a transient oxygen species other than the hydroxyl radical formed through MFR reactions was likely responsible for degradation (Watts et al., 1999a; Watts et al., 2005a). The species responsible for this destruction may be superoxide or hydroperoxide (Watts and Teel, 2005), both of which are produced by propagation reactions in MFR systems using relatively high concentrations of hydrogen peroxide. The propagation reactions are followed by termination reactions:

$$OH^{\cdot} + Fe^{2+} \rightarrow OH^{-} + Fe^{3+} \tag{2.21}$$

$$2OH^{\cdot} \rightarrow H_2O_2 \tag{2.22}$$

$$HO_2^{\cdot} + Fe^{3+} \rightarrow O_2 + Fe^{2+} + H^{+} \tag{2.23}$$

$$HO_2^{\cdot} + Fe^{2+} \rightarrow HO_2^{-} + Fe^{3+} \tag{2.24}$$

$$R^{\cdot} + OH^{\cdot} \rightarrow ROH \tag{2.25}$$

$$R^{\cdot} + Fe^{3+} \rightarrow R^{+} + Fe^{2+} \tag{2.26}$$

$$R^{\cdot} + Fe^{2+} \rightarrow R^{-} + Fe^{3+} \tag{2.27}$$

$$2R^{\cdot} \rightarrow R-R \tag{2.28}$$

2.2.2.3. *Catalyst*

The acidic properties of Fe(II) lower the pH to 3–4 in classical Fenton's system, when Fe(II) is added to deionised water. In soil and ground water the effects of pH are more complicated since most of these systems possess a strong buffering capacity which will maintain the pH neutral and result in iron precipitation and ineffective catalysis:

$$Fe^{3+} + nOH^{-} \rightarrow \text{Amorphous iron oxide/hydroxides (precipitate)} \tag{2.29}$$

As an alternative to lowering the pH in the entire treatment area to 2–4 with, e.g. sulphuric acid, other methods have been used to catalyse the reaction in MFR. These include soluble iron (Murphy et al., 1989; Tyre et al., 1991; Watts and Dilly, 1996), iron minerals (Tyre et al., 1991; Ravikumar and Gurol, 1994; Khan and Watts, 1996; Valentine and Wang, 1998; Huang et al., 2001; Teel et al., 2001; Kwan and Voelker, 2003), and chelated iron (Sun and Pignatello, 1992; Pignatello and Baehr, 1994; Luzzatto et al., 1995; Pignatello and Day, 1996; Wang and Brusseau, 1998; Seol and Javandel, 2008).

For concentrations of hydrogen peroxide greater than 30 mM (>1 g/l), which are often used for ISCO, iron (III) is a more effective catalyst compared to iron (II) and the reactions proceed through formation of superoxide/perhydroxyl radicals (Murphy et al., 1989; Watts and Dilly, 1996):

$$H_2O_2 + Fe^{3+} \rightarrow Fe - OOH^{2+} + H^+ \rightarrow Fe^{2+} + HO_2^{\bullet} + H^+ \qquad (2.30)$$

The main advantages of using chelated catalysts are that the process can be conducted at neutral pH and that chelated iron may travel farther in the subsurface compared to soluble iron (Watts and Teel, 2005). However, the dosage of oxidants has to be increased since MFR also oxidises the chelating agents. It has been documented that contaminant oxidation can take place in MFR systems containing soil with naturally occurring iron oxides but no soluble iron catalyst (Tyre et al., 1991). Crystalline iron oxides like goethite are more effective catalysts compared to amorphous ones such as ferrihydrite (Valentine and Wang, 1998; Huang et al., 2001; Watts et al., 1999c). Also soluble manganese (II) and other transition metals can catalyse the Fenton's reaction and the presence of manganese oxides (amorphous and crystalline) can catalyse the formation of reductants (Watts et al., 2005b).

2.2.2.4. *Decomposition and Transport*

Besides all of above-mentioned reactions a large part of the hydrogen peroxide decomposes to molecular oxygen in contact with aquifer materials resulting in waste of oxidant when applied for ISCO:

$$2H_2O_2 \rightarrow 2H_2O + O_2 \qquad (2.31)$$

The successful application of hydrogen peroxide-based ISCO is strongly limited by the instability of hydrogen peroxide (Watts and Teel, 2006; Pignatello et al., 2006). Several studies have evaluated the impact of NOM and minerals on decomposition rates (Watts et al., 1999; Valentine and Wang, 1998; Teel et al., 2001; Kwan and Voelker, 2003; Watts et al., 1999; Watts et al. 2007; Teel et al., 2007; De Laat et al., 2004; Kong et al., 1998). Attempts have been done to minimise this decomposition by stabilisation of hydrogen peroxide using stabilising agents such as citrate, malonate, etc. (Seol and Javandel, 2008; Watts et al., 2007). Stabilisation is important in order to enhance transport of hydrogen peroxide and thereby create contact between oxidants and contaminants farther from the injection point. The one-dimensional transport of hydrogen peroxide with dispersion and advection can be described as:

$$\frac{\partial[H_2O_2]}{\partial t} = D\frac{\partial^2[H_2O_2]}{\partial x^2} - v\frac{\partial[H_2O_2]}{\partial x} - k_{obs}[H_2O_2] \qquad (2.32)$$

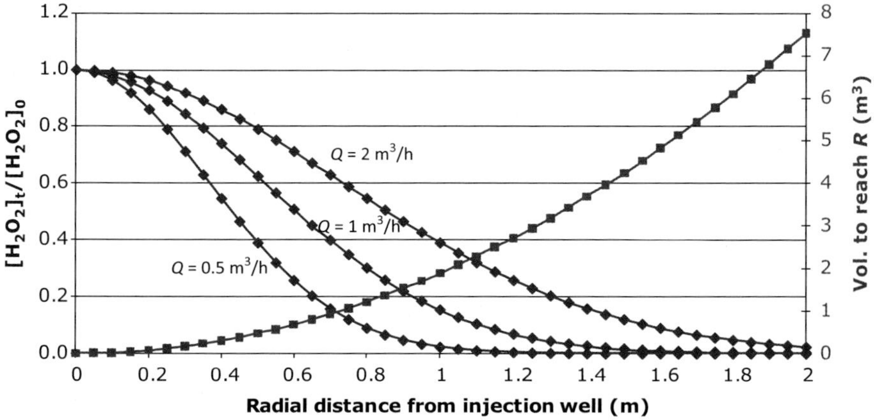

FIGURE 2.4 **Example of reactive transport of hydrogen peroxide in soil at different injection flow rates.** $k_{obs} = 1/h$, $Z = 3$ m, $n = 0.2$, $Q = 0.5–2\ m^3/h$. (For colour version of this figure, the reader is referred to the online version of this book.)

where t (s) is the time, D (m^2/s) is the dispersion coefficient, x (m) is the distance, v (m/s) is the pore water velocity and k_{obs} (1/s) is the rate of hydrogen peroxide decomposition/reaction. At steady-state conditions and negligible dispersion the solution for the equation will be:

$$x = -\frac{v}{k_{obs}} \ln\left(\frac{[H_2O_2]}{[H_2O_2]_0}\right) \tag{2.33}$$

showing that the transport distance greatly depends on decomposition/reaction and pore water velocity. If site-specific stability data of hydrogen peroxide is collected it is possible to use these data to determine injection rates and injection point spacing. By rearranging Eqn (2.33), the radial flow from an injection well can be calculated:

$$\frac{[H_2O_2]}{[H_2O_2]_0} = \exp\left(\frac{-k_{obs} \cdot R^2 \cdot \pi \cdot Z \cdot n}{Q}\right) \tag{2.34}$$

where R (m) is the radial distance, Z (m) is the vertical injection interval, n is the porosity and Q (m^3/s) is the flow rate. Figure 2.4 shows an example of using Eqn (2.34) to calculate the radial distribution of hydrogen peroxide in a homogeneous soil. From the figure it is seen that higher injection rates can result in a greater radius of influence (ROI).

2.2.3. Persulphate

Persulphate (peroxodisulphate) salts dissociate in water to persulphate anions ($S_2O_8^{2-}$), which are strong oxidising agents with an oxidation potential of 2.01 V (House, 1962), Reaction (2.8). However, the kinetics involved in degrading many organic compounds is very slow (House, 1962). To enhance the oxidation strength, persulphate needs to be activated in order to produce free sulphate radicals ($SO_4^{\bullet -}$) with an oxidation potential of 2.6 V (Huie et al., 1991), Reaction (2.10). As with hydroxyl radicals, these radicals are short lived and highly reactive. Sulphate radicals have a reported half-life of about 4 s at 40 °C (Banerjee and Konar, 1984). The primary limitation of activated persulphate in ISCO is the minimal knowledge of its reaction pathways in the subsurface (Furman et al., 2010).

2.2.3.1. Activation

Activation of persulphate can be accomplished by heat (House, 1962; Huang et al., 2002; Huang et al., 2005; Liang et al., 2003), metals (Kolthoff and Miller, 1951; Anipsitakis and Dionysiou, 2004), or ultraviolet (UV) light (Neta et al., 1977). These mechanisms are well described. However, in full-scale ISCO applications more recent activation methods include strongly alkaline pH (Buxton et al., 1988) or addition of hydrogen peroxide (Crimi and Taylor, 2007).

UV activation is obviously not an option for ISCO. Both UV and heat activation occur via homolysis of the oxygen−oxygen bond (Waldemer et al., 2007), Reaction (2.35). Thermal activation has been demonstrated at temperatures ranging from 35 to 130 °C (House, 1962; Huang et al., 2002; Peyton, 1993; Tsitonaki et al., 2008; Tsitonaki et al., 2010) and the relation between temperature and reaction rate obeys the Arrhenius equation (Huang et al., 2002; Liang et al., 2003; Liang et al., 2007), Eqn (2.36), where A is the pre-exponential factor, E_a is the activation energy, R is the universal gas constant and T is the temperature.

$$S_2O_8^{2-} \xrightarrow{\text{Heat or UV}} 2SO_4^{\bullet -} \tag{2.35}$$

$$k = Ae^{-E_a/RT} \tag{2.36}$$

Metal activation occurs via oxidation and reduction processes analogous to the Fenton's reaction with electron donors including e^- from radiolysis of water (Neta et al., 1977) or low-valent metal ions such as Fe^{2+} (Anipsitakis and Dionysiou, 2004), Reaction (2.37). Because of the similarities to Fenton's chemistry, the knowledge about this activation technique and potential problems, such as the unwanted Reaction (2.38), are well known. Chelating agents are often used to keep iron in solution (Crimi and Taylor, 2007; Dahmani et al., 2006; Killian et al., 2007; Liang et al., 2004b). It has also been indicated that naturally occurring iron oxides (Liang et al., 2003, 2004a) and zero valent iron (Brown et al., 2004) can activate persulphate.

$$S_2O_8^{2-} + Fe^{2+} \rightarrow SO_4^{\bullet -} + SO_4^{2-} + Fe^{3+} \quad (2.37)$$

$$SO_4^{\bullet -} + Fe^{2+} \rightarrow SO_4^{2-} + Fe^{3+} \quad (2.38)$$

Base is the most commonly used activator for persulphate in full-scale ISCO (Furman et al., 2010). Several mechanisms have been proposed for the alkaline activation (House, 1962; Kolthoff and Miller, 1951), but the mechanisms have not been evaluated and confirmed. A recent thorough study by Furman et al. (Furman et al., 2010) documented the reactions involved in base-catalysed hydrolysis of persulphate to hydroperoxide anion and sulphate followed by the reduction of another persulphate molecule by hydroperoxide. Reduction by hydroperoxide causes decomposition of persulphate into a sulphate radical and sulphate anion, and hydroperoxide is oxidised to superoxide. Summing the proposed reactions yields the following net Reaction (2.39) and under highly alkaline conditions, the sulphate radical reacts with hydroxide to form hydroxyl radicals according to Reaction (2.40). Hence, the hydroxyl radical is likely the dominant radical available for organic oxidation using alkaline-activated persulphate. Furthermore, at high pH, hydroxyl radicals themselves may dissociate into oxide radicals, Reaction (2.41), which can impact their reactivity (Buxton et al., 1988).

$$2S_2O_8^{2-} + 2H_2O \rightarrow 3SO_4^{2-} + SO_4^{\bullet -} + O_2^{\bullet -} + 4H^+ \quad (2.39)$$

$$SO_4^{\bullet -} + OH^- \rightarrow SO_4^{2-} + OH^{\bullet} \quad (2.40)$$

$$OH^{\bullet} \rightleftarrows O^{\bullet -} + H^+ \quad pK_a = 11.9 \quad (2.41)$$

Hydrogen peroxide has been shown to be able to activate persulphate even without addition of transition metal catalysts (Crimi and Taylor, 2007; Block et al., 2004). The mechanisms involved are unknown, but it might be due to activation of hydrogen peroxide by naturally occurring metal oxides resulting in the formation of radicals able to interact with persulphate to form sulphate radicals through propagation reactions. Another suggestion is that activation happens because of the heat generated through the exothermic hydrogen peroxide reactions (Robinson et al., 2004).

It has also been demonstrated that ketones, primary alcohols and low-carbon-chain aldehydes can activate persulphate to generate reactive species (Ocampa, 2009):

$$S_2O_8^{2-} + RH \rightarrow SO_4^{\bullet -} + R^{\bullet} + H^+ + SO_4^{2-} \quad (2.42)$$

2.2.3.2. Propagation and Termination Reactions

Once persulphate is activated via the initiating reactions, the sulphate radical can initiate a series of propagation and terminating reactions involving other radicals and oxidants (Anipsitakis and Dionysiou, 2004; Waldemer et al., 2007; Peyton, 1993; Berlin, 1986) including the reaction involving hydroxyl radicals described in Section 2.2.2.2. Examples of these reactions are:

$$SO_4^{\bullet -} + H_2O \rightleftarrows OH^{\bullet} + H^+ + SO_4^{2-} \tag{2.43}$$

$$SO_4^{\bullet -} + RH \rightarrow R^{\bullet} + H^+ + SO_4^{2-} \tag{2.44}$$

$$R^{\bullet} + S_2O_8^{2-} \rightarrow SO_4^{\bullet -} + SO_4^{2-} + R \tag{2.45}$$

$$H_2O_2 + S_2O_8^{2-} \rightarrow 2H^+ + 2SO_4^{2-} + O_2 \tag{2.46}$$

$$SO_4^{\bullet -} + OH^{\bullet} \rightarrow HSO_5^- \tag{2.47}$$

$$2SO_4^{\bullet -} \rightarrow S_2O_8^{2-} \tag{2.48}$$

$$2OH^{\bullet} \rightarrow H_2O_2 \tag{2.49}$$

Because different reactive species are generated in activated persulphate systems, the kinetics can be complicated, but COC removal should be described with a pseudo-first-order rate constant, k_{obs} (Waldemer et al., 2007):

$$\frac{d[COC]}{dt} = -k_{obs}\,[COC] \tag{2.50}$$

$$k_{obs} = k'_{SO_4^{\bullet -}}\left[SO_4^{\bullet -}\right] + k'_{OH^{\bullet}}[OH^{\bullet}] + k'_{S_2O_8^{2-}}\left[S_2O_8^{2-}\right] + k'_{other}\,[other] \tag{2.51}$$

Under most conditions the dominant term will be the one involving sulphate radicals (Kolthoff and Miller, 1951; Peyton, 1993), but as shown in the previous section, the activation method, pH and possible other factors may affect the kinetics and types of reactive species generated and other terms in the rate Eqn (2.51) may be dominant. The type of reactive species generated also decides which contaminants can be degraded by activated persulphate.

Besides generating reactive species able to degrade contaminants, the application of persulphate can cause acidification of aquifers with low buffering capacity (Huang et al., 2005; Liang et al., 2004a). The low pH can potentially increase the mobilisation of metals, see Section 2.5. On the other hand, well buffered systems often contain high concentrations of carbonates that will act as radical scavengers (Buxton et al., 1988; Haag and Yao, 1992; Neta et al., 1977) and thereby reduce the efficiency of the persulphate, see Section 2.6. Decomposition and transport of persulphate is similar to what is described in Section 2.2.2.4 for hydrogen peroxide.

One of the main advantages of persulphate is that persulphate can be present for relatively long time in the subsurface, but its fundamental chemistry in the subsurface is only now being elucidated.

2.2.4. Permanganate

Permanganate is a selective ISCO oxidant that has been used almost exclusively for the treatment of chloroalkenes, such as trichloroethylene (TCE) and perchloroethylene (PCE). For ISCO two forms of permanganate are used: potassium permanganate ($KMnO_4$) typically supplied as a crystalline solid and prepared at the site to concentrations up to 4% and sodium permanganate ($NaMnO_4$) typically supplied as 40% liquid solution that is diluted at the site before injection. Both forms are strong oxidisers and participate in quite complex reactions in the subsurface due to its multiple valence state and mineral forms. Some of the primary redox reactions taking place are:

$$MnO_4^- + 8H^+ + 5e^- \rightarrow Mn^{2+} + 4H_2O \quad (pH < 3.5) \tag{2.52}$$

$$MnO_4^- + 2H_2O + 3e^- \rightarrow MnO_2 + 4OH^- \quad (3.5 < pH < 12) \tag{2.53}$$

$$MnO_4^- + e^- \rightarrow MnO_4^{2-} \quad (pH > 12) \tag{2.54}$$

Under common environmental conditions Reaction (2.53) leading to the precipitation of manganese dioxide is the most pronounced reaction. This reaction involved a three electron transfer and based on stoichiometric calculations the amount of oxidant needed to completely mineralise a known amount of contaminants can be calculated. If PCE is used as an example, the complete mineralisation will be:

$$4MnO_4^- + 3C_2Cl_4 + 4H_2O \rightarrow 6CO_2 + 2MnO_2 + 12Cl^- + 8H^+ \tag{2.55}$$

In this case 0.96 kg of MnO_4^- is required to mineralise 1 kg of PCE. When using permanganate the oxidation of contaminants occurs through direct electron transfer rather than through radical formation as in the case with hydrogen peroxide and persulphate.

As showed in Section 2.3.3, oxidants are nonselective and will therefore react with other compounds in the subsurface besides the contaminants and the needed amount of oxidants cannot be calculated only by the stiochiometric demand. The NOM and other reduced species present in the soil can make up a large oxidant demand.

Compared to other oxidants used for ISCO, the main advantages of permanganate are that the oxidant is relatively stable and can persist in the subsurface for several months, permitting both advective and diffusive transport, and can result in good distribution of the oxidant and treatment of low-permeability materials (Struse et al., 2002). The purple colour of the permanganate makes it easy to identify visually and simple to measure quantitatively. One of the main disadvantages is that mass transport and mass transfer limitations can occur when MnO_2 precipitates near the injection wells and at a nonaqueous phase liquid (NAPL) interface, respectively.

2.2.5. Ozone

Ozone-based ISCO has a high degree of potential for treating ground water contaminants, but is limited by its stability in the subsurface and the need for capital equipment at the site. Because ozone (O_3) is a gas it differs a lot from the other oxidants used for ISCO. To treat contaminated soil and ground water, gaseous ozone can be injected in the vadose zone or purged below the ground water table. Also, the ozone can be injected as ozonated water.

Degradation of organic contaminants happens either by direct oxidation by ozone itself or indirectly through formation of hydroxyl radicals. These radicals can be created from reactions with hydroxide or hydrogen peroxide in a process sometimes referred to as perozone:

$$2O_3 + 3H_2O_2 \rightarrow 2OH^{\bullet} + 4O_2 + 2H_2O$$

2.2.6. Site Characteristics

Site characterisation is a critical component of hazardous site remediation and should provide information that is required for conducting risk assessments and for designing and implementing remediation systems. It is often stated that the completion of thorough site characterisation is the most important factor in the success of any remediation project. In recent years there have been many advances in site characterisation, and technologies such as direct push technologies can result in valuable information acquired at a lower cost compared to traditional methods. In the following two sections some of the important parameters are outlined.

2.2.6.1. Geology

Naturally occurring subsurface heterogeneities such as zones of high permeability as well as subsurface utility corridors and other anthropogenic subsurface disturbances can act as preferential pathways and unpredictable flow patterns for the injected oxidant. If low-permeability zones are present, it will be difficult to obtain contact between oxidant and contaminant present in these zones.

2.2.6.2. Hydrogeology

Due to fast reaction rates of hydrogen peroxide in the subsurface, ground water flow rate and direction only have a minimum of impact on the oxidant transport. Depending on the activation techniques used for persulphate, this oxidant might be transported significantly from the injection point and will undergo advective and diffusive transport as described in Section 2.2.2.4. The greater the hydraulic conductivity and hydraulic gradient of the aquifer, the farther the transport distance will be.

2.2.6.3. Geochemistry

Some of the important geochemical parameters to characterise are redox conditions, pH, buffering capacity, iron concentration, alkalinity, temperature and NOM. Reduced conditions favour the presence of soluble Fe^{2+} and Mn^{2+} that can activate hydrogen peroxide and persulphate. However, the reduced conditions are also a source of treatment inefficiency due to presence of reduced species that will consume oxidant, which is further described in Section 2.2.9. pH is important, especially if iron activation is desired since this requires a low pH (or use of chelating agents). The natural buffering capacity of soils will in most cases possess significant acid neutralisation capacity and thereby maintain neutral pH. To overcome this, large amounts of acid are needed to lower pH or base to obtain alkaline activation of persulphate. High alkalinity can result in scavenging of the radicals and high concentrations of NOM can potentially consume large amounts of oxidants.

2.2.6.4. Contaminants

Types, distribution, concentrations and presence of NAPL contaminants are critical parameters to determine before ISCO is evaluated as a potential technology for remediation. Based on the characterisation, oxidant type and dosage can be estimated. This characterisation is also very important in order to evaluate the effectiveness of ISCO, since this should be done based on total mass and not dissolved phase evaluations, which would be inaccurate and misleading. This is because ISCO will alter the equilibrium partitioning of total mass distribution by converting sorbed and free phase mass to dissolved phase both during and for a period of time after the application of ISCO.

2.2.7. Contaminants Amenable

The reactive species generated when the peroxygens are activated determines which compounds can be degraded. For hydrogen peroxide it is mainly hydroxyl radicals and for persulphate it is mainly sulphate radicals. However, as shown in the previous sections other reactive species can be generated under specific condition and by applying specific activation methods.

The reaction mechanisms involved in the oxidation of organic compounds by hydroxyl radicals and persulphate radicals can be significantly different. The sulphate radical prefers to remove electrons from the organic molecule thereby producing organic radicals (Reaction (2.53)), whereas the hydroxyl radical adds to unsaturated bonds of alkenes (or alkynes) or abstracts hydrogen from the C–H bonds (Reaction (2.57)) (Ocampa, 2009; Forsey, 2004; Norman, 1979). Example with the carboxylate ion:

$$CH_3CO_2^- + SO_4^{\bullet -} \rightarrow CH_3CO_2^{\bullet} + SO_4^{2-} \rightarrow {}^{\bullet}CH_3 + CO_2 + SO_4^{2-} \quad (2.56)$$

$$CH_3CO_2^- + OH^{\bullet} \rightarrow {}^{\bullet}CH_2CO_2^- + H_2O \quad (2.57)$$

Reactions with aromatic or heterocyclic compounds also differ since the sulphate radical reacts by an electron transfer mechanism to produce a radical cation, Reaction (2.58), whereas the hydroxyl radical reacts via a mechanism analogous to an electrophilic substitution reaction (Anbar et al., 1966) in which the attack on the aromatic ring results in the formation of the cyclohexadienyl radical (Forsey, 2004), Reaction (2.59):

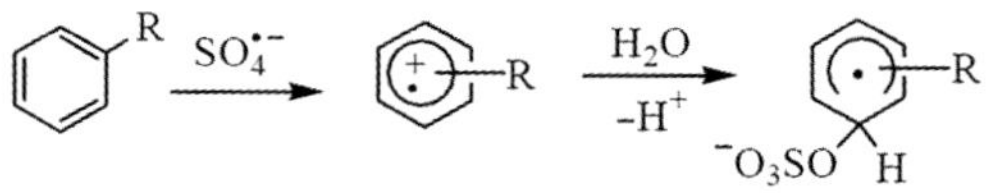

(2.58)

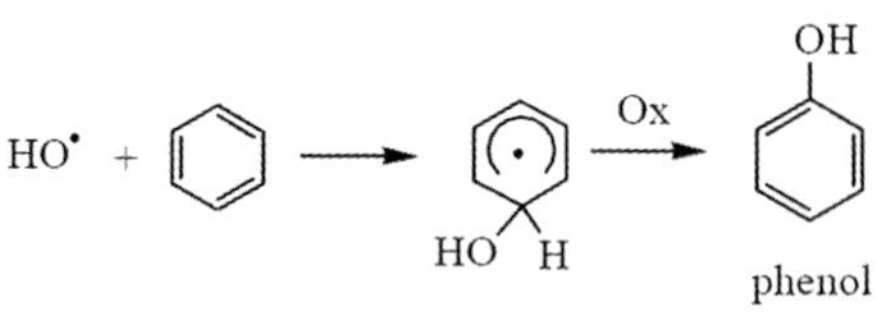

(2.59)

The reaction rates between hydroxyl radicals and aromatic compounds are close to diffusion controlled (Buxton et al., 1988), but because of the electrophilic substitution reactions, electron-withdrawing substituents such as nitro (–NO_2) or carbonyl (C=O) will decrease reaction rates and electron-donating substituents such as hydroxyl ($-OH$), amino ($-NH_2$) or alkoxy ($-OR$) will accelerate the reactions (Anbar et al., 1966). The presence of halogens in organic compounds has been shown to slow down the reaction rates for hydroxyl radicals (Haag and Yao, 1992) and this will possibly also apply for sulphate radicals (Tsitonaki et al., 2010).

Besides the hydroxyl radical and sulphate radicals other reactive species formed with MFR and ASP, such as superoxide, can significantly enhance degradation of recalcitrant organic contaminants.

It is important to notice that the reactive species generated in MFR and ASP will react with a wide range of different reactants present in the system besides the contaminant and to describe this, competition kinetics has to be applied.

Some general rules have been established in Watts et al., 2006 for the reactivity of hydroxyl radical with organic contaminants:

$k \approx 10^{10}$/M s Rate of diffusion.

$k > 10^9$/M s ISCO can be used.

$k < 10^8$/M s ISCO cannot be used.

Second-order rate constants for the reactivity of contaminants with hydroxyl and sulphate radicals are listed in Tables 2.4 and 2.5. Again, it should be noticed that, for example, carbon tetrachloride which is

TABLE 2.4 Second-Order Rate Constants for Contaminants and Hydroxyl Radicals

Compound	$k_{OH^\bullet}$ **(1/M s)**
2,4,5-Trichlorophenoxy acetic acid	4.0×10^9
Pentachlorophenol	4.0×10^9
Benzene	7.8×10^9
Toluene	3.0×10^9
Nitrobenzene	2.9×10^9
2,4,6-Trinitrotoluene	4.3×10^8
Polycyclic aromatic hydrocarbons	1.0×10^{10}
Carbon tetrachloride	$<2 \times 10^6$
Chloroform	$\approx 5 \times 10^6$
PCE	2.8×10^9
TCE	4.0×10^9
2,3,7,8-Tetrachlorodibenzo-*p*-dioxin	4.0×10^9
Methyl *tert*-butyl ether	$(3.9 \pm 0.73) \times 10^9$

Modified from Watts and Teel (2006). Sources are Haag and Yao (1992), Chang and Young (2000) and Schmelling et al. (1998).

TABLE 2.5 Second-Order Rate Constants for Contaminants and Sulphate Radicals

Compound	$k_{SO_4^{\cdot-}}$ **(1/M s)**
Methanol	1.0×10^7
Ethanol	4.3×10^7
2-Propanol	8.2×10^7
t-Butyl alcohol	$<10^6$
1-Hexanol	1.6×10^8
1-Octanol	3.2×10^8
Anisole	4.9×10^9
Acetanilide	3.6×10^9
Benzene	3.0×10^9
Benzoic acid	1.2×10^9
Nitrobenzene	$<10^6$

Modified from Watts and Teel (2006). Sources are Buxton et al. (1988), Neta et al. (1977) and Padmaja et al. (1993). More rate constants are found in Tsitonaki et al. (2010).

unreactive with hydroxyl radical is rapidly degraded in MFR systems through reactions with superoxide in the presence of sufficient hydrogen peroxide to provide a solvent effect (Smith et al., 2004; Teel et al., 2001; Watts and Stanton, 1999; Smith et al., 2006).

2.2.8. Destruction of Free Phase Contaminants

A critical issue associated with hazardous waste sites is the potential presence of immiscible liquid contamination in the subsurface. These NAPLs become trapped in the pore space due to capillary forces and once present, they are difficult to remove and will serve as a long-term source of contamination as they dissolve into the ground water under mass transfer-controlled conditions. Since the rate-limiting step in NAPL remediation is mostly dissolution, NAPLs can be present for decades even when treated actively with traditional pump-and-treat techniques (Lemke et al., 2004; Johnson and Pankow, 1992). It has been estimated that the cost to cleanup sites contaminated by dense nonaqueous phase liquid in the United States with traditional techniques could be $100 billion or more (EPA, 2003) and even though treatment of DNAPLs by traditional methods as pump and treat have been proved to be ineffective (Kavanaugh et al., 2003), this technique is still widely used at DNAPL sites (Kovalick, 2008). This clearly suggests that development or implementation of existing and emerging effective source-depletion technologies is needed. Modelling and limited data suggest that even removal of more than 90% of the DNAPL source is not enough to reduce concentrations in ground water (Stroo et al., 2003). The most prevalent types of DNAPLs are the halogenated organic solvents including TCE and PCE, but many sites are contaminated with other types of DNAPLs including coal tar and creosotes, polychlorinated biphenyls and pesticides.

Historically, ISCO has been rejected at sites with large amounts of DNAPL (Blanchard, 2010), but there is a theoretical basis and there are experimental results suggesting that DNAPL can be degraded by ISCO (Crimi and Siegrist, 2005).

The rate-limiting step in treating NAPLs is the mass transfer of contaminant into the aqueous phase (dissolution), which can generally be modelled based on the stagnant film model utilising a first-order linear driving force (Siegrist et al., 2006):

$$\frac{dc_{\text{bulk}}}{dt} = -k_{\text{La}}(c_{\text{bulk}} - c_{\text{sat}}) \tag{2.60}$$

describing the rate of change of aqueous concentration per time (g/m^3/s), k_{La} is the NAPL dissolution rate (1/s), c_{bulk} is the aqueous solute concentration (g/m^3), and c_{sat} is the solubility of the solute (g/m^3). The model is based on Fick's law of diffusion:

$$J_{\text{a}} = -D_{\text{m}}\frac{dC}{dz} \tag{2.61}$$

stating that the mass flux J_{a} (g/m^2/s) is proportional to the concentration gradient at the surface (g/m^4). The linear concentration gradient can be approximated by $(c_{\text{sat}} - c_{\text{bulk}})/\delta$, where δ (m) is the thickness of the film. The model is illustrated in Figure 2.5 and also shows the accelerated mass transfer occurring when oxidants are introduced to the system as

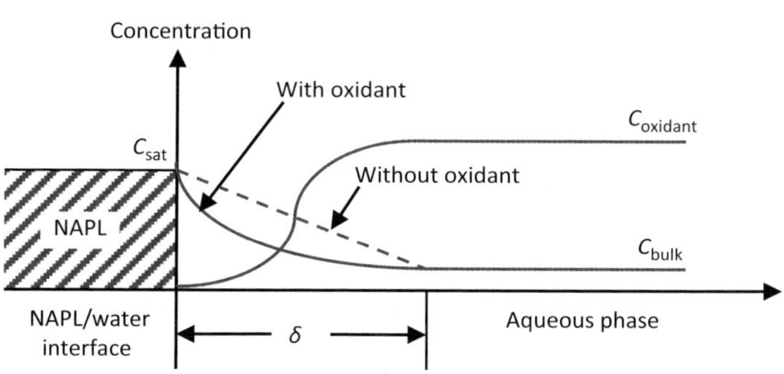

FIGURE 2.5 Thin film model with ISCO. *Adapted from Schnarr et al. (1998).* (For colour version of this figure, the reader is referred to the online version of this book.)

suggested by Schnarr et al. (1998). The concentration gradient is increased by reactions in the aqueous phase and reactions within the diffusion film layer. All ISCO technologies (and other mass removal technologies) will increase mass transfer of soluble organic contaminants, but the treatment is not very efficient, since it is still limited by diffusion.

In recent years it has been demonstrated in several laboratory experiments that MFR can be used for destruction of NAPL and that contaminants are degraded more rapidly than they dissolved into the aqueous phase (Watts et al., 1999; Watts et al., 2006; Watts and Stanton, 1999; Crimi and Siegrist, 2005; Siegrist et al., 2006; Kang et al., 2006; Watts et al., 2002; Yeh et al., 2003). By applying hydroxyl radical scavengers and reductant scavengers in experiments with MFR treatments of sorbed phase contaminant, it has been demonstrated that a reductive pathway in MFR is responsible for enhanced contaminant desorption (Watts et al., 1999). The experiments suggest that treatment of NAPL with MFR occurs by an enhanced desorption caused by reductants followed by oxidation and reduction of the contaminants in the aqueous phase. For destruction of carbon tetrachloride DNAPL, superoxide is the reactive species in MFR (Smith et al., 2006).

Until very recently there was no documentation showing if ASP could enhance desorption and NAPL destruction. In 2009, laboratory tests showed that treatment of sorbed phase PCE by Fe(II) citrate and alkaline-activated persulphate was equal or less than the rates of gas-purge desorption indicating that enhanced treatment of the sorbed contaminants did not occur in the ASP reactions (Teel et al., 2009). However, in 2010 it was demonstrated that alkaline-activated persulphate generates superoxide (Furman et al., 2010), Reaction (2.39). This discovery may lead to ASP applications for enhanced NAPL treatment in the future, but further studies are needed to evaluate this.

Only few well-documented field studies have demonstrated the destruction of NAPL. In 1997 at the 'Savannah river project' in South Carolina, MFR was demonstrated to reduce the mass of DNAPL chlorinated solvents from 269 to 17 kg (94% reduction) over 6 days of MFR treatment (Jerome et al., 1997).

2.2.9. Oxidant Demand

Proper design of a field-scale implementation of ISCO requires data on target contaminant levels as well as quantitative estimates of other oxidant sinks. There are a number of chemical and physical factors that contribute to the total oxidant demand (TOD) of a subsurface environment. These include (Haselow et al., 2003):

1. Dissolved, sorbed and free phase contaminant.
2. Dissolved, sorbed and solid phase reduced minerals.

3. Dissolved and sorbed phase NOM.
4. Thermal and chemical decomposition.

There are no accepted standard methods for conducting TOD tests and several methods have been reported in the literature. Overall, these are based on tests where varying ratios of oxidant mass to soil mass are prepared in separate vials and the oxidant consumption is monitored to describe the kinetics of the oxidation and the total consumption of oxidants after 48 or 72 h. This approach works rather well for permanganate, and TOD tests with persulphate are very similar to permanganate tests, whereas additional considerations have to be taken into account when using hydrogen peroxide, since this oxidant undergoes autocatalytic decomposition and also minerals can catalyse the decomposition resulting in an infinite oxidant demand. This is also partly true for persulphate and makes TOD measurements difficult for activated peroxygens.

2.2.10. Coupling ISCO with Biological Remediation

ISCO has traditionally been used as a stand-alone treatment technology, but in recent years ISCO is often used as integrated parts of sequential remediation systems, where ISCO is used to rapidly reduce high concentrations of contaminants or to destroy NAPL. Typical ISCO is coupled with traditional mass transfer technologies and monitored natural attenuation (MNA).

Several laboratory tests and field tests have documented that it is extremely difficult for ISCO technologies to render a site biologically inactive, even in cases involving reductive dechlorination (anaerobic) bacteria (Rowland et al., 2001; EPA, 2004). Posttreatment bacterial populations have been demonstrated to grow rapidly within short time after even the most aggressive treatments.

Applying bioremediation as a polishing step after ISCO can be an effective technology since ISCO in most cases is only cost-effective for bringing down concentrations to a certain level. For treating the lower concentrations after ISCO, bioremediation can be an effective technology, and has several advantages:

- High concentrations that might be toxic for microorganisms are reduced.
- Nonbiodegradable compounds can be degraded by ISCO.
- Some compounds will be more biodegradable after ISCO pretreatment due to partial oxidation (Lee and Hosomi, 2001; Kao et al., 2003).
- ISCO will increase the bioavailability of contaminants (Bittkau et al., 2004; Nadarajah et al., 2002).
- ISCO will create an oxidised environment leaving oxygen for the aerobic microbial degradation (Chapelle et al., 2005).

- Organic matter released from the soil particles as a result of ISCO can act as electron donor for some processes (Miller et al., 1996).
- Increased temperatures after ISCO can enhance the bioavailability of contaminants (Kosegi et al., 2000) and the microbial activity (Friis et al., 2007).

Challenges:

- Oxidants are toxic to microorganisms (Sahl and Munakata-Marr, 2006).
- Changes in water chemistry after ISCO: low or high pH.
- Changes in water chemistry after ISCO: high sulphate concentrations after ASP.

Results from Tsitonaki et al. (2008) suggest that ASP may be less damaging to microorganisms than MFR.

In the Kærgaard Plantation case study MFR was followed by MNA at one subsite and enhanced reductive dechlorination (ERD) at another subsite.

2.3. CASE STUDY: KÆRGAARD PLANTATION

The case study considers remediation of a very large contaminated site located in the Western part of Denmark, see Figure 2.6. The site located in Kærgaard Plantation has a varying geology as well as a very complex contaminant mixture and distribution. It represents one of the most difficult remediation tasks in Scandinavia and is considered the largest contaminated site in Denmark. The case study involved laboratory tests with soil and ground water from the site combined with extensive pilot testing at the site. The results and general issues will be presented and interesting topics will be dealt with in more details.

2.3.1. History

The contamination in Kærgaard Plantation originates from disposal of about 286.000 m^3 of pharmaceutical wastewater that was legally discharged by the former Grindstedværk in six separate waste pits in the plantation from 1956 to 1973. These waste pits were located 400–800 m from the shore and after the disposal the pits were covered by sand. The wastewater originated from the plant's production of vitamins and medicinal products, and it was transported to the site in tank trucks. The disposed wastewater was estimated to include (COWI, 2006a):

- 40,000 tons of inorganic salt.
- 15,000 tons of organic acid salts.

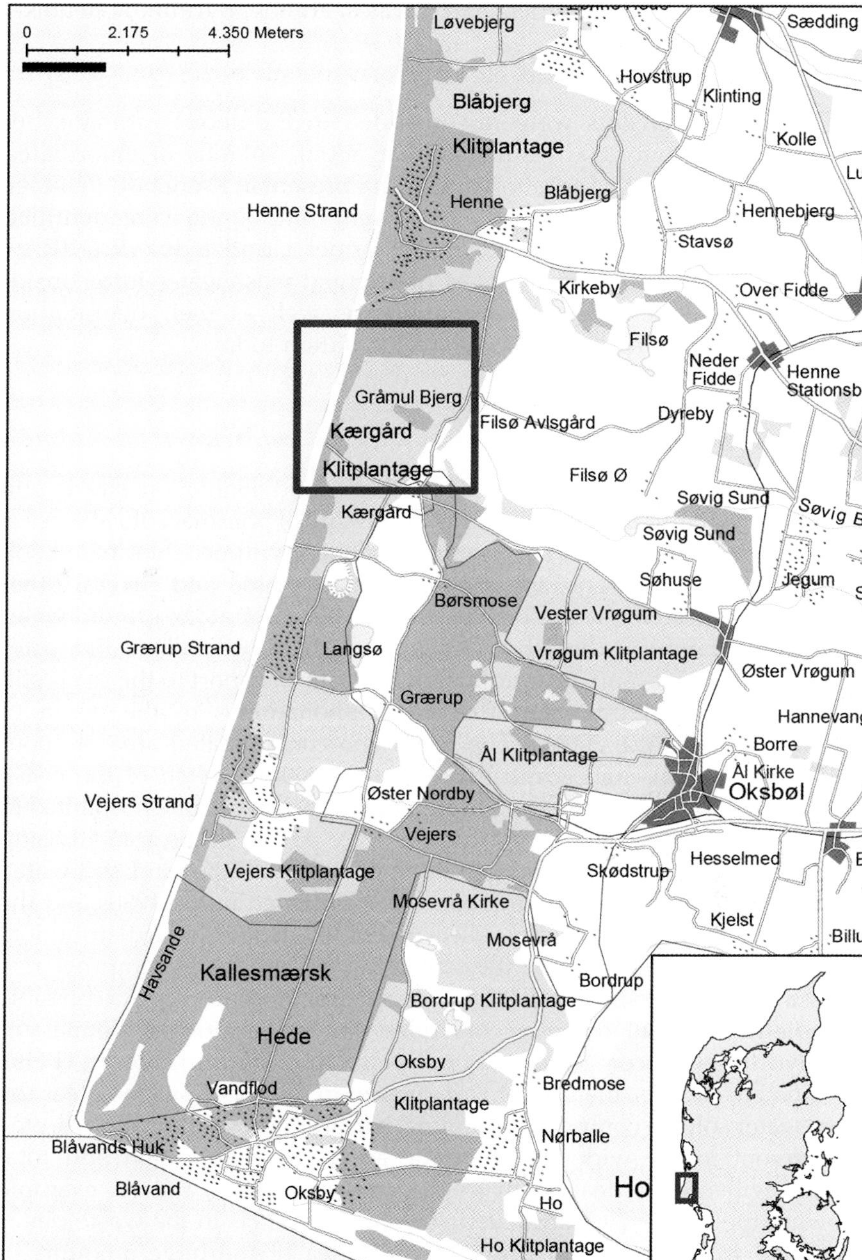

FIGURE 2.6 **The location of Kærgaard Plantation in the western part of Denmark.** *From COWI (2006b).* (For colour version of this figure, the reader is referred to the online version of this book.)

- 1500 tons of sulphonamides, barbiturates, aniline, pyridine, phenols, benzene and toluene.
- Lithium, mercury, cyanide, etc.

Chlorinated solvents were not included in the above estimate, but investigations from 2006 estimated that about 340 tons of chlorinated solvents and 230 tons of sulphonamides are present in Kærgaard Plantage today (COWI, 2006a). Recently, 1160 different compounds were identified to have been produced or used at the former Grindstedværk (NIRAS, 2009). Taking the huge amounts of discharged wastewater into consideration, a number of these compounds as well as their degradation products may be present in Kærgaard Plantation today.

2.3.2. Contamination

Soil and ground water beneath the Kærgaard Plantation site is still extremely contaminated with a complex mixture of pharmaceutical wastes, including sulphonamides, barbiturates, aniline, pyridine, chlorinated solvents (DNAPL), benzene, toluene, mercury, cyanide and several other compounds. From the waste pits located just above or at the ground water table, the contaminants have been transported to the soil and ground water underneath the pits and a continuous significant transport to the sea is still occurring with an estimated flux of sulphonamides to the ocean of 4000 kg/year (COWI, 2006a). The result is a contaminated area of about 500.000 m^2 and discharge into the sea at an 800-m stretch of the coast. Because of the contamination, swimming in the ocean became prohibited in 1964 and this still applies. In 2002, signs were put up in the area advising against staying in the area and getting in contact with the sand and water.

Investigations of the contamination were initiated in the 1980s and the site is still being investigated to collect more knowledge on the distribution, spreading and composition of the contamination and thereby improve risk assessments and increase the basis for decision concerning a possible remediation. Overall, it can be concluded that the majority of the mass of the hydrophobic compounds (total petroleum hydrocarbons (TPHs), chlorinated solvents, metals, etc.) are present in the waste pits, whereas the more water soluble compounds (aniline, barbiturates, sulphonamides, etc.) are present in the widespread plume area. The distribution and total masses will not be dealt with in the present chapter, but an example showing the masses of chlorinated solvents and TPHs in the waste pits is presented in Figure 2.7 to provide some idea of the huge amounts of contaminants present at the site. More details can be found in COWI, 2006a.

A conceptual model showing the spreading of contaminants from the waste pits is presented in Figure 2.8. From this model it is seen that DNAPL has penetrated the sand/gravel layer (layer 2) and the fine-grained

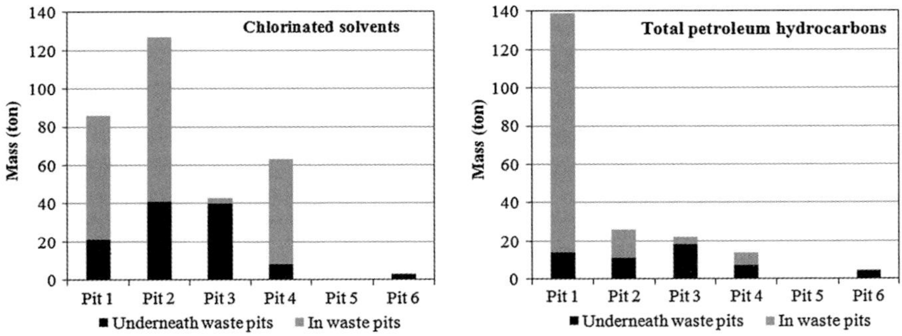

FIGURE 2.7 Calculated masses of chlorinated solvents and TPHs in the waste pits based on investigations performed in 2006, before waste pits 1 and 2 were excavated in 2008. *Data from COWI (2006a).*

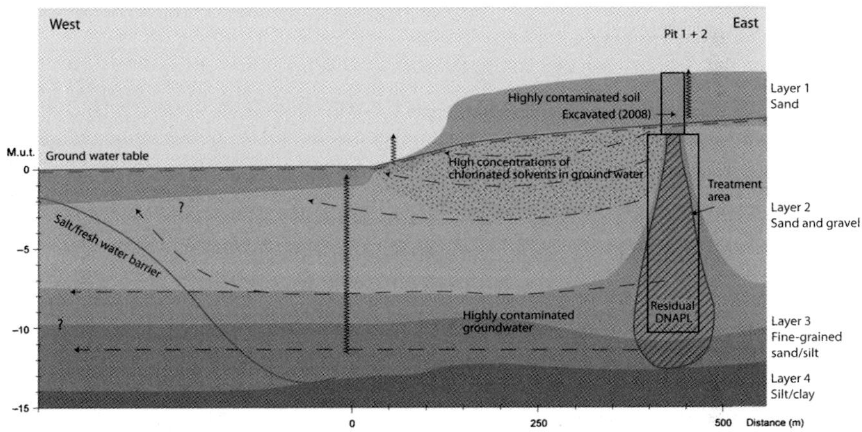

FIGURE 2.8 **Conceptual model of the site showing why the contamination poses a risk.** ?: Uncertainty on geology in the area. (For colour version of this figure, the reader is referred to the online version of this book.)

sand/silt layer (layer 3) all the way down to an impermeable silt/clay layer (layer 4). Today, residual DNAPL is still present in this area and continues to result in a significant contaminant flux towards the beach and the sea. The DNAPL is mainly present in the top and bottom of layer 2 and in layer 3, which is due to a higher hydraulic conductivity in the middle of layer 2 causing a more significant leaching of the chemicals present.

A more detailed conceptual model for waste pit 1 is presented in Figure 2.9, indicating the location of the waste pit (excavated in 2008) and the assumed presence of residual DNAPL PCE beneath the pit. Typical contaminant concentrations in water and soil as well as estimated mass

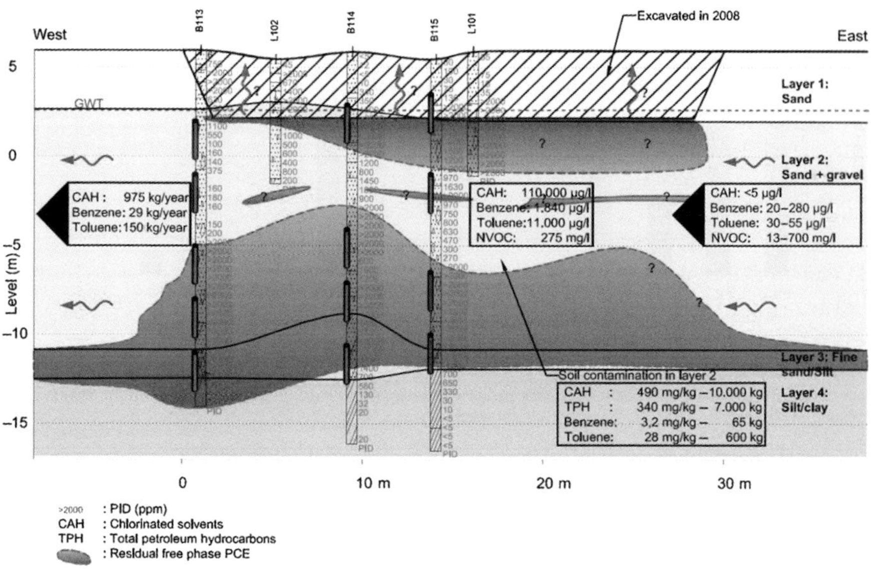

FIGURE 2.9 **Conceptual model of waste pit 1.** GWT: Ground water table ?: Uncertainty on distribution of the contamination in the area. (For colour version of this figure, the reader is referred to the online version of this book.)

fluxes of contaminants leaving the pit area are inserted in the model. Work in the case study was carried out with water and soil from layer 2 with focus on the top of layer 2 (layer 2.1).

2.3.3. Remediation

In 2007, a working group with the purpose of giving recommendations to a phase 1 for the future work in Kærgaard Plantation was appointed by the Minister of Environment and the chairman of the regional council. The aim was to improve conditions for staying on the beach and in the dynes. The recommendations were divided into three:

1. Excavation of highly contaminated soil in pit 1 and 2
2. Test, selection and design of remediation technique for contamination located below the ground water table
   a. Laboratory experiments
      - Chemical: MFR
      - Chemical: ASP
      - Thermal: Steam
      - Biological: ERD
   b. Pilot tests
3. Full-scale treatment of contamination in pit 1 and 2

The excavation was finished in 2009 and from there work focused on finding a technique or combination of techniques to treat the contamination present under the ground water table. As there was no precedent for treating this type of mixture, the feasibility of a variety of in situ remediation technologies was being evaluated in bench tests at this stage, including biological, chemical and thermal techniques. The presented results will focus on laboratory experiments with MFR and ASP followed by pilot tests with MFR combined with ERD in waste pit 1 at the site. Based on the results from the bench and pilot tests a full-scale remediation technique was chosen.

2.4. BENCH AND PILOT TESTS

Even though ISCO has been used at thousands of sites throughout the world and several peer-reviewed papers describe the involved processes, there is still a lack of published, well-documented field applications to back up the results observed in laboratories.

The objectives of the work presented in this section were to collect knowledge on the performance of MFR and ASP, and evaluate if the processes were applicable for treating complex contaminant mixtures including the ability to destroy NAPL. This was done by:

- Batch test with MFR and ASP.
- Pilot test with MFR followed by ERD.

Specific purposes are described in the following sections summarising the results obtained.

As an introductory remark, it should be noticed that the sandy soil from Kærgaard Plantation almost had no buffering capacity, Figure 2.10, since titration curves for this soil were almost identical to those of titration of demineralised water. It is believed that the large amounts of acids deposited at the site dissolved and removed carbonates.

2.4.1. Kærgaard Plantation: MFR and ASP (Bench)

As a part of selecting and designing a remediation approach for contamination located below the ground water table in waste pits 1 and 2 in Kærgaard Plantation, bench-scale tests with MFR and ASP were conducted. The overall objective of the laboratory tests was to determine the optimal treatment strategy for pilot tests using MFR and ASP and the specific objectives for the laboratory tests were to:

- Estimate the buffering capacity of the soil.
- Determine best suited activation method for persulphate.

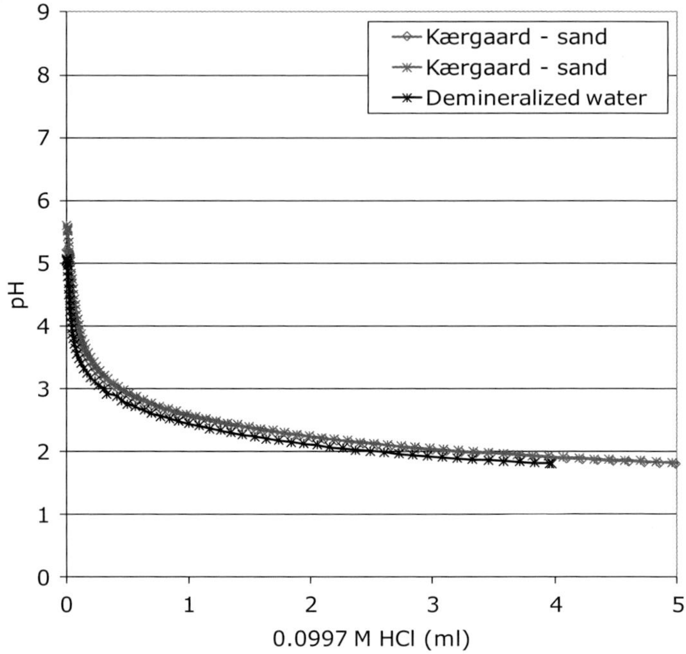

FIGURE 2.10 pH in Kærgaard Plantation soil used for bench and pilot tests as a function of acid added. (For colour version of this figure, the reader is referred to the online version of this book.)

- Determine the stability of the hydrogen peroxide and persulphate in the presence of subsurface solids.
- Determine oxidation effectiveness towards chlorinated solvents, BTEX (benzene, toluene, ethylbenzene and xylenes), barbiturates, sulphonamides, aniline and pyridine.
- Evaluate the potential for mobilising cyanide, mercury and other metals during treatment; see Section 2.5.
- Evaluate potential by-product formation.

Because of a very low buffering capacity of the soil and ground water from Kærgaard Plantation, pH dropped to below 2 in all tests and all the way down to 1, when high dosages of persulphate were used. As described in Section 2.2.3.1 there are different methods to activate persulphate. To determine the optimal methods for the Kærgaard Plantation matrix, four different methods were tested and the results from closed batch reactors based on PCE degradation in soil slurries are presented in Figure 2.11. All activation methods worked, but surprisingly, the natural activation (no activation) was more effective (75% PCE removal) compared to the alkaline activation, which was only able to oxidise 49% of

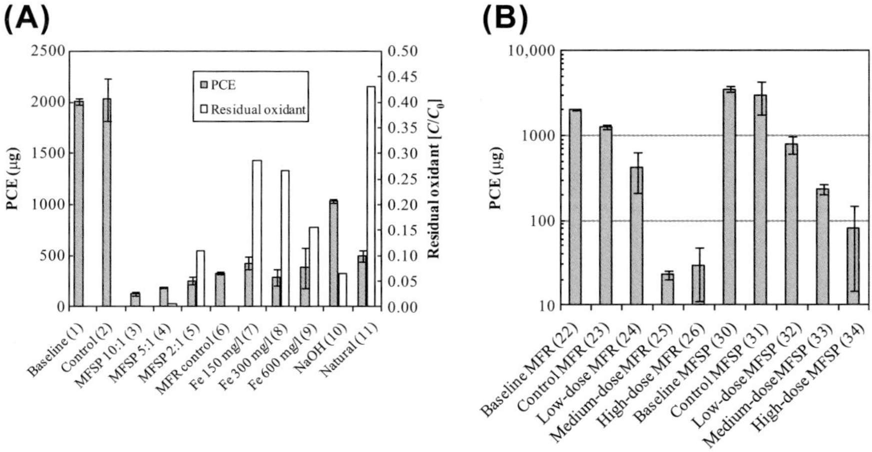

FIGURE 2.11 (A) Total amount of PCE in the reactors (soil, water, and air) after 9 days of reaction compared to baseline (day 0) and control (day 9) as a function of activation method. Residual oxidant present in the reactors after 9 days is also presented. (B) Total amount of PCE in the reactors after reaction with different concentrations of MFR-only and MFSP.

the PCE. The most effective removal of contaminants was obtained when MFR (H_2O_2 and chelated iron) was added to activate persulphate; this will be referred to as MFSP. MFR will activate sodium persulphate via two mechanisms – heat activation resulting from the exothermic nature of the MFR (modified Fenton's activated sodium persulfate (MFSP)), and chelated iron activation. Furthermore, the effervescence via decomposition of hydrogen peroxide to oxygen and the heat generated by the MFR are expected to promote the rates of desorption of adsorbed contaminants and the dissolution of NAPL contaminants, increasing their availability for concurrent aqueous phase oxidation and subsequent polishing treatment by persulphate (Teel et al., 2009). Based on these considerations and given the aggressive treatment goals at the Kærgaard Plantation, it was decided to continue with MFSP for further experiments to combine the strengths of MFR and ASP.

After having determined the activation technique for persulphate, the MFR-only and MFSP could be compared by setting up soil slurry batch reactors and treating these with different dosages of MFR-only (5, 15, 25 g H_2O_2/kg soil) and MFSP (5, 15, 25 g $Na_2S_2O_8$/kg soil + 1, 6, 11 g H_2O_2/kg soil for activation). Some of the results are presented in Figure 2.11 and MFR-only and MFSP were proved to be effective with 98% and 89% degradation of PCE, respectively.

The degradation of several other contaminants present in the water phase by MFSP was also investigated with two different dosages of persulphate (21 and 83 mM $S_2O_8^{2-}$) activated by MFR (63 and 315 mM H_2O_2

and 1.8 mM chelated Fe). The results presented in Figure 2.11 clearly demonstrate that MFSP was able to degrade the wide range of contaminants present in the ground water. For all the organic nitrogen compounds, sulphonamides and barbiturates, the removal was greater than 99.9% or below detection limit (0.02–0.1 μg/l) with the high oxidant dose except for sulfanilic acid, which was reduced by 96%. The petroleum hydrocarbons were also degraded, but long carbon chains (C_{25}–C_{40}) seem more difficult to degrade compared to the smaller molecules (C_5–C_{10}). BTEX were also easily degraded and the high dosage also degraded the chlorinated ethenes completely. Similar tests with ground water from Kærgaard Plantation for MFR-only were conducted in 2006 by ISOTEC (In-Situ Oxidative Technologies Inc.) and similar results were obtained (Geosyntec Consultants, 2006; Durant et al., 2007) and there was no doubt that the MFR chemistry, when used to activate persulphate played an important and probably superior role compared to persulphate. No other studies concerning ISCO for remediation of these pharmaceutical exists in the peer reviewed literature (Figure 2.12).

The presented data and other test results showed that MFR and MFSP were comparable in many ways. Overall, both techniques and the dosages applied have been proved to be effective towards the contaminants in the Kærgaard Plantation. Reaction rates, heat and gas production were controlled and acceptable in all experiments.

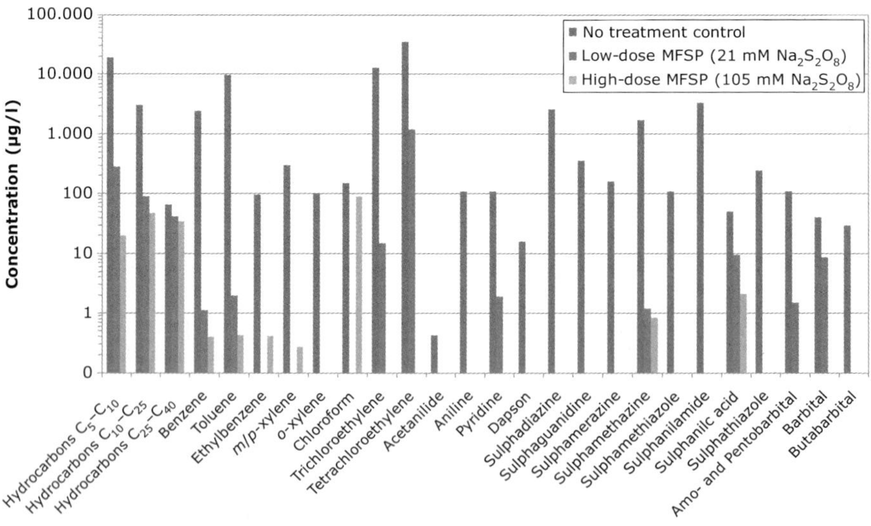

FIGURE 2.12 **Example of degradation of different contaminants in the water phase with different dosages of MFSP.** (For colour version of this figure, the reader is referred to the online version of this book.)

However, some differences were demonstrated. Compared to MFR-only, the MFSP reaction (oxidant consumption) was faster, resulted in lower pH and produced less gas. As a result of the different activation mechanisms included in the MFSP, this technique will potentially be able to treat a wider range of contaminants. On the other hand, MFSP will be more expensive to apply in the field and leave high concentrations of residual sulphate and acidity in the ground water after treatment. On this basis, it was decided to test MFR-only in a field pilot study, where ERD was to be included as a polishing step.

2.4.2. Kærgaard Plantation – MFR Followed by ERD (Pilot)

The objective of the proposed remedial action for Kærgaard Plantation was to significantly reduce the concentrations of COCs, including DNAPL PCE, in soil and ground water. More specifically, the objectives of the pilot test were to demonstrate the effectiveness of MFR followed by ERD as a polishing step towards the COCs within a test cell and to collect design data for the full-scale remedial design. The following were to be clarified by this pilot test:

- Quantify the effect of the treatment.
- Determine the ROI.
- Determine the consumption of oxidant and catalyst.
- Determine the stability of oxidant.
- Evaluate different injection techniques.
- Determine the number of injection events to achieve cleanup goals.
- Determine how heat and gas formation can be controlled and managed.
- Evaluate extent of potential metal and cyanide mobilisation out of the treatment zone.
- Determine optimal injection pressure and flow.
- Determine if clean soil above the treatment area is recontaminated.
- Determine the effectiveness of MFR for destruction of BTEX and chlorinated ethenes.
- Demonstrate whether reducing and pH-neutral conditions can be reestablished after MFR injection and, if so, determine the time frame to achieve these conditions.
- Demonstrate whether it is possible to achieve reductive dechlorination of PCE to ethene after MFR injection.

The pilot test setup is illustrated in Figure 2.13. A total of five nested injection wells screened separately through Layer 2.1 and 2.2 were utilised for the pilot test. Monitoring results from only two central wells (M3 and M4) and a well (M105) placed 2.5 m down-gradient from the

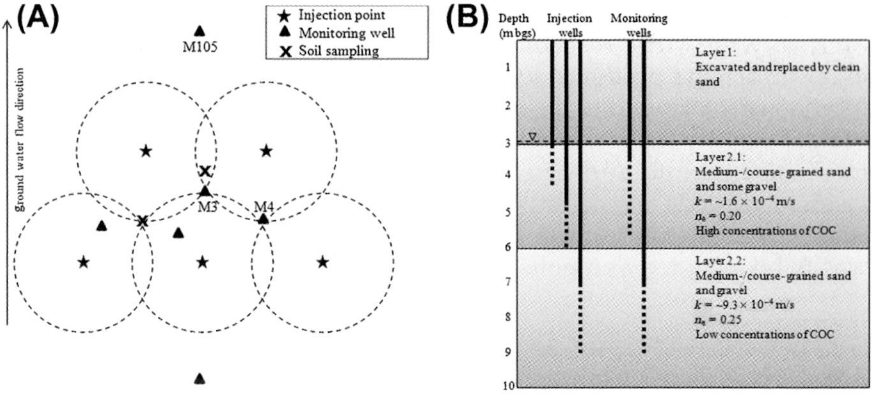

FIGURE 2.13 (A) Site plan showing the test setup with assumed 1.5 m ROI of injection wells. (B) A conceptual model showing the geology, hydrogeology and degree of contamination as a function of depth as well as construction of wells and the two treatment zones, layers 2.1 and 2.2. The figures are not to scale.

nearest injection well are included in this thesis. To accomplish the objectives, a number of injections were made in the two layers, see Table 2.6.

Photos from the pilot test showing the setup of mixing tanks, pumps, well heads etc. are seen in Figure 2.14.

Before, during and after MFR injections, water samples were collected from all the monitoring wells. The changes in pH and ORP over time in a few selected wells is presented in Figure 2.15 and as indicated by the

TABLE 2.6 Injected Volumes of Reagents in the Two Treatment Zones

Day	Layer 2.1 (3–6 m bgs)	Layer 2.2 (7–9 m bgs)
0	MFR: 5550 l	MFR: 6450 l
56	MFR: 7200 l	MFR: 4800 l
104	MFR: 9000 l	–
146	MFR: 9000 l	–
157	–	Buffer: 62 kg $NaHCO_3$ and 100 kg Na_2CO_3 diluted to 10,000 l
161	–	Donor: 675 l Newman zone diluted to 18,000 l
225	–	Buffer + donor: 18 kg $NaHCO_3$ and 450 l Newman zone diluted to 18,000 l Bacteria: 5 l KB1®

All volumes were distributed equally in the injection wells.

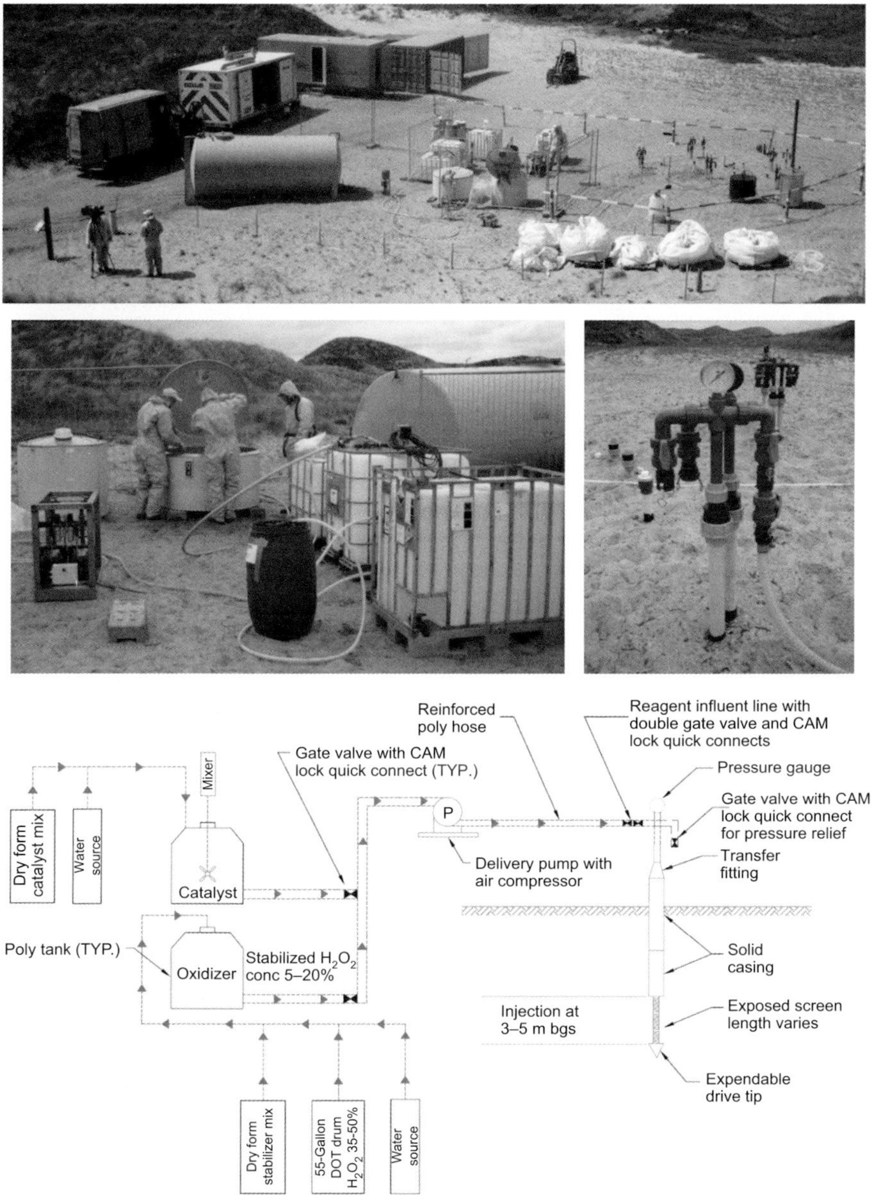

FIGURE 2.14 Photos from the test site showing the setup, the mixing and injection equipment, an injection well with pressure gauge, and a process block diagram supplied by ISOTEC. (For colour version of this figure, the reader is referred to the online version of this book.)

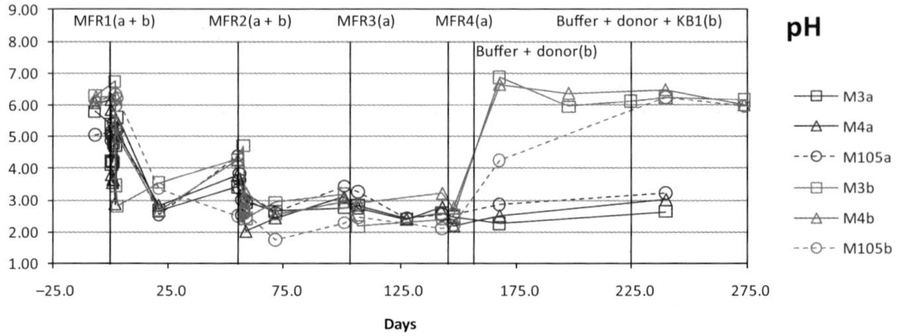

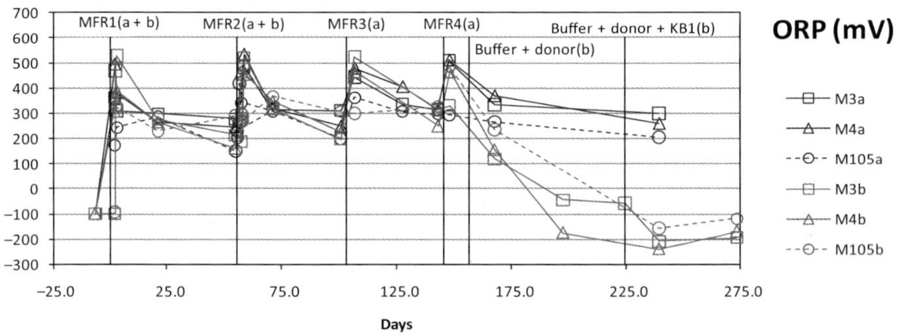

FIGURE 2.15 pH and ORP are selected monitoring wells screened in layer 2.1 (red) and layer 2.2 (blue) in the treatment area (full line) and 2.5 m down-gradient the nearest injection well (stippled line). (For interpretation of the references to colour in this figure legend, the reader is referred to the online version of this book.)

laboratory tests, pH dropped significantly as a result of the MFR injections due to low buffering capacity and presumably formation of carboxylic acids as no strong acids were utilised as part of MFR. Both pH and ORP were expected to rebound quickly because of a relative high influx of reduced ground water with a neutral pH, but this did not happen and therefore it was necessary to neutralise the pH by the addition of buffer and create reducing conditions by the addition of an electron donor into layer 2.2, where ERD was to be applied. These injections quickly resulted in conditions favourable for ERD (ORP < -50 mV, pH 6.5–8.2).

Several other parameters were monitored in the wells: temperature, water level, dissolved iron, dissolved oxygen, conductivity, chloride, and formation of hydroxyl radicals (ability of ground water samples to decolourise methylene blue (Satoh et al., 2007)). All parameters indicated an ROI of minimum 1.5 m (data not shown). Soil samples collected before and after treatment, see Figure 2.16, were used to document if contaminant levels were reduced. Based on average results, the concentration of residual free phase PCE in layer 2.1 was reduced from 3100 to 1600 mg/kg

FIGURE 2.16 Examples of intact soil cores collected from the treatment area before MFR injections (A) and after two injection events (B, with depth intervals in m bgs). The cores also illustrates the layered geology. (For colour version of this figure, the reader is referred to the online version of this book.)

with a hydrogen peroxide dosage of 14 g/kg soil and in layer 2.2 the concentration of PCE was reduced from 160 to 1 mg/kg with a dosage of 8 g/kg soil. In layer 2.1, the concentration decreased in the treatment area corresponding to destruction of about 290 kg out of 600 kg PCE. Also, the colour of the soil changed after treatment from completely black to a white/yellow colour, which presumable was due to oxidation of reduced sulphur compounds.

High contaminant concentrations in the soil were removed in layer 2.2 and ERD was applied as described in Table 2.6. Even though pH and redox conditions had been extreme for a month, concentration of the degradation product *cis*-dichloroethylene was observed to increase to almost a factor of 40 above baseline concentrations before MFR only 1 month after the first donor addition (Newman zone, emulsified soybean oil) indicating biological activity. Monitoring 15 days after bioaugmentation with KB1® culture showed concentration of 52,000 μg/l (factor 430 above baseline concentration) and small increases in vinyl chloride and ethene. The KB1®culture is a natural microbial culture used to introduce *Dehalococcoides* bacteria, since these are the only known organisms capable of dechlorinating PCE all the way to nontoxic ethene.

2.4.3. Implications and Perspectives

ISCO with MFR or ASP has only been tested at a limited number of sites in Denmark. This case study has provided more information and hopefully more acceptance of the technique as a safe (if handled correct) and effective technique for source zone treatment of both fuel oil and chlorinated solvents including NAPL contamination and most other organic contaminants. It was also shown that MFR and MFSP both were able to treat a wide range of different compounds in ground water including different pharmaceuticals. Until recently, ISCO with aggressive methods (MFR and ASP) was not very accepted in Denmark and the rest of Scandinavia compared to, e.g. North America, but it seems as if this is slowly changing and ASP is currently being tested by others too at a few sites in Denmark. In the future, aggressive ISCO technologies will be a supplement or alternative to traditional techniques and can be used as part of a sequential treatment strategy. Especially, ISCO followed by bioremediation will most likely be an attractive cost-effective combination. There are still many unknown processes involved and reactions taking place in situ when these technologies are applied, but these years a lot of effort is put into understanding these details and thereby be able to control the reactions and predict the applicability of the techniques. However, it is still essential to conduct bench and pilot studies to collect important design data for full-scale applications.

2.5. METAL MOBILISATION

The coexisting occurrence of organic contaminants and heavy metals in soil is a significant factor complicating remediation of contaminated sites, as the enhanced desorption of organic contaminants by the MFR and ASP reactions may lead to mobilisation of heavy metals (Monahan et al., 2005; Teel and Watts, 2008). Mobilisation can potentially be caused by changes in pH, changes in redox conditions, degradation of NOM or the use of chelating agents.

In soils and ground water the effects of pH during ISCO can be complicated since most of these systems possess a strong buffering capacity that will maintain pH neutral and result in iron precipitation and ineffective catalysis. As an alternative to lowering pH in the entire treatment area to 2–4 with, e.g. sulphuric acid, which was the traditional custom practice, many other methods have been used to activate hydrogen peroxide and persulphate. These include soluble iron (Fe^{2+} and Fe^{3+}) (Murphy et al., 1989; Tyre et al., 1991; Watts and Dilly, 1996), iron minerals (Tyre et al., 1991; Ravikumar and Gurol, 1994; Khan and Watts, 1996; Valentine and Wang, 1998; Huang et al., 2001; Teel et al., 2001; Kwan and Voelker, 2003), and chelated iron (Sun and Pignatello, 1992; Pignatello and Baehr, 1994; Luzzatto et al., 1995; Pignatello and Day, 1996; Wang and Brusseau, 1998; Seol and Javandel, 2008).

The main advantages of using chelates as catalysts are that the process can be conducted at neutral pH and that chelates may travel farther in the subsurface compared to soluble iron (Watts and Teel, 2005). However, the dosage of oxidants has to be increased, since MFR or ASP will also oxidise the chelating agents that are typically organic compounds. Studies have shown that the widespread use of chelating agents for consumer products and industrial processes combined with poor biodegradability of these chemicals have led to accumulation of chelating agents in the environment and ethylenediaminetetraacetic acid (EDTA) is now among the highest concentrated anthropogenic compounds in European surface waters (Schmidt et al., 2004). Chelating agents may enhance the mobility and transport of heavy metals once released into the environment and high concentrations are also able to remobilise metals out of soils (Yuan and VanBriesen, 2006).

In this section, results of an investigation of the mobilisation of different metals from the Kærgaard Plantation soil treated with MFR and MFSP is presented in order to illustrate typical observations. This was done by:

- Batch tests with MFR.
- Column studies with MFR and MFSP.
- Pilot tests with MFR.

Before the results are presented, the existing knowledge on metal mobilisation and the chemistry of chelating agents will be reviewed in order to understand the mechanisms causing the mobilisation.

2.5.1. Existing Knowledge

Only few studies have investigated metal mobilisation during treatment of contaminated soils with MFR. In the first study in 2005 (Monahan et al., 2005), the mobilisation of Cd, Cu, Pb, Ni and Zn during MFR treatment was investigated employing kaolinite as a model sorbent. Using soluble Fe(III) as catalyst at pH 3 resulted in displacement of Cd, Cu and Zn and using Fe(III)–NTA (Nitrilotriacetic acid) as catalyst at pH 6 resulted in displacement of Cd, Cu, Pb and Zn. Also, the concentrations of the metals released into the aqueous phase were higher in the Fe(III)–NTA-catalysed systems. The observed differences were suggested to have been mediated by transient oxygen species, such as superoxide, which may reduce sorbed metals and facilitate their release. In the second study in 2008 (Teel and Watts, 2008) the release of Pb was investigated during treatment of soil with MFR using the same catalysts as in the first study, only the pH was changed from 6 to 7. At pH 3, Pb release was minimal and at pH 7 no Pb was released. When hydroxyl radicals were scavenged from the reactions by adding isopropanol, leaving superoxide as the primary reactant, Pb desorption increased significantly. Again, it was suggested that superoxide played a role in Pb desorption.

Chelating agents can participate in different processes, where the most important related to potential metal mobilisation during ISCO are metal exchange in solution, adsorption to mineral surfaces, dissolution of minerals, remobilisation of adsorbed metals, degradation by oxidation, and redox reactions involving the coordinated metal ion. Equilibrium of metal exchange is determined by stability constants. However, it is very important to include knowledge on kinetics, since ligands containing multiple bonding groups such as EDTA are often kinetically hindered in their reaction (Nowack, 2002). Also, especially the reactions involving Fe(III)–EDTA that have shown to be a catalyst for MFR (Luzzatto et al., 1995) are very slow (Margerum et al., 1978), and half-life of Fe(III)–EDTA, when exchanging with Zn^{2+}, has been shown to be about 20 days in natural waters (Xue et al., 1995):

$$\mathrm{Fe(III) - EDTA^{-} + Zn^{2+} \rightarrow Zn(II) - EDTA^{2-} + Fe^{3+}_{(precipitates)}} \quad (2.62)$$

Sorption of metals changes significantly when chelating agents are used. Most interesting is the fact that metals in general show increasing adsorption at increasing pH, while the opposite is true for chelated metals. This means that chelating agents are able to immobilise metals through adsorption, not at high pH, but especially at low pH. However, this is mostly true for low concentrations of chelating agents, whereas high concentrations used for technical purposes such as ISCO in many cases will decrease adsorption by forming dissolved complexes (Nowack, 2002).

Remobilisation of adsorbed metals can be accomplished by chelating agents, which is a process that has been used for years for characterising bioavailability (Beckett, 1988), for extraction of heavy metals from contaminated soils (Ehsan et al., 2007; Hong et al., 2002; Lo and Yang, 1999; Norvell, 1984; Qiu et al., 2010; Wang et al., 2007; Wuana et al., 2010) and to enhance phytoextraction of heavy metals from contaminated soils (Firdaus-E-Bareen and Tahira, 2010; Kim and Lee, 2010; Muhammad et al., 2009). These fields of application indirectly suggest that using chelating agents for ISCO may increase metal mobilisation. In addition, the chelating agents can enhance dissolution of minerals via ligand exchange reactions (Stumm, 1997).

2.5.2. Batch Study

As shown earlier, the buffering capacity of the Kærgaard Plantation soil was very low and the ISCO treatment resulted in significant pH changes. To learn more about potential metal mobilisation a batch study with soil from the Kærgaard Plantation was performed to investigate the influence of chelating agents, pH and oxidation/reduction processes on metal mobilisation. Ten different reactors with soil and water were used, all in duplicate. The study included a control reactor with water (#1); a reactor with H_2O_2 (#2); a reactor with H_2O_2 and Fe^{2+} (#3); reactors with chelated Fe^{2+} using EDTA, citrate and pyrophosphate (#4–6); reactors with H_2O_2 and chelated Fe^{2+} (#7–9) and finally, a control reactor where pH was lowered with sulphuric acid (#10). pH was significantly affected in all reactors containing hydrogen peroxide, as shown in Figure 2.17.

The low pH could increase mobilisation, but as shown in Figure 2.18, metal mobilisation in the pH control reactor (#10) was only slightly above the concentrations in the water control (#1) for most of the investigated metals. It is also seen that the chelated iron catalyst (#4–6) itself does not increase the mobilisation. A high degree of mobilisation was observed in all reactors containing hydrogen peroxide (#2,3,7–9) and for As, Cu, Pb and Zn up to 38–129% of the metals analysed in the soil were mobilised to the aqueous phase. Results above 100% mobilisation are caused by a standard deviation in the analysis results of the soil samples of up to 30% because of the heterogeneous nature of the samples. Less than 0.5% of Ni was mobilised. One suggestion is that the increased mobilisation observed, when hydrogen peroxide is present, is caused by superoxide that has the potential to rapidly reduce and mobilise transition metals (Zafiriou et al., 1998) which is in accordance with suggestions from

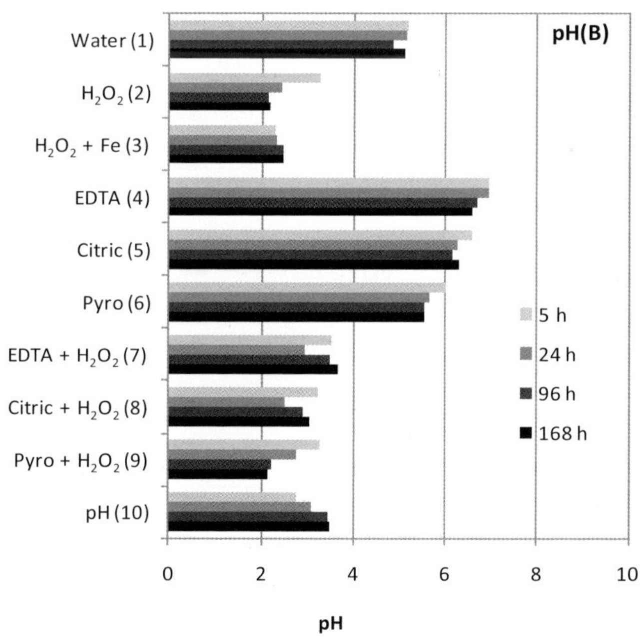

FIGURE 2.17 **Average pH in the reactors after 5, 24, 96 and 168 h** (Bennedsen et al., 2011).

Monahan et al. (2005). Another suggestion for the increased mobilisation in the presence of H_2O_2 is the fact that oxalic acid and other small carboxylic acids are produced when aromatic rings are oxidised by hydroxyl radicals (Kavitha and Palanivelu, 2005). Aromatic structures are abundant in soil organic matter. These acids can strongly chelate some metal cations. The two suggestions along with the acidity might explain the observed mobilisation.

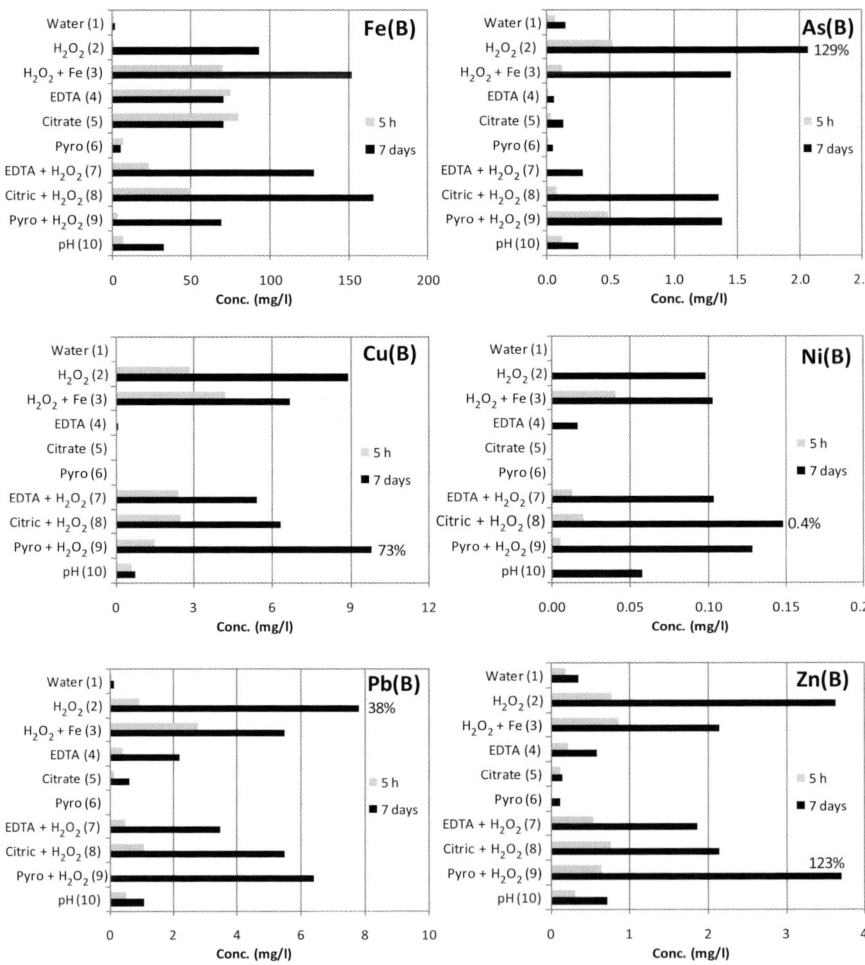

FIGURE 2.18 **Average metal concentrations in reactors after 5 h and 7 days of reaction.** Note the different scales. The highest degree of mobilisation compared to a maximal theoretical mobilisation (corresponding to 100% mobilisation of the metals analysed in the soil) is showed for each metal except Fe (Bennedsen et al., 2012).

2.5.3. Column Study

Results from the batch test were further evaluated by column studies. Columns were filled with 700 g Kærgaard soil and demineralised water was flushed through the columns until stable conditions (pH and metals) were measured in the outlet. At this point flushing with MFSP or MFR-only was started at a flow rate of 7–10 ml/h. Chelated iron catalysts from ISOTEC were used for the tests. Concentrations of As, Cu, Pb, Zn, Cr and Hg measured in the outlet over time are presented in Figure 2.19. The results clearly showed that mobilisation occurred immediately after addition of MFSP or MFR-only to the columns. Concentrations reached levels at or above the concentrations observed in the batch study and then started to decrease as a function of time (or pore volumes exchanged). Hg was included in the analysis, but could not be observed at any time. However, Cr was mobilised in concentrations up to about 120 and 75 μg/l for MFSP and MFR-only, respectively. Cr and to a lesser extent Pb, were released over a longer period of time compared to the other metals indicating that As, Cu and Zn were more easily mobilised compared to Cr and Pb. This also corresponds to the results from the batch tests in Section 2.5.2, where a larger parts of As, Cu and Zn were mobilised compared to Pb. As(III) and AS(V) form oxyanion complexes and thereby differ from the other metals investigated. pH was about 1.8 and 2.3, respectively, in the outlets of the MFSP and MFR-only reactor during oxidant addition.

2.5.4. Field Tests

Based on the results obtained in the laboratory batch and column studies, monitoring in the field during MFR pilot and full-scale

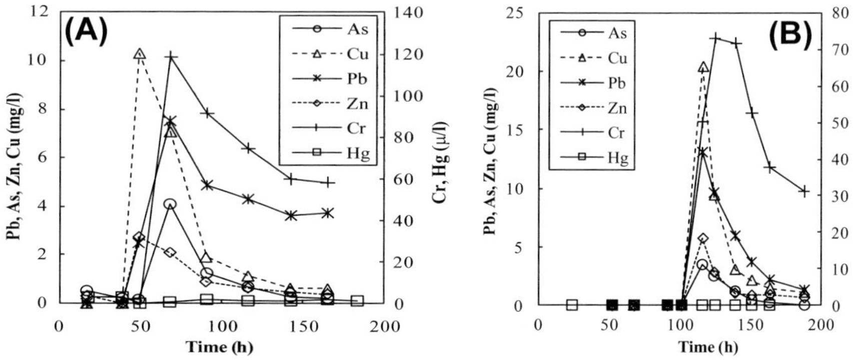

FIGURE 2.19 **Concentrations of metals in the outlet of the columns treated with MFSP (A) and MFR-only (B).**

TABLE 2.7 Results in Milligrams Per Liter from the Pilot Study at Kærgaard Plantation

Time (days)	0	29	79	135	175	247
CENTRE OF TREATMENT AREA						
pH	5.78	2.66	2.65	2.41	2.26	2.64
Pb	2.20	39.00	52.20	9.47	8.60	2.40
Cu	1.40	16.00	33.70	28.20	12.00	1.60
Ni	0.01	0.35	0.25	0.14	0.14	0.02
Fe	16.00	63.00	60.00	130.00	64.60	26.00
Mn	0.66	1.80	1.70	3.90	1.38	0.40
2.5 M DOWN-GRADIENT THE NEAREST INJECTION WELL						
pH	5.04	2.54	2.61	2.43	2.86	3.44
Pb	0.03	14.00	n.a.	9.50	5.31	n.a.
Cu	0.01	21.00	n.a.	28.10	11.80	n.a.
Ni	0.01	0.21	n.a.	0.14	0.19	n.a.
Fe	12.00	71.00	n.a.	160.00	62.80	n.a.
Mn	0.70	2.60	n.a.	4.40	1.37	n.a.

A total of four MFR injection events were performed after 7, 63, 111 and 153 days as described in Section 2.4.2.
n.a.: Not analysed.

treatments was performed. At the Kærgaard Plantation MFR pilot project, metals were monitored in the treatment area, 2.5 m down-gradient the nearest injection well and 25 m down-gradient. The results are presented in Table 2.7, from where it is seen that metals were mobilised in concentrations even higher than those measured in the bench-scale tests. Again, it was especially Cu and Pb measured in concentrations up to 52 and 34 mg/l in the treatment area and 14 and 28 mg/l 2.5 m down-gradient the injection wells. No metals above background concentrations were observed 25 m down-gradient. Except the very high concentrations of metals, these trends are typical for ISCO, where metals are released in the treatment area, but precipitate when transported outside the reactive zone.

2.5.5. Implications and Perspectives

The results presented from the case study in this section demonstrate that metals are mobilised during MFR (and MFSP) applications, but precipitate again when pH and redox conditions are neutralised outside the treatment area. Lowering of pH will cause some mobilisation, but the most significant release is caused by reactions based on hydrogen peroxide, and according to the suggestions in the peer-reviewed literature, reductants such as superoxide play an important role. But formation of, e.g. oxalic acid as a result of oxidation reactions, might also cause increased mobilisation. Even though chelating agents are used to mobilise metals, it seems as if they do not significantly impact the release, when used to chelate Fe catalyst in ISCO. Only few studies have dealt with this issue and more research within this area is needed in order to fully understand the processes and thereby minimise mobilisation in future ISCO applications. However, the problem might not be very significant, especially not in well-buffered soils.

2.6. SCAVENGING OF HYDROXYL RADICALS

Radicals produced in oxidations systems react rapidly with other chemical species in solution at near diffusion-controlled rates (Buxton et al., 1988; Haag and Yao, 1992; Neta et al., 1977). Besides contaminant degradation these reactions include nonproductive reactions with other compounds and radicals resulting in reduced reaction efficiency and effectiveness, often referred to as scavenging reactions. In this section the processes of radical scavenging are reviewed with focus on carbonates and chloride. As an example, high concentrations up to 4600 mg/l (130 mM) of chloride and 6900 mg/l (113 mM) of hydrogen carbonate have been observed in the source zone in the Kærgaard Plantation. Also very high concentration of chloride can occur when DNAPL chlorinated solvents are treated with ISCO.

The scavenging from chloride and carbonates in different ASP systems was investigated too in the case study by:

- Batch tests with heat-activated persulphate.
- Batch test with alkaline-activated persulphate.
- Batch test with iron-activated persulphate.

Specific objectives are described in the following sections summarising the results obtained. Before the results are presented, the existing knowledge on radical scavenging will be reviewed in order to understand the mechanisms causing the scavenging.

2.6.1. Existing Knowledge

Scavenger ions are one of the most common factors limiting the oxidation efficiency. Especially chloride and hydrogen carbonate/carbonate have the potential to impact pathways, kinetics, and efficiency of oxidation reactions both as radical scavengers and as metal complexing agents (Valentine and Wang, 1998; De Laat et al., 2004; Lipczynskakochany et al., 1995; Beltran et al., 1998). Also, phosphate is often added to stabilise hydrogen peroxide, since phosphate is believed to adsorb to and inactivate catalytic and scavenging sites on, e.g. iron oxides. Through addition of phosphate, oxidation efficiency can be enhanced and oxidant decomposition reduced significantly, which is assumed to be through reduction of surface scavenging sites in the soil (Miller, 1995). Carbonates and chloride can act in the same way and form aqueous complexes or solid precipitates with metals or mineral surfaces (Valentine and Wang, 1998; De Laat et al., 2004). In the peer-reviewed literature, experiments with MFR showed that these complexation processes are decreasing the rate of contaminant degradation, oxidant decomposition and the overall oxidation efficiency (De Laat et al., 2004; Lipczynskakochany et al., 1995; Beltran et al., 1998; Sung and Morgan, 1980), whereas others have shown that the overall efficiency was not decreased, even though the reaction rates for degradation of contaminants were decreased (Valentine and Wang, 1998). Carbonates have also been shown to reduce rates of contaminant degradation as well as oxidant decomposition rates for ASP (Huang et al., 2002; Peyton, 1993; Liang et al., 2006). The adverse impact of the carbonates in ASP increases with increasing pH (Peyton, 1993; Liang et al., 2006).

Besides the complexation processes, chloride and carbonates can also form radicals on their own via reaction with the reactive species in MFR and ASP systems (Buxton et al., 1988; Huie et al., 1991; Liang et al., 2006; Yu and Barker, 2003; Zuo et al., 1999; Kiwi et al., 2000):

$$OH^{\bullet} + HCO_3^- \rightarrow OH^- + HCO_3^{\bullet} \tag{2.63}$$

$$OH^{\bullet} + CO_3^{2-} \rightarrow OH^- + CO_3^{\bullet-} \tag{2.64}$$

$$OH^{\bullet} + Cl^- \rightleftarrows ClHO^{\bullet-} \tag{2.65}$$

$$SO_4^{\bullet-} + CO_3^{2-} \rightarrow SO_4^{2-} + CO_3^{\bullet-} \tag{2.66}$$

$$SO_4^{\bullet-} + HCO_3^- \rightarrow SO_4^{2-} + HCO_3^{\bullet} \tag{2.67}$$

$$SO_4^{\bullet-} + Cl^- \rightleftarrows SO_4^{2-} + Cl^{\bullet} \tag{2.68}$$

The formed carbonate and chloride radicals can react further (Kiwi et al., 2000). The chloride radicals react as follows:

$$ClHO^{\bullet-} + H^+ \rightleftarrows Cl^{\bullet} + H_2O \tag{2.69}$$

$$Cl^{\bullet} + Cl^{-} \rightleftarrows Cl_2^{\bullet -} \tag{2.70}$$

$$2Cl_2^{\bullet -} \rightarrow 2Cl^{-} + Cl_2 \tag{2.71}$$

$$Cl_2^{\bullet -} + H_2O \rightarrow ClHO^{\bullet -} + H^{+} + Cl^{-} \tag{2.72}$$

The carbonate ions and radicals react as follows (Zuo et al., 1999):

$$HCO_3^{\bullet} \rightleftarrows CO_3^{\bullet -} + H^{+} \quad ; pK_a = 9.5 \tag{2.73}$$

$$HCO_3^{-} \rightleftarrows CO_3^{2-} + H^{+} \quad ; pK_a = 10.45 \tag{2.74}$$

$$H_2CO_3 \rightleftarrows HCO_3^{-} + H^{+} \quad ; pK_a = 6.33 \tag{2.75}$$

However, the reactivity of these radicals with organic contaminants or participation in chain propagation reactions are not widely investigated or well understood. It is traditionally assumed that they have an overall negative impact on the performance (Huang et al., 2002; Liang et al., 2006). The carbonate radical ($CO_3^{\bullet -}$) generated by Reaction (2.62) was reported to yield reduction potential ($CO_3^{\bullet -}$, H^{+}/HCO_3^{-}) of 1.63 V at pH 8.4 (Zuo et al., 1999) and it has been shown to be strong enough to degrade PAHs (Umschlag and Herrmann, 1999). Based on these findings, carbonate radicals might play an important role and should not strictly be considered a scavenger. It has also been suggested that the chloride radicals can participate in propagation reactions with oxidant and thereby not be entirely unproductive (Yu et al., 2004) and that the dichloride radical catalyses perhydroxyl radical formation from hydrogen peroxide and the propagation reaction (Yu and Barker, 2003).

The concentration of chloride needed to decrease oxidation efficiency of heat-activated persulphate at 130 °C was shown to be more than 20 mM Cl^{-} (Aiken, 1992), and other studies with persulphate at 20 °C at neutral pH showed that only chloride levels above 200 mM affected the degradation of TCE and that hydrogen carbonate up to 9.2 mM had no influence (Liang et al., 2006).

High concentrations (>100 mM) of chloride have also been reported to result in formation of halogenated by-products (Aiken, 1992) and even hexachloroethane has been observed to be formed during treatment of PCE as a result of the reactive chloride species (Waldemer et al., 2007).

It is worth noticing that because hydrogen peroxide is generally present at high concentrations in MFR systems and has a moderate rate constant for reaction with hydroxyl radicals (2.7×10^{7}/M s (Buxton et al., 1988), Reaction (2.14)), hydrogen peroxide is itself a primary source of inefficiency in MFR systems (Huling et al., 1998). Also, chelating agents used for keeping catalyst in solution represents a sink for the radicals (Voelker and Sulzberger, 1996) and the catalyst itself might consume significant amounts of radicals (Liang et al., 2004a).

2.6.2. Experimental Results with *p*-Nitrosodimethylaniline

The organic dye *p*-nitrosodimethylaniline (RNO) was used as a model compound, since it is easily detectable by measuring absorbance at 440 nm, thermally stable, stable at high pH (Bennedsen et al., 2012), and bleached by oxidation with hydroxyl radicals (Baxendale and Khan, 1969), sulphate radicals (Simonsen et al., 2010) and other strong oxidants. It is not reactive with singlet oxygen, superoxide anions or other peroxygens (Comninellis, 1994; Holt et al., 2003; Kraljic and Trumbore, 1965).

The effects of chloride and carbonates were investigated for heat-activated persulphate at 65 °C at neutral pH, for alkaline-activated persulphate at pH 12.4 and for chelated iron-activated persulphate at neutral pH. Initial concentrations of persulphate and RNO were 5 and 0.05 mM, respectively. Some results are presented in Figure 2.20.

Heat-activated persulphate was not affected by hydrogen carbonate based on RNO bleaching, but chloride significantly affected the bleaching by enhancing the initial reaction rate. However, when the concentration of RNO was decreased during the run, high (200−400 mM) and low (1 mM) concentrations of chloride seemed to inhibit the bleaching reaction compared to the control. Only the 10 mM chloride enhanced the bleaching over the entire 60-min reaction period. Less than 5% of the persulphate was consumed after 60 min.

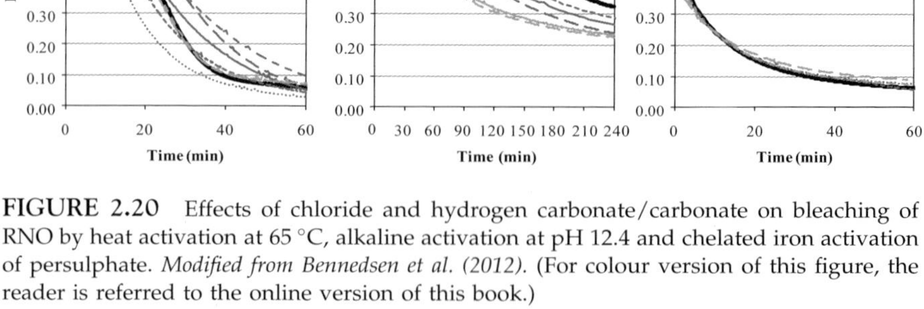

FIGURE 2.20 Effects of chloride and hydrogen carbonate/carbonate on bleaching of RNO by heat activation at 65 °C, alkaline activation at pH 12.4 and chelated iron activation of persulphate. *Modified from Bennedsen et al. (2012).* (For colour version of this figure, the reader is referred to the online version of this book.)

Alkaline-activated persulphate at pH 12.4 was affected by the presence of carbonate that enhanced the bleaching rate significantly. Chloride, in all of the tested concentrations 1–400 mM, also enhanced the bleaching of RNO compared to control reactors containing no chloride. Less than 5% of the persulphate was consumed after 240 min.

Chelated iron-activated persulphate was not affected significantly by chloride or hydrogen carbonate. This activation mechanism caused the fastest bleaching of RNO with a very rapid initial reaction rate resulting in 50% bleaching within seconds. This was also the only activation mechanism consuming measurable amounts of persulphate since about 25% of the persulphate reacted within the first 2 min and about 30% after 60 min.

Chloride and carbonates are generally assumed to have an overall negative impact on the performance in ISCO because of scavenging of hydroxyl and sulphate radicals, but the results presented in Figure 2.20 indicated that chloride and carbonates may have some positive effects enhancing oxidation reactions. From these results it was not evident if chlorine and carbonate species produced according to Reactions (2.61–2.70) are directly involved in degradation of the target compounds or if they participate in propagation reactions with persulphate causing more sulphate radicals to be produced.

2.6.3. Experimental Results with PCE

A test similar to the RNO test was conducted to investigate PCE destruction by chelated iron-activated persulphate at ambient temperature with molar concentrations of $S_2O_8^{2-}/Fe^{2+}$/citric acid/PCE of 4/2/0.4/0.2 mM. No effects were observed in the RNO tests at neutral pH, but in these PCE tests no phosphate buffer was used and pH was therefore 2.5–2.6 in all tests. The results shown as degraded PCE and the consumed persulphate are presented in Figure 2.21. The conclusion from these results was that increased chloride concentration decreased the rate of oxidation, but the overall oxidation efficiency was not affected.

2.6.4. Implications and Perspectives

Reviewing the various processes involved in scavenging of radicals related to MFR and ASP reveals a system with numerous parameters to be optimised involving oxidant concentration, catalyst concentrations, use of chelating agents, naturally occurring ions, NOM, unknown processes etc. In well-controlled laboratory systems it might be possible to optimise these processes to minimise scavenging, but most studies only focused on optimising a single parameter, e.g. oxidant/catalyst ratio. When taking all aspects into account it is always a question of compromises since

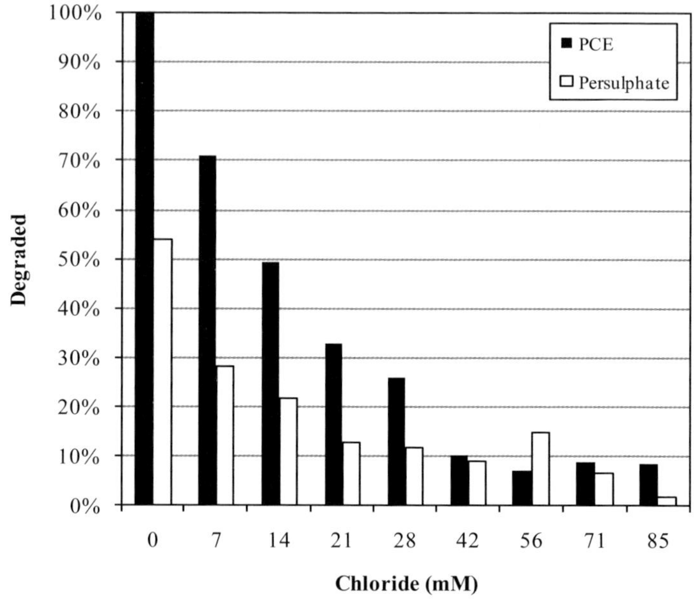

FIGURE 2.21 **Degradation of PCE by chelated iron-activated persulphate as a function of chloride concentration after 60 min of reaction.** Consumption of persulphate is also shown (Bennedsen et al., 2012).

changing one parameter to enhance a process in most cases will interfere with another process. When applying MFR or ASP for full-scale treatments more unknowns are introduced and optimisation becomes more difficult and relies to a great extent on experience from similar sites and rules of thumb.

As demonstrated from the results presented in this section, the effects of scavenger ions are not fully understood or consistent for different activation mechanisms. Further research may improve the understanding of the role of chloride and carbonate radicals in oxidation systems, and potentially allow for optimisation of process chemistry that would minimise their negative impacts and maybe even enhance their potential positive aspects.

2.7. CONCLUSION

Today soil and ground water contamination is a widespread and challenging problem threatening ground water resources throughout the world. As with all other technologies, ISCO with activated peroxygens is not a universal solution solving all the problems, but the technologies are important supplements to the toolbox of remediation technologies available. In particular these techniques are suited for permeable geology, complex contaminant mixtures, toxic contaminants or high concentrations of contaminants including NAPLs.

Some of the most significant specific conclusions drawn from the research presented in this chapter are:

- MFR can degrade >1000 mg/kg levels of fuel oil and PCE in soils.
- MFR and ASP reactions can be controlled. The oxidants are relatively stable in the presence of contaminated soil and ground water and observed temperature increases during MFR treatments were only 5–10 °C.
- MFSP can efficiently degrade aniline, pyridine, sulphonamides, barbiturates, petroleum hydrocarbons, BTEX, PCE, and TCE in the aqueous phase.
- ERD can be used as a second step in a treatment train in source zones after aggressive ISCO with MFR. Redox and pH conditions need to be changed by injecting buffer and electron donor.
- MFR and MFSP can cause mobilisation of metals (Cu, Pb, Fe, As, Mn, Ni, Zn and Mg) in the treatment area, but these are most likely precipitated when leaving the treatment area. Cyanide and Hg were not mobilised. Results indicate that low pH and chelating agents cause some of the mobilisation, but the most significant mobilisation is caused by redox reaction.
- Chloride and carbonates should not be strictly considered as scavengers in ASP systems. For some contaminants they may enhance oxidation rates and for other they may decrease reaction rates, but not affect the overall oxidation efficiency.

2.8. PERSPECTIVES AND FUTURE RESEARCH

Recent advantages in the use of activated peroxygens show that the involved processes are complex and still not fully understood. The techniques are widely used in full scale to treat contaminated soil and ground water, but to increase the chances for successful ISCO application with activated peroxygen more knowledge is needed. Further research within the following areas will help improve the understanding of ISCO with activated peroxygens:

- In general more published well-documented field applications are needed to support and validate the observation from small scale laboratory tests.

- MFR has been investigated in most details and hence more research on persulphate is needed, e.g. knowledge on activation mechanisms and interactions with subsurface soils.
- Optimisation of coupled ISCO and bioremediation and other combinations.
- Mobilisation of metals are still very often questioned when ISCO is applied, and more studies investigating the transport and fate of the mobilised metals are needed. Most research suggests that metals are precipitated outside of the treatment area, but well-documented field tests could elucidate this issue.
- Is it possible to enhance the positive aspects of chloride and carbonates in ISCO? Which contaminants can be degraded by carbonate or chloride radicals? Can the ions catalyse the propagation reactions?
- Development of decision and design tools that incorporate process conditions specific to both the site and the contaminants to enable site-specific engineering of ISCO.

The application of activated peroxygens in combination with other technologies as treatment trains (Sutton et al., 2011) of direct coupling of, e.g. surfactants/co-solvents with ISCO, to enhance NAPL removal (Dugan et al., 2010) attracts increasingly attention and holds promise for enhanced treatment of contaminated soil and ground water in the future when the involved complex processes are more understood and optimised.

ABBREVIATIONS

AOP Advanced oxidation processes
ASP Activated sodium persulphate
Bgs Below ground surface
BTEX Benzene, toluene, ethylbenzene and xylenes
CHP Catalysed hydrogen peroxide propagation
***cis*-DCE** *cis*-1,2-Dichloroethylene
COC Contaminants of concern
DO Dissolved oxygen
DNAPL Dense nonaqueous phase liquids
ERD Enhanced reductive dechlorination
ISCO In situ chemical oxidation
m bgs Metres below ground surface
MFR Modified Fenton's reagent
MFSP Modified Fenton's activated sodium persulphate
MNA Monitored natural attenuation
NAPL Nonaqueous phase liquid
NOM Natural organic matter
ORP Oxidation/reduction potential
PAH Polycyclic aromatic hydrocarbons

PCE Perchloroethylene
PCB Polychlorinated biphenyl
RNO *p*-Nitrosodimethylaniline
ROI Radius of influence
TCE Trichloroethylene
TOD Total oxidant demand
TPH Total petroleum hydrocarbons
VC Vinyl chloride

References

Aiken, G.R., 1992. Chloride interference in the analysis of dissolved organic carbon by the wet oxidation method. Environ. Sci. Technol. 26 (12), 2435–2439.

Anbar, M., Meyerste, D., Neta, P., 1966. Reactivity of aromatic compounds toward hydroxyl radicals. J. Phys. Chem. 70 (8), 2660.

Anipsitakis, G.P., Dionysiou, D.D., 2004. Radical generation by the interaction of transition metals with common oxidants. Environ. Sci. Technol. 38 (13), 3705–3712.

Banerjee, M., Konar, R.S., 1984. Polymerization of acrylonitrile initiated by $K_2S_2O_8$-Fe(II) redox system. J. Polym. Sci. A Polym. Chem. 22 (5), 1193–1195.

Baxendale, J.H., Khan, A.A., 1969. The pulse radiolysis of p-nitrosodimethylaniline in aqueous solution. Int. J. Radiat. Phys. Chem. 1 (1), 11–24.

Beckett, P.H.T., 1988. The use of extractants in studies on trace metals in soils sewage sludges, and sludge-treated soils. Adv. Soil Sci. 9, 143.

Beltran, F.J., Gonzalez, M., Rivas, F.J., Alvarez, P., 1998. Fenton reagent advanced oxidation of polynuclear aromatic hydrocarbons in water. Water Air Soil Pollut. 105 (3–4), 685–700.

Bennedsen, L.R., Krischker, A., Jørgensen, T.H., Søgaard, E., 2011. Mobilization of metals during treatment of contaminated soils by modified Fenton's reagent using different chelating agents. J. Hazard. Mater. 199–200, 128–134.

Bennedsen, L.R., Muff, J., Søgaard, E.G., 2012. Influence of chloride and carbonates on the reactivity of activated persulfate. Chemosphere 86 (11), 1092–1097.

Berlin, A.A., 1986. Kinetics of radical-chain decomposition of persulfate in aqueous-solutions of organic-compounds. Kinet. Catal. 27 (1), 34–39.

Bittkau, A., Geyer, R., Bhatt, M., Schlosser, D., 2004. Enhancement of the biodegradability of aromatic groundwater contaminants. Toxicology 205 (3), 201–210.

Blanchard, C., 2010. Applying chemical oxidation technologies at NAPL-impacted sites to facilitate risk-based closure. Remediation Spring, 77–94.

Block, P.A., Brown, R.A., Robinson, D., 2004. Novel activation technologies for sodium persulfate in situ chemical oxidation. In: Proceedings of the Fourth International Conference on the Remediation of Chlorinated and Recalcitrant Compounds Monterrey, California.

Brown, R.A., Robinson, D., Block, P.A., 2004. Simultaneous reduction and oxidation: combining sodium persulfate with zero valent iron.

Buxton, G.V., Greenstock, C.L., Helman, W.P., Ross, A.B., 1988. Critical review of rate constants for reactions of hydrated electrons, hydrogen atoms and hydroxyl radicals ($^{\bullet}OH/^{\bullet}O^-$ in aqueous solution). J. Phys. Chem. Ref. Data 17 (2), 513–886.

Chang, P.B.L., Young, T.M., 2000. Kinetics of methyl tert-butyl ether degradation and by-product formation during UV/hydrogen peroxide water treatment. Water Res. 34 (8), 2233–2240.

Chapelle, F.H., Bradley, P.M., Casey, C.C., 2005. Behavior of a chlorinated ethene plume following source-area treatment with Fenton's reagent. Ground Water Monit. Remediat. 25 (2), 131–141.

Comninellis, C., 1994. Electrocatalysis in the electrochemical conversion/combustion of organic pollutants for waste-water treatment. Electrochim. Acta 39 (11–12), 1857–1862.

COWI, 2006a. Delrapport nr. 1. Jord – Og Grundvandsforurening Ved Kærgård Plantage, Arbejdsgruppen Vedrørende Kærgård Plantage.

COWI, 2006b. Samlerapport – Undersøgelser 2005/2006, Arbejdsgruppen vedrørende Kærgård Plantage.

Crimi, M.L., 2010. Lessons learned from ISCO applications: towards an improved systematic approach for site specific engineering, ATV ISCO theme day, Vingstedcentret, Denmark.

Crimi, M.L., Siegrist, R.L., 2005. Factors affecting effectiveness and efficiency of DNAPL destruction using potassium permanganate and catalyzed hydrogen peroxide. J. Environ. Eng. 131 (12), 1724–1732.

Crimi, M.L., Taylor, J., 2007. Experimental evaluation of catalyzed hydrogen peroxide and sodium persulfate for destruction of BTEX contaminants. Soil & Sediment Contam. 16 (1), 29–45.

Dahmani, M.A., Huang, K., Hoag, G.E., 2006. Sodium persulfate oxidation for the remediation of chlorinated solvents (USEPA superfund innovative technology evaluation program). Water Air Soil Pollut. 6, 127–141.

De Laat, J., Le, G.T., Legube, B., 2004. A comparative study of the effects of chloride, sulfate and nitrate ions on the rates of decomposition of H_2O_2 and organic compounds by Fe(II)/H_2O_2 and Fe(III). Chemosphere 55 (5), 715–723.

Dugan, P.J., Siegrist, R.L., Crimi, M.L., 2010. Coupling surfactants/cosolvents with oxidants for enhanced DNAPL removal: a review. Rem. J. 20 (3), 27–49.

Durant, N.D., Mackinnon, L., Cox, E., Dworatzek, S., Jørgensen, T.H., 2007. Chemical oxidation of sulfa drugs, barbiturates and chlorinated solvents in groundwater: a bench test evaluation. In: The 23rd Annual International Conference on Soils, Sediments and Water, University of Massachusetts October 15, 2007, p. 34.

Ehsan, S., Prasher, S.O., Marshall, W.D., 2007. Simultaneous mobilization of heavy metals and polychlorinated biphenyl (PCB) compounds from soil with cyclodextrin and EDTA in admixture. Chemosphere 68 (1), 150–158.

EPA, 2004. Evaluates Microbial Responses to Groundwater Remediation Technologies. EPA/54-N-04-002.

EPA, 2003. The DNAPL Remediation Challenge: Is There a Case for Source Depletion? U.S. Environmental Protection Agency, Cincinnati, Ohio.

European Environmental Agency, 2007. Progress in Management of Contaminated Sites (CSI 015). Assessment Published Aug 2007. Homepage of European Environmental Agency. Available from: http://www.eea.europa.eu/data-and-maps/indicators/progress-in-management-of-contaminated-sites/progress-in-management-of-contaminated-1 (21.02.11.).

Fenton, H.J.H., 1894. Oxidation of tartaric acid in presence of iron. J. Chem. Soc. 65, 899.

Firdaus-E-Bareen, Tahira, S.A., 2010. Efficiency of seven different cultivated plant species for phytoextraction of toxic metals from tannery effluent contaminated soil using EDTA. Soil Sediment Contam. 19 (2), 160–173.

Forsey, S.P., 2004. In situ Chemical Oxidation of Creosote/Coal Tar Residuals: Experimental and Numerical Investigation.

Friis, A.K., Heimann, A.C., Jakobsen, R., Albrechtsen, H.J., Cox, E., Bjerg, P.L., 2007. Temperature dependence of anaerobic TCE-dechlorination in a highly enriched dehalococcoides-containing culture. Water Res. 41 (2), 355–364.

Furman, O.S., Teel, A.L., Watts, R.J., 2010. Mechanism of base activation of persulfate. Environ. Sci. Technol. 44 (16), 6423–6428.

Furman, O., Laine, D.F., Blumenfeld, A., Teel, A.L., Shimizu, K., Cheng, I.F., Watts, R.J., 2009. Enhanced reactivity of superoxide in water-solid matrices. Environ. Sci. Technol. 43 (5), 1528–1533.

Geosyntec Consultants, 2006. Derapport 4, Laboratory Treatability tests of chemical oxidation and bioremediation technologies for groundwater remediation, Kærgård Plantage (Grindstedværkets gruber), Arbejdsgruppen vedrørende Kærgård Plantage.

Haag, W.R., Yao, C.C.D., 1992. Rate constants for reaction of hydroxyl radicals with several drinking-water contaminants. Environ. Sci. Technol. 26 (5), 1005–1013.

Haber, F., Weiss, J., 1934. The catalytic decomposition of hydrogen peroxide by iron salts. Proc. R. Soc. Lond. A Math. Phys. Sci. 147 (861), 332–351.

Haselow, J.S., Siegrist, R.L., Crimi, M., Jarosch, T., 2003. Estimating the total oxidant demand for in situ chemical oxidation design. Remediation Autumn, 5–16.

Holt, K.B., Forryan, C., Compton, R.G., Foord, J.S., Marken, F., 2003. Anodic activity of boron-doped diamond electrodes in bleaching processes: effects of ultrasound and surface states. J. Chem. 27 (4), 698–703.

Hong, P.K.A., Li, C., Jiang, W.M., Chen, T.C., Peters, R.W., 2002. Chelating agents for extraction of heavy metals from soil. In: Tedder, D.W., Pohland, F.G. (Eds.), Emerging Technologies in Hazardous Waste Management 8. Springer, US, Boston, pp. 9–20.

House, D.A., 1962. Kinetics and mechanism of oxidations by peroxydisulfate. Chem. Rev. 62 (3), 185–200.

Huang, H.H., Lu, M.C., Chen, J.N., 2001. Catalytic decomposition of hydrogen peroxide and 2-chlorophenol with iron oxides. Water Res. 35 (9), 2291–2299.

Huang, K.C., Couttenye, R.A., Hoag, G.E., 2002. Kinetics of heat-assisted persulfate oxidation of methyl tert-butyl ether (MTBE). Chemosphere 49 (4), 413–420.

Huang, K.C., Zhao, Z.Q., Hoag, G.E., Dahmani, A., Block, P.A., 2005. Degradation of volatile organic compounds with thermally activated persulfate oxidation. Chemosphere 61 (4), 551–560.

Huie, R.E., Clifton, C.L., Neta, P., 1991. Electron-transfer reaction-rates and equilibria of the carbonate and sulfate radical-anions. Radiat. Phys. Chem. 38 (5), 477–481.

Huling, S.G., Arnold, R.G., Sierka, R.A., Miller, M.R., 1998. Measurement of hydroxyl radical activity in a soil slurry using the spin trap alpha-(4-pyridyl-1-oxide)-N-tert-butylnitrone. Environ. Sci. Technol. 32 (21), 3436–3441.

Huling, S.G., Pivetz, B.E., 2006. Engineering Issue: In-situ Chemical Oxidation EPA/600/R-06/072. U.S. Environmental Protection Agency.

ITRC, 2005. Technical and Regulatory Guidance for In situ Chemical Oxidation of Contaminated Soil and Groundwater. Interstate Technology & Regulatory Council.

Jerome, K.M., Riha, B., Looney, B.B., 1997. Final report for demonstration of in situ oxidation of DNAPL using the Geo-Clense Technology. WSRC-TR-97-00283.

Johnson, R.L., Pankow, J.F., 1992. Dissolution of dense chlorinated solvents into groundwater 2. Source functions for pools of solvent. Environ. Sci. Technol. 26 (5), 896–901.

Kang, N., Hua, I., Rao, P.S.C., 2006. Enhanced Fenton's destruction of non-aqueous phase perchloroethylene in soil systems. Chemosphere 63 (10), 1685–1698.

Kao, C.M., Chen, S.C., Wang, J.Y., Chen, Y.L., Lee, S.Z., 2003. Remediation of PCE-contaminated aquifer by an in situ two-layer biobarrier: laboratory batch and column studies. Water Res. 37 (1), 27–38.

Kavanaugh, M.C., Rao, P.S.C., Abriola, L., Cherry, J., Destouni, G., Falta, R., 2003. The DNAPL remediation challenge: source removal or long term management. EPA/600/R-03/143. Report of an Expert Panel to the U.S. EPA National Risk Management Laboratory and Technology Innovation Office.

Kavitha, V., Palanivelu, K., 2005. Destruction of cresols by Fenton oxidation process. Water Res. 39 (13), 3062–3072.

Khan, A.J., Watts, R.J., 1996. Mineral-catalyzed peroxidation of tetra chloroethylene. Water Air Soil Pollut. 88 (3–4), 247–260.

Killian, P.F., Bruell, C.J., Liang, C., Marley, M.C., 2007. Iron (II) activated persulfate oxidation of MGP contaminated soil. Soil Sediment Contam. 16 (6), 523–537.

Kim, S., Lee, I., 2010. Comparison of the ability of organic acids and EDTA to enhance the phytoextraction of metals from a multi-metal contaminated soil. Bull. Environ. Contam. Toxicol. 84 (2), 255–259.

Kiwi, J., Lopez, A., Nadtochenko, V., 2000. Mechanism and kinetics of the OH-radical intervention during Fenton oxidation in the presence of a significant amount of radical scavenger (Cl^-). Environ. Sci. Technol. 34 (11), 2162–2168.

Kolthoff, I.M., Miller, I.K., 1951. The chemistry of persulfate. I. The kinetics and mechanism of the decomposition of the persulfate ion in aqueous medium. J. Am. Chem. Soc. 73 (7), 3055–3059.

Kong, S.H., Watts, R.J., Choi, J.H., 1998. Treatment of petroleum-contaminated soils using iron mineral catalyzed hydrogen peroxide. Chemosphere 37 (8), 1473–1482.

Kosegi, J.M., Minsker, B.S., Dougherty, D.E., 2000. Feasibility study of thermal in situ bioremediation. J. Environ. Eng. 126 (7), 601–610.

Kovalick Jr, W.W., 2008. Review of characterization and remediation technologies for NAPL's in groundwater. In: Annable, M.D., Teodorescu, M., Hlavinek, P., Diels, L. (Eds.), Methods and Techniques for Cleaning-up Contaminated Sites. Springer, Netherlands, pp. 165–175.

Kraljic, I., Trumbore, C.N., 1965. P-Nitrosodimethylaniline as an OH radical scavenger in radiation chemistry. J. Am. Chem. Soc. 87 (12), 2547–2550.

Krembs, F.J., Siegrist, R.L., Crimi, M.L., Furrer, R.F., Petri, B.G., 2010. ISCO for groundwater remediation: analysis of field applications and performance. Ground Water Monit. Remediat. 30 (4), 42–53.

Kwan, W.P., Voelker, B.M., 2003. Rates of hydroxyl radical generation and organic compound oxidation in mineral-catalyzed Fenton-like systems. Environ. Sci. Technol. 37 (6), 1150–1158.

Lee, B.D., Hosomi, M., 2001. Clean-up of benz(a) anthracene-contaminated soils by fenton oxidation-microbial treatment. Kagaku Kogaku Ronbunshu 27 (3), 411–415.

Lemke, L.D., Abriola, L.M., Goovaerts, P., 2004. Dense nonaqueous phase liquid (DNAPL) source zone characterization: influence of hydraulic property correlation on predictions of DNAPL infiltration and entrapment. Water Resour. Res. 40 (1), W01511.

Liang, C.J., Bruell, C.J., Marley, M.C., Sperry, K.L., 2004a. Persulfate oxidation for in situ remediation of TCE. I. Activated by ferrous ion with and without a persulfate-thiosulfate redox couple. Chemosphere 55 (9), 1213–1223.

Liang, C.J., Bruell, C.J., Marley, M.C., Sperry, K.L., 2004b. Persulfate oxidation for in situ remediation of TCE. II. Activated by chelated ferrous ion. Chemosphere 55 (9), 1225–1233.

Liang, C.J., Bruell, C.J., Marley, M.C., Sperry, K.L., 2003. Thermally activated persulfate oxidation of trichloroethylene (TCE) and 1,1,1-trichloroethane (TCA) in aqueous systems and soil slurries. Soil Sediment Contam. 12 (2), 207–228.

Liang, C.J., Wang, Z.S., Mohanty, N., 2006. Influences of carbonate and chloride ions on persulfate oxidation of trichloroethylene at 20 degrees C. Sci. Total Environ. 370 (2–3), 271–277.

Liang, C., Wang, Z., Bruell, C.J., 2007. Influence of pH on persulfate oxidation of TCE at ambient temperatures. Chemosphere 66 (1), 106–113.

Lipczynskakochany, E., Sprah, G., Harms, S., 1995. Influence of some groundwater and surface waters constituents on the degradation of 4-chlorophenol by the Fenton reaction. Chemosphere 30 (1), 9–20.

Lo, I.M.C., Yang, X.Y., 1999. EDTA extraction of heavy metals from different soil fractions and synthetic soils. Water Air Soil Pollut. 109 (1–4), 219–236.

Luzzatto, E., Cohen, H., Stockheim, C., Wieghardt, K., Meyerstein, D., 1995. Reactions of low-valent transition-metal complexes with hydrogen-peroxide – are they "Fenton-like" or not? 4. The case of Fe(II)L, L=edta hedta and tcma. Free Radical Res. 23 (5), 453–463.

Margerum, D.W., Cayley, G.R., Weatherburn, D.C., Pagenkopf, G.K., 1978. Kinetics and mechanism of complex formation and ligand exchange. In: Martell, A. (Ed.), Coordination Chemistry, vol. 2, ACS Monograph 174, vol. 2. ed., pp. 1–220.

Mckillop, A., Sanderson, W.R., 1995. Sodium perborate and sodium percarbonate – cheap, safe and versatile oxidizing-agents for organic-synthesis. Tetrahedron 51 (22), 6145–6166.

Miller, C.M., 1995. Hydrogen peroxide decomposition and contaminant degradation in the presence of sandy aquifer materials.

Miller, C.M., Valentine, R.L., Roehl, M.E., Alvarez, P.J.J., 1996. Chemical and microbiological assessment of pendimethalin-contaminated soil after treatment with Fenton's reagent. Water Res. 30 (11), 2579–2586.

Monahan, M.J., Teel, A.L., Watts, R.J., 2005. Displacement of five metals sorbed on kaolinite during treatment with modified Fenton's reagent. Water Res. 39 (13), 2955–2963.

Morris, B.L., Lawrence, A.R.L., Chilton, P.J.C., Adams, B., Calow, R.C., Klinck, B.A., 2003. Groundwater and Its Susceptibility to Degradation: A Global Assessment of the Problem and Options for Management. United Nations Environment Programme, Nairobi, Kenya.

Muff, J., Bennedsen, L.R., Søgaard, E.G., 2011. Study of electrochemical bleaching of p-nitrosodimethylaniline and its role as hydroxyl radical probe compound. J. Appl. Electrochem. 41 (5), 599–607. http://dx.doi.org/10.1007/s10800-011-0268-1.

Muhammad, D., Chen, F., Zhao, J., Zhang, G., Wu, F., 2009. Comparison of EDTA- and citric acid-enhanced phytoextraction of heavy metals in artificially metal contaminated soil by *Typha angustifolia*. Int. J. Phytoremediation 11 (6), 558–574.

Murphy, A.P., Boegli, W.J., Price, M.K., Moody, C.D., 1989. A Fenton-like reaction to neutralize formaldehyde waste solutions. Environ. Sci. Technol. 23 (2), 166–169.

Nadarajah, N., Van Hamme, J., Pannu, J., Singh, A., Ward, O., 2002. Enhanced transformation of polycyclic aromatic hydrocarbons using a combined Fenton's reagent, microbial treatment and surfactants. Applied Microbiol. Biotechnol. 59 (4–5), 540–544.

Neta, P., Madhavan, V., Zemel, H., Fessenden, R.W., 1977. Rate constants and mechanism of reaction of sulfate radical anion with aromatic-compounds. J. Am. Chem. Soc. 99 (1), 163–164.

NIRAS, 2009. Redegørelse over anvendte kemikalier på Grindstedværket og deres potentielle trussel i forhold til miljøet, Region of Southern Denmark, Region Syddanmark.

Norman, R.O.C., 1979. Application of E.S.R. spectroscopy to kinetics and mechanism in organic chemistry. Chem. Soc. Rev. 1, 1–27.

Norvell, W.A., 1984. Comparison of chelating-agents as extractants for metals in diverse soil materials. Soil Sci. Soc. Am. J. 48 (6), 1285–1292.

Nowack, B., 2002. Environmental chemistry of aminopolycarboxylate chelating agents. Environ. Sci. Technol. 36 (19), 4009–4016.

Ocampa, A.M., 2009. Persulfate Activation by Organic Compounds. Washington State University.

Padmaja, S., Alfassi, Z.B., Neta, P., Huie, R.E., 1993. Rate constants for reactions of SO_4 – radicals in acetonitrile. Int. J. Chem. Kinet. 25 (3), 193–198.

Peyton, G.R., 1993. The free-radical chemistry of persulfate-based total organic-carbon analyzers. Mar. Chem. 41 (1–3), 91–103.

Pignatello, J.J., Baehr, K., 1994. Ferric complexes as catalysts for Fenton degradation of 2,4-D and metolachlor in soil. J. Environ. Qual. 23 (2), 365–370.

Pignatello, J.J., Day, M., 1996. Mineralization of methyl parathion insecticide in soil by hydrogen peroxide activated with iron(III)-NTA or -HEIDA complexes. Hazard. Waste Hazard. Mater. 13 (2), 237–244.

Pignatello, J.J., Oliveros, E., MacKay, A., 2006. Advanced oxidation processes for organic contaminant destruction based on the Fenton reaction and related chemistry. Crit. Rev. Environ. Sci. Technol. 36 (1), 1–84.

Qiu, R., Zou, Z., Zhao, Z., Zhang, W., Zhang, T., Dong, H., Wei, X., 2010. Removal of trace and major metals by soil washing with Na(2)EDTA and oxalate. J. Soils Sediments 10 (1), 45–53.

Ravikumar, J.X., Gurol, M.D., 1994. Chemical oxidation of chlorinated organics by hydrogen-peroxide in the presence of sand. Environ. Sci. Technol. 28 (3), 394–400.

Robinson, D., Brown, R.A., Dablow, J., Rowland, K., 2004. Chemical oxidation of MGP residuals and dicyclopentadiene at a former MGP site. In: The Fourth International Conference on Remediation of Chlorinated and Recalcitrant Compounds, Monterey, California, Paper 2A-03.

Rowland, M.A., Brubaker, G.R., Kohler, K., Westray, M., Morris, D., 2001. Effects of potassium permanganate oxidation on subsurface microbial activity. Anaerobic Degrad. Chlorinated Solvents 6 (7), 1–12.

Sahl, J., Munakata-Marr, J., 2006. The effects of in situ chemical oxidation on microbiological processes: a review. Remediat. J. 16 (3), 57–70.

Satoh, A.Y., Trosko, J.E., Masten, S.J., 2007. Methylene blue dye test for rapid qualitative detection of hydroxyl radicals formed in a Fenton's reaction aqueous solution. Environ. Sci. Technol. 41 (8), 2881–2887.

Schmelling, D.C., Gray, K.A., Kamat, P.V., 1998. Radiation-induced reactions of 2,4,6-trinitrotoluene in aqueous solution. Environ. Sci. Technol. 32 (7), 971–974.

Schmidt, C.K., Fleig, M., Sacher, F., Brauch, H.E., 2004. Occurrence of aminopolycarboxylates in the aquatic environment of Germany. Environ. Pollut. 131 (1), 107–124.

Schnarr, M., Truax, C., Farquhar, G., Hood, E., Gonullu, T., Stickney, B., 1998. Laboratory and controlled field experiments using potassium permanganate to remediate trichloroethylene and perchloroethylene DNAPLs in porous media. J. Contam. Hydrol. 29 (3), 205–224.

Seol, Y., Javandel, I., 2008. Citric acid-modified Fenton's reaction for the oxidation of chlorinated ethylenes in soil solution systems. Chemosphere 72 (4), 537–542.

Siegrist, R.L., Crimi, M., Munakata-Marr, J., Illangasekare, T.L.K., Cuyk, S.V., Dugan, P., Heiderscheidt, J., Jackson, S., Petri, B., Sahl, J., Seitz, S., 2006. Reaction and Transport Processes Controlling In situ Chemical Oxidation of DNAPLs. SERDP Project CU-1290. Colorado School of Mines.

Simonsen, M.E., Muff, J., Bennedsen, L.R., Kowalski, K.P., Sogaard, E.G., 2010. Photocatalytic bleaching of p-nitrosodimethylaniline and a comparison to the performance of other AOP technologies. J. Photochem. Photobiol. A 216 (2–3), 244–249.

Smith, B.A., Teel, A.L., Watts, R.J., 2004. Identification of the reactive oxygen species responsible for carbon tetrachloride degradation in modified Fenton's systems. Environ. Sci. Technol. 38 (20), 5465–5469.

Smith, B.A., Teel, A.L., Watts, R.J., 2006. Mechanism for the destruction of carbon tetrachloride and chloroform DNAPLs by modified Fenton's reagent. J. Contam. Hydrol. 85 (3–4), 229–246.

Stroo, H.F., Unger, M., Ward, C.H., Kavanaugh, M.C., Vogel, C., Leeson, A., Marqusee, J.A., Smith, B.P., 2003. Remediating chlorinated solvent source zones. Environ. Sci. Technol. 37 (11), 224A–230A.

Struse, A.M., Siegrist, R.L., Dawson, H.E., Urynowicz, M.A., 2002. Diffusive transport of permanganate during in situ oxidation. J. Environ. Eng. 128 (4), 327–334, 10.1061/(ASCE)0733-9372(2002)128:4(327).

Stumm, W., 1997. Reactivity at the mineral-water interface: dissolution and inhibition. Colloids Surf., A 120 (1–3), 143–166.

Sun, Y.F., Pignatello, J.J., 1992. Chemical treatment of pesticide wastes – evaluation of Fe(III) chelates for catalytic hydrogen-peroxide oxidation of 2,4-D at circumneutral pH. J. Agri. Food Chem. 40 (2), 322–327.

Sung, W., Morgan, J.J., 1980. Kinetics and product of ferrous iron oxygenation in aqueous systems. Environ. Sci. Technol. 14 (5), 561–568.

Sutton, N.B., Grotenhuis, J.T.C., Langenhoff, A.A.M., Rijnaarts, H.H.M., 2011. Efforts to improve coupled in situ chemical oxidation with bioremediation: a review of optimization strategies. J. Soils Sediments 11 (1), 129–140.

Teel, A.L., Warberg, C.R., Atkinson, D.A., Watts, R.J., 2001. Comparison of mineral and soluble iron Fenton's catalysts for the treatment of trichloroethylene. Water Res. 35 (4), 977–984.

Teel, A.L., Cutler, L.M., Watts, R.J., 2009. Effect of sorption on contaminant oxidation in activated persulfate systems. J. Environ. Sci. Health A Tox. Hazard. Subst. Environ. Eng. 44 (11), 1098–1103.

Teel, A.L., Finn, D.D., Schmidt, J.T., Cutler, L.M., Watts, R.J., 2007. Rates of trace mineral-catalysed decomposition of hydrogen peroxide. J. Environ. Eng. 133 (8), 853–858.

Teel, A.L., Watts, R.J., 2008. Fate of sorbed lead during treatment of contaminated soils by catalyzed H_2O_2 propagations (modified Fenton's reagent). Soil Sediment Contam. 17 (6), 654–664.

Tsitonaki, A., Petri, B., Crimi, M., Mosbaek, H., Siegrist, R.L., Bjerg, P.L., 2010. In situ chemical oxidation of contaminated soil and groundwater using persulfate: a review. Crit. Rev. Environ. Sci. Technol. 40 (1), 55–91.

Tsitonaki, A., Smets, B.F., Bjerg, P.L., 2008. Effects of heat-activated persulfate oxidation on soil microorganisms. Water Res. 42 (4–5), 1013–1022.

Tyre, B.W., Watts, R.J., Miller, G.C., 1991. Treatment of 4 biorefractory contaminants in soils using catalyzed hydrogen-peroxide. J. Environ. Qual. 20 (4), 832–838.

Umschlag, T., Herrmann, H., 1999. The carbonate radical (HCO_3 center dot/CO_3-center dot) as a reactive intermediate in water chemistry: kinetics and modelling. Acta Hydrochim. Hydrobiol. 27 (4), 214–222.

Valentine, R.L., Wang, H.C.A., 1998. Iron oxide surface catalyzed oxidation of quinoline by hydrogen peroxide. J. Environ. Eng. 124 (1), 31–38.

Voelker, B.M., Sulzberger, B., 1996. Effects of fulvic acid on Fe(II) oxidation by hydrogen peroxide. Environ. Sci. Technol. 30 (4), 1106–1114.

Waldemer, R.H., Tratnyek, P.G., Johnson, R.L., Nurmi, J.T., 2007. Oxidation of chlorinated ethenes by heat-activated persulfate: kinetics and products. Environ. Sci. Technol. 41 (3), 1010–1015.

Walling, C., 1975. Fentons reagent revisited. Acc. Chem. Res. 8 (4), 125–131.

Wang, G., Koopmans, G.F., Song, J., Temminghoff, E.J.M., Luo, Y., Zhao, Q., Japenga, J., 2007. Mobilization of heavy metals from contaminated paddy soil by EDDS, EDTA, and elemental sulfur. Environ. Geochem. Health 29 (3), 221–235.

Wang, X.J., Brusseau, M.L., 1998. Effect of pyrophosphate on the dechlorination of tetrachloroethene by the Fenton reaction. Environ. Toxicol. Chem. 17 (9), 1689–1694.

Watts, R.J., Bottenberg, B.C., Hess, T.F., Jensen, M.D., Teel, A.L., 1999a. Role of reductants in the enhanced desorption and transformation of chloroaliphatic compounds by modified Fenton's reactions. Environ. Sci. Technol. 33 (19), 3432–3437.

Watts, R.J., Foget, M.K., Kong, S.H., Teel, A.L., 1999b. Hydrogen peroxide decomposition in model subsurface systems. J. Hazard. Mater. 69 (2), 229–243.

Watts, R.J., Udell, M.D., Kong, S.H., Leung, S.W., 1999c. Fenton-like soil remediation catalyzed by naturally occurring iron minerals. Environ. Eng. Sci. 16 (1), 93–103.

Watts, R.J., Dilly, S.E., 1996. Evaluation of iron catalysts for the Fenton-like remediation of diesel-contaminated soils. J. Hazard. Mater. 51 (1–3), 209–224.

Watts, R.J., Finn, D.D., Cutler, L.M., Schmidt, J.T., Teel, A.L., 2007. Enhanced stability of hydrogen peroxide in the presence of subsurface solids. J. Contam. Hydrol. 91 (3–4), 312–326.

Watts, R.J., Frank, J.L., Teel, A.L., 2006. Improved Understanding of Fenton-like Reactions for the In situ Remediation of Contaminated Groundwater Including Treatment of Sorbed Contaminants and Destruction of DNAPLs. SERDP Project No. CU-1288.

Watts, R.J., Howsawkeng, J., Tee, A.L., 2005a. Destruction of a carbon tetrachloride dense nonaqueous phase liquid by modified Fenton's reagent. J. Environ. Eng. 131 (7), 1114–1119.

Watts, R.J., Sarasa, J., Loge, F.J., Teel, A.L., 2005b. Oxidative and reductive pathways in manganese-catalyzed Fenton's reactions. J. Environ. Eng. 131 (1), 158–164.

Watts, R.J., Stanton, P.C., 1999. Mineralization of sorbed and NAPL-phase hexadecane by catalyzed hydrogen peroxide. Water Res. 33 (6), 1405–1414.

Watts, R.J., Stanton, P.C., Howsawkeng, J., Teel, A.L., 2002. Mineralization of a sorbed polycyclic aromatic hydrocarbon in two soils using catalyzed hydrogen peroxide. Water Res. 36 (17), 4283–4292.

Watts, R.J., Teel, A.L., 2006. Treatment of contaminated soils and groundwater using ISCO. Pract. Period. Hazard. Toxic Radioact. Waste Manage. 10, 2–9.

Watts, R.J., Teel, A.L., 2005. Chemistry of modified Fenton's reagent (catalyzed H_2O_2 propagations-CHP) for in situ soil and groundwater remediation. J. Environ. Eng. 131 (4), 612–622.

Wuana, R.A., Okieimen, F.E., Imborvungu, J.A., 2010. Removal of heavy metals from a contaminated soil using organic chelating acids. Int. J. Environ. Sci. Technol. 7 (3), 485–496.

Xue, H.B., Sigg, L., Kari, F.G., 1995. Speciation of EDTA in natural-waters – exchange kinetics of Fe-EDTA in river water. Environ. Sci. Technol. 29 (1), 59–68.

Yeh, C.K.J., Wu, H.M., Chen, T.C., 2003. Chemical oxidation of chlorinated non-aqueous phase liquid by hydrogen peroxide in natural sand systems. J. Hazard. Mater. 96 (1), 29–51.

Yu, X.Y., Bao, Z.C., Barker, J.R., 2004. Free radical reactions involving Cl-center dot, Cl-2(-center dot), and SO_4-center dot in the 248 nm photolysis of aqueous solutions containing $S_2O_8^{2-}$ and Cl^-. J. Phys. Chem. A 108 (2), 295–308.

Yu, X.Y., Barker, J.R., 2003. Hydrogen peroxide photolysis in acidic aqueous solutions containing chloride ions. I. Chemical mechanism. J. Phys. Chem. A 107 (9), 1313–1324.

Yuan, Z.W., VanBriesen, J.M., 2006. The formation of intermediates in EDTA and NTA biodegradation. Environ. Eng. Sci. 23 (3), 533–544.

Zafiriou, O.C., Voelker, B.M., Sedlak, D.L., 1998. Chemistry of the superoxide radical (O-2(−)) in seawater: reactions with inorganic copper complexes. J. Phys. Chem. A 102 (28), 5693–5700.

Zuo, Z.H., Cai, Z.L., Katsumura, Y., Chitose, N., Muroya, Y., 1999. Reinvestigation of the acid-base equilibrium of the (bi)carbonate radical and pH dependence of its reactivity with inorganic reactants. Radiat. Phys. Chem. 55 (1), 15–23.

CHAPTER

3

Electrochemical Oxidation – A Versatile Technique for Aqueous Organic Contaminant Degradation

Assistant Professor Jens Muff

Section of Chemical Engineering, Department of Biotechnology, Chemistry and Environmental Engineering, Aalborg University Esbjerg, Niels Bohrs Vej 8, DK 6700 Esbjerg, Denmark

3.1. INTRODUCTION

For some decades, a large number of companies have been marketing electrochemical devices for metal ion removal and metal recovery, treatment of liquors containing dissolved chromium, flue-gas desulphurisation, desalination and salt splitting (Panizza and Cerisola, 2009a). However, so far the role of electrochemical technologies for the treatment of organic pollutants has been relatively small. But, nowadays thanks to intensive investigations that have improved the electrocatalytic activity and stability of electrode materials and optimised reactor geometry, electrochemical technologies have reached a promising state of development and can be effectively used for disinfection and purification of water polluted with organic compounds (Comninellis et al., 2008).

The electrochemical oxidation (EO) method for full mineralisation or part conversion of organic pollutants is a technology considered very interesting for the treatment of particularly dilute waste water (chemical oxygen demand (COD) < 5 g/l) (Comninellis et al., 2008) and is in competition with the processes of chemical oxidation with mixtures of strong oxidants, photocatalysis, etc. Some authors do, however, claim EO

Chemistry of Advanced Environmental Purification Processes of Water
http://dx.doi.org/10.1016/B978-0-444-53178-0.00003-1

to be feasible up to COD levels of 30 g/l (Serikawa et al., 2009) or even 100 g/l (Fryda et al., 2003). The main advantage of the EO technology is that chemicals are not used. In fact, only electrical energy is consumed for the oxidation of organic pollutants and electrochemistry is often referred to as an environmentally friendly approach, since the electron itself is considered a 'clean', safe and very effective reagent. The main pros as often stressed in the literature are (Comninellis et al., 2008; Rajeshwar and Ibanez, 1997) as follows:

1. Versatility: EO is generally not selective and can deal with many pollutants and treat from microlitres to millions of litres.
2. Energy efficiency: The electrochemical processes generally require lower temperature with respect to equivalent nonelectrochemical counterparts (e.g. thermal incineration), the potential can be easily controlled and operational parameters can be designed to minimise power loss.
3. Amenability to automation: The electrical parameters used in the process are well suited for data acquisition, process automation and control.

For the above reasons, electrochemistry and EO is a very interesting and promising alternative technology in environmental engineering for the prevention of pollution problems, especially within water treatment, but other fields of application, as, e.g. soil remediation, do exist.

3.2. THE PRINCIPLE AND MECHANISMS IN EO

The underlying principles of the oxidation mechanisms in EO of organics have been subject to comprehensive research within the last decade. These works have resulted in a somehow generally accepted theoretical understanding on how the different anode materials, the composition of the electrolyte and the nature and structure of the organics influence the efficiency of the organic oxidation and evolution in common parameters as COD and total organic carbon (TOC) during electrolysis.

The use of electricity in order to study chemical reactions was one of the cornerstone tools in the development of modern chemistry from the mid-1700s. Despite the long history of electrochemistry originating from great notabilities as Galvani (1791) and Volta (1800) in the late eighteenth century, and Faraday (1849) and Daniell (1836) in the early nineteenth century among many others (see information box: Historic electrochemistry), the feasibility of electrochemical conversion/destruction of organics in waste water first attracted attention in the mid-1970s (Dabrowski et al., 1975). This interest increased steadily during the 1980s and early 1990s and much more rapidly in the late 1990s and forward.

Within the last 15 years, many research groups around the globe have been and still are involved in studying the potentials of EO due to the obvious importance in developing new technology to preserve our water resources and environment. The main contributors on the theoretical side are groups focussed around C. Comninellis (Kapalka et al., 2007, 2008, 2008d; Fierro et al., 2007, 2008; Duo et al., 2004; Panizza et al., 2001, Simond et al., 1997; Comninellis and De Battisti, 1996; Comninellis and Nerini, 1995; Comninellis, 1994; Foti et al., 1999), A. De Battisti (Martinez-Huitle et al., 2005; Martinez-Huitle et al., 2004; Ferro and De Battisti, 2002; Bonfatti et al., 1999, 2000a, 2000b), O. Scialdone (Scialdone, 2009; Scialdone et al., 2009), M. Panizza and G. Ceriscola (Panizza et al., 2001, 2008, 2009a,b), P. Cañizares (Canizares et al., 2003, 2004), A.M. Polcaro (Polcaro et al., 2002, 2003, 2009) and E. Brillas (Brillas et al., 2009), but several others could be mentioned as well. On the application side, the number of research groups is even higher, and a hunch of the number of groups involved can be obtained by going through the references at the end of the chapter. However, compared to other advanced oxidation processes (AOPs) as photocatalysis, the field is still in a growing phase and especially in Scandinavia it has so far attracted only little attention.

This section presents the main principles in EO of organics in aqueous solution. Electrochemistry offers different oxidation pathways, and the direct electrolysis, the electrochemical oxygen transfer reaction and the indirect mediated oxidation processes will all be discussed. In the published literature, a large effort has been dedicated to development of theoretical models in order to predict process efficiency and optimise the performance of the electrochemical reactors and the basic principles in these models will be presented.

HISTORIC ELECTROCHEMISTRY

Four of the great pioneers in the birth of electrochemistry

Luigi Galvani (1737–1798): In the mid-1780s, anatomist Luigi Galvani (Bologna, Italy) was studying the effects of atmospheric electrical discharge. One day, in his garden, he fastened brass hooks between the spinal cord of a dissected frog and an iron railing. To his amazement the frog's legs began twitching wildly, not only when lightning flashed, but also when the sky was calm. Galvani interpreted his results in terms of animal electricity (incorrect). Galvani proclaimed that the muscle retained a nerveo-electrical fluid similar to that of an electric eel. While his papers (Galvani, 1791) on the subject ignited research among scientists of Europe, the most significant consequence of Galvani's discovery was the concept of Galvanism which refers to the production of electrical

(Continued)

HISTORIC ELECTROCHEMISTRY (*cont'd*)

current from the contact of two metals in a moist environment. Shortly before he died, Galvani was dismissed from his professorship at the University of Bologna, because he refused to swear allegiance to Napoleon's Republic.

Alessandro Giuseppe Volta (1745–1827): Volta was born in Como, Italy (near Milan). In 1774, he began his first academic position as principal of the state Gymnasium in Como. In 1777, he was appointed Professor of Physics at the University of Pavia. Here he began to repeat Galvani's famous experiments with decapitated frogs. He observed that Galvani had connected brass hooks between the frog's spinal cord and an iron railing. According to Volta's interpretation, the muscle twitches were induced by current flowing between two dissimilar metals connected by the moist flesh of the frog's leg. This led him to develop the first device which demonstrated chemical production of electric current. In 1799, Volta arranged a vertical pile of metal discs (zinc with copper or silver) and separated them from each other with paperboard discs that had been soaked in saline solution. This stack became known as the voltaic pile and was the first electric battery (Volta, 1800).

In his pursuit of the current generated by his primitive batteries, Volta developed several new devices. He invented the electrophore, a forerunner of the capacitor; the condensatore, a device that detected weak electrical current and the straw electrometer, a meteorology tool which measured atmospheric electricity. The term volt, a unit of electrical measurement, is named in his honour. In 1801, Volta was summoned to Paris to demonstrate his discovery before the Academy of Sciences. Napoleon declared his presentation a triumph, awarded him a gold medal and initiated the annual Volta Prize in his honour.

Michael Faraday (1791–1867): Although Faraday came from very humble beginnings, left school at the age of 12 years, and was essentially self-taught, he is acknowledged as one of the greatest of all scientists. His mentor in his early years of physics and electrochemistry research was Sir Humphrey Davy. For a time, in fact, Faraday extended and developed the research begun by Davy at the Royal Institution in London, where Faraday began his career in 1813 as Davy's Laboratory Assistant. Most of Faraday's early experiments and published papers bore the stamp of Davy's involvement.

Faraday achieved scientific prominence of his own for the First Law of Electrochemistry, developed in 1834: The chemical power of a current of electricity is in direct proportion to the absolute quantity of electricity which passes. The Second Law of Electrochemistry, also defined by Faraday, states electrochemical equivalents coincide, and are the same,

HISTORIC ELECTROCHEMISTRY *(cont'd)*

with ordinary chemical equivalents. The work that led to these two laws also resulted in many of the modern electrochemical terms such as electrode, electrolyte, and ion (Faraday, 1849).

But Faraday did not consider himself an electrochemist; he preferred the title of natural philosopher and devoted his life to proving the interconnection of natural forces. His electrochemical research was one outcome of this effort, exploring the connection between the chemical and electrical forces of the voltaic battery.

John Frederic Daniell (1790–1845): Daniell was born in London and was appointed professor of chemistry at King's College, London. Daniell's research into development of constant current cells took place at the same time (late 1830s) that commercial telegraph systems began to appear. Early telegraph messages were brief and travelled short distances. Daniell's copper battery (1836) (Daniell, 1836) became the standard for British and American telegraph systems. In 1839, Daniell experimented on the fusion of metals with a 70-cell battery. He produced an electric arc so rich in ultraviolet rays that it resulted in instant, artificial sunburn. These experiments caused serious injury to Daniell's eyes as well as the eyes of spectators. Ultimately, Daniell showed that the ion of the metal, rather than its oxide, carries an electric charge when a metal-salt solution is electrolysed.

Text compiled from the 1997 History of Electrochemistry calendar by Bioanalytical Systems, Inc.

3.2.1. Short Introduction to Fundamental Electrochemistry

The field of electrochemistry concerns either utilisation of energy from spontaneous chemical reactions in the form of electricity or the other way around, to supply energy in order to get non-spontaneous chemical reactions running. Electrochemistry is redox chemistry, where electrons are transferred from one species to another. A species is in this terminology said to be oxidised when it loses electrons and reduced when it gains electrons.

When chemical reactants capable of spontaneous redox reaction are physically separated and the electrons from the reaction flow through an electric circuit by means of electrodes and wires, it produces a current that can be utilised as an energy source (battery or fuel cell). The electric current is proportional to the reaction rate, and the cell voltage is proportional to the free energy change for the electrochemical reaction (see information box: Nernst and the electromotive force).

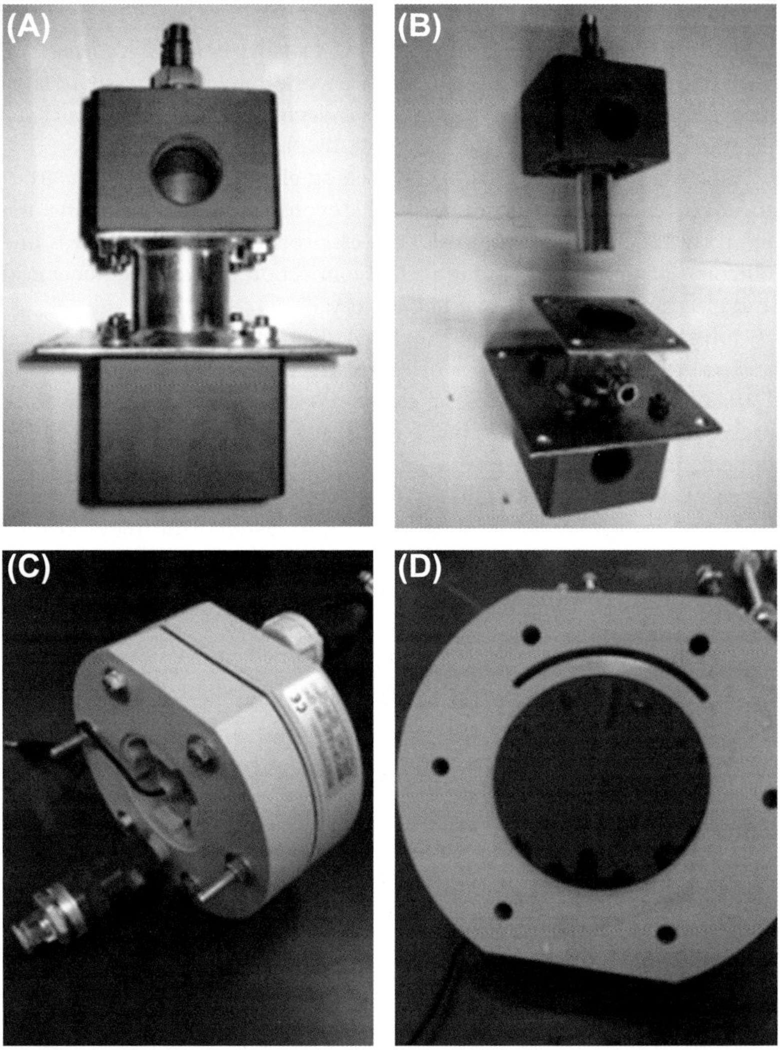

FIGURE 3.1 Examples of small-scale commercial electrochemical cell for wastewater treatment (A) cell in tubular design with Ti/Pt-Ir anode from the company Watersafe S.A., Chrania, Greece (capacity ≈ 6 m^3/h). (B) The Watersafe cell dismantled. (C) Cell (Diacell) with plate design with two Si/boron-doped diamond (BDD) electrodes for polarity reversal. From the company Adamant Technologies (capacity up to 1 m^3/h). (D) The Diacell dismantled. (For colour version of this figure, the reader is referred to the online version of this book.)

In electrolysis, non-spontaneous chemical reactions can be forced to occur at the surface of an electrode by an imposed potential difference (energy) high enough to overcome the activation energy for a reaction at the specific electrode and the resistance in the reagent media. Two electrodes, most commonly solid materials, are connected via a power supply and submerged in a solution containing the chemical reagents, termed the electrolyte. If the supplied energy is sufficient, electrons will be removed from the compounds in the electrolyte (they become oxidised) at the positive electrode (anode), while electrons will be delivered to compounds in the electrolyte (they become reduced) at the negative electrode (cathode). Electrons cannot travel freely in solutions, so the charge passed through an electrochemical cell originates from continuous simultaneous oxidation and reduction of chemical species and electrostatic and hydrodynamic transport of ions and molecules.

This system of electrodes and electrolyte, where electrochemical reactions occur, is termed the electrochemical cell. In order to avoid reduction of the oxidised species at the cathode, the anode compartment (anolyte) and cathode compartment (catholyte) can be divided by a membrane, but just as frequently undivided cells are used due to their more simple construction and maintenance. At the electrodes both organic and inorganic molecules and ions can be involved in electron transfer in the electrolysis process. It is utilisation of the EO at the anode that is the subject of this chapter (Figure 3.1).

NERNST AND THE ELECTROMOTIVE FORCE

Hermann Walther Nernst (1864–1941): Walther Nernst was the son of a civil servant. He studied in Zürich, Würzburg, and Graz. At Graz, he worked with von Ettinghausen, with whom he published work in 1886 which formed part of the experimental foundation of the modern electronic theory of metals. He took his Ph.D. at Würzburg, beginning his career as a physicist. His transition to chemistry began in Leipzig but developed fully in his subsequent position as an associate professor of physics at Göttingen. In 1905, he was appointed professor of physical chemistry in Berlin. He became director of the Physikalisch-Technisches Reichsanstalt in 1922 and finally professor of physics at Berlin in 1924 before retiring in 1933 (Central Electrochemical Research Institute). (1997 History of Electrochemistry calendar by Bioanalytical Systems, Inc.)

Nernst's first outstanding work was his theory of *the electromotive force* of the voltaic cell (Nernst, 1889).

The cell potential for a voltaic cell under standard conditions can be calculated from the standard electrode potentials. But real voltaic cells will typically differ from the standard conditions. The Nernst equation relates the cell potential to its standard cell potential.

(Continued)

NERNST AND THE ELECTROMOTIVE FORCE *(cont'd)*

$E_{cell} = E^0_{cell} - \frac{RT}{nF} \ln Q$

R = gas constant
T = temperature in Kelvin
Q = thermodynamic reaction quotient
F = Faraday's constant
n = number of electrons transferred

The quantity Q, the thermodynamic reaction constant, is like a dynamic version of the equilibrium constant in which the concentrations and gas pressures are the instantaneous values in the reaction mixture.

3.2.2. Direct Electrolysis and Thermodynamics of Electrochemical Mineralisation

In the process termed direct electrolysis, pollutants are oxidised after adsorption on the anode surface, without involvement of any substances other than the removed electron, producing an adsorbed organic radical.

$$R_{ads} \rightarrow P^{\bullet}_{ads} + ze^- \quad (3.1)$$

Conversely, the process of electrochemical mineralisation, where the organic compound is completely degraded to carbon dioxide, water and inorganic minerals, involves oxygen transferred from water molecules to the organic pollutant by help of the electrical energy imposed. This mechanism is usually in the literature termed the electrochemical oxygen transfer reaction (EOTR) (Panizza and Cerisola, 2009) and is presented in detail in Section 3.2.3. An example of EOTR is electrochemical mineralisation of bisphenol A (BPA) (3.2), a known endocrine disrupting chemical that is widely used in the production of polycarbonate plastics and epoxy resins, and found frequently in surface waters contaminated with industrial effluents (Cui et al., 2009).

$$C_{15}H_{16}O_2 + 28H_2O \rightarrow 15CO_2 + 72H^+ + 72e^- \quad (3.2)$$

In this reaction, water is the source of oxygen atoms for complete oxidation of BPA to 15 carbon dioxide molecules at the anode of the electrolytic cell by removal of 72 electrons. The corresponding reaction feeding electrons to the solution in order to maintain charge neutrality is primarily hydrogen evolution on the cathode from the protons liberated in the oxidation reaction (3.2) or from reduction of water.

$$72H^+ + 72e^- \rightarrow 36H_2 \quad (3.3)$$

$$36H_2O + 72e^- \rightarrow 36H_2 + 72OH^- \quad (3.4)$$

From a thermodynamic view point, electrochemical mineralisation is achievable also through direct electrolysis of any soluble organic compound in water at potentials comfortable below the thermodynamic potential of oxidation of water to molecular oxygen ($E^0 = 1.23$ V). This oxygen evolution reaction given by (3.5) is in EO of organics mostly considered an energy-consuming side reaction, which is the optimal goal to avoid.

$$2H_2O \rightarrow O_2 + 4H^+ + 4e^- \tag{3.5}$$

As an example, the thermodynamic reduction potential (E^0) of the mineralisation of BPA (3.2) under standard conditions is 0.17 V, calculated using the standard Gibbs energy of reaction ($\Delta_r G^0$) based on the standard Gibbs free energy of formation ($\Delta_f G^0$) for reactants and products (−454.7 kJ/mol for BPA[1], −237.13 kJ/mol for H_2O and −394.4 kJ/mol for CO_2), the number of exchanged electrons, $n = 72$, and Faraday's constant, F. $\Delta_r G^0 = 15\cdot(-394.4 \text{ kJ/mol}) - (-454.7 \text{ kJ/mol} + 28\cdot(-237.13 \text{ kJ/mol}))$

$$\Delta_r G^0 = 1178 \ \text{kJ/mol} \tag{3.6}$$

$$E^0_{\text{ox}} = \frac{-\Delta_r G^0}{nF} = \frac{-(1178\cdot 10^3)\ \text{J/mol}}{72\cdot 96485\ \text{C/mol}} = -0.17\ \text{V}$$

$$E^0_{\text{red}} = -E^0_{\text{ox}} = -(\,-0.17\ \text{V}) = 0.17 \tag{3.7}$$

In this sense, it would theoretically be possible to treat a stream with an aqueous organic pollutant such as BPA as a fuel cell with cogeneration of electrical energy (fuel cell regime), if the anodic oxidation of the organic was counterbalanced by reduction of oxygen at the cathode, the reverse reaction of (3.5). The total reaction of this hypothetic fuel cell is given in (3.8).

$$C_{15}H_{16}O_2 + 18O_2 \rightarrow 8H_2O + 15CO_2 \tag{3.8}$$

The standard thermodynamic cell potential of such a hypothetical fuel cell would be 1.04 V (1.23–0.17 V). However, even though the thermodynamics seems favourable, the kinetics of these kinds of reactions is in practise reported to be very slow. In the actual state of the art, the fuel cell operated electrochemical mineralisation is not feasible due to the lack of active electrocatalytic anode materials (Kapalka et al., 2010).

[1]Calculated by thermodynamic data from NIST webbook database (U.S. National Institute of Standards and Technology) ($\Delta H^0_f = -369$ KJ/mol and $S_f = 287.7$ J/mol/K) at $T = 298$ K.

3.2.2.1. *Poisoning Effect*

The main problem of direct anodic electrolysis of organics at a fixed potential below oxygen evolution is a decrease in catalytic activity due to formation of a polymer layer on the anode surface, a process termed poisoning effect (Panizza and Cerisola, 2009). The extent of this deactivation depends on the adsorption properties of the anode surface, the concentration and nature of the organic compounds, and their oxidative intermediates. In particular it has been observed that the poisoning effect is less pronounced with anodes with weak adsorption properties (further explained in the next section), while it is more evident in the presence of high organic concentrations and presence of aromatic compounds such as phenol, chlorophenols, nitrophenols, aniline, etc.

An example of poisoning is seen in the direct oxidation of 2-naphthol, which was studied by Panizza and Cerisola (2003) and Panizza et al. (2001) using different anode materials. Cyclic voltammetry (CV) studies showed that the anodic peak for naphthol oxidation rapidly decreased to lower potentials until a steady state was reached after around six scans. This behaviour was attributed to the formation of a polymeric film as a consequence of the polymerisation of naphthyloxy radicals, formed by the initial step in the oxidation mechanism.

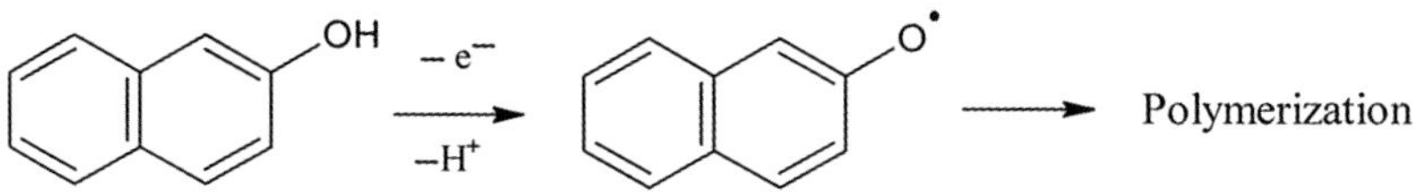

(3.9)

The nature of the brown organic film was confirmed by infrared analysis, which showed the presence of C–H, C=C, C=C–H and C–OH bonds on the anode surfaces (Panizza et al., 2001). The organic film was strong and resistant to washing with organic solvents and the anode fouling was observed for $TiRuSnO_2$, PbO_2 and BDD anodes. However, regeneration of the electrochemical activity in the same solution was possible through polarisation applying higher potentials in the region of oxygen evolution, but only using PbO_2 and BDD (Panizza and Cerisola, 2003). These anode materials are characterised as having a high overpotential for oxygen evolution with adsorbed hydroxyl radicals as intermediates that oxidised the polymeric film and restored the electrode activity. The PbO_2 and BDD electrodes belong to a class of anodes termed as so-called nonactive anodes, which also will be further explained in the next section.

As this example demonstrates, the poisoning effect can be avoided by performing oxidation in the region of water discharge. Operated under these conditions simultaneous to the oxygen evolution, the organics can be oxidised electrochemically on the anode surface through EOTR or by a process termed indirect electrolysis, where redox reagents are electrochemically generated in situ in solution as a chemical reactant to oxidise the organics.

3.2.3. The Electrochemical Oxygen Transfer Reaction

EO of organic compounds in aqueous solution can be obtained without electrode fouling by performing electrolysis at high anodic potentials in the region of water discharge due to the participation of intermediates of oxygen evolution (Comninellis et al., 2008). This process results in part conversion or full mineralisation of the organics, does not need to add oxidation catalysts to the solution and does not in principle produce any by-products. However, current efficiencies are diminished by the secondary reaction of oxygen evolution during oxidation, and generally, removal efficiencies are related to operating conditions and selected electrode materials.

In 1994, Comninellis published a paper on 'Electrocatalysis in the electrochemical conversion/combustion of organic pollutants for wastewater treatment' (Comninellis, 1994), based on earlier work by Feng and Johnson (1990), which set the foundation for exploring and documenting the mechanisms involved in EO of organics in water. Prior to this paper, EO for treatment of water was recognised as a useful method for oxidation for some troublesome organics like phenols. However, the mechanisms involved were not clear, but in general believed to be rather simple in nature. The papers by Comninellis and several others published from the mid-1990s to the early 2000s showed that the nature of the electrode material strongly influences both the selectivity and the efficiency of the process and, in particular, several anodes favoured the partial and selective oxidation of pollutants (conversion), while others favoured complete mineralisation to carbon dioxide (Panizza and Cerisola, 2009). In order to interpret these observations, a comprehensive model was proposed for oxidation of organics at metal oxide electrodes with simultaneous oxygen evolution, which currently largely is considered as globally accepted. A recent paper from Panizza and Cerisola reviews a lot of these results obtained by different anode materials for anodic oxidation of organics (Panizza and Cerisola, 2009).

3.2.3.1. *The Reaction Scheme*

The first step in the EOTR is the discharge of water molecules at the anode surface to form adsorbed hydroxyl radicals, where MO_x denotes the metal oxide anode.

$$MO_x + H_2O \rightarrow MO_x(^{\bullet}OH) + H^+ + e^- \tag{3.10}$$

The following steps have been shown to depend on the nature of the electrode materials and make it possible to distinguish between limiting classes of electrodes defined as 'active' or 'nonactive' anodes. The active anodes have higher oxidation states available on the electrode surface and may interact stronger with the adsorbed hydroxyl radicals by further oxidation, forming higher oxides:

$$MO_x(^{\bullet}OH) \rightarrow MO_x + H^+ + e^- \quad (3.11)$$

The surface redox couple MO_{x+1}/MO_x is often denoted chemisorbed active oxygen and can act as mediator in the conversion or partial oxidation of organics on active anodes:

$$MO_{x+1} + R \rightarrow MO_x + RO \quad (3.12)$$

In contrast to the active anodes, the nonactive electrodes are characterised by a weak electrode–hydroxyl radical interaction forming physisorbed active oxygen (hydroxyl radicals), which can assist in a complete mineralisation of the organics to carbon dioxide:

$$MO_x\ (^{\bullet}OH) + R \rightarrow MO_x + CO_2 + H_2O + H^+ + e^- \quad (3.13)$$

Both the chemisorbed and physisorbed active oxygen undergo a competitive oxygen evolution side reaction resulting in decreased anodic process efficiency and loss of energy:

$$MO_{x+1} \rightarrow MO_x + \tfrac{1}{2}O_2 \quad (3.14)$$

$$MO_x(^{\bullet}OH) \rightarrow MO_x + \tfrac{1}{2}O_2 + H^+ + e^- \quad (3.15)$$

The reaction scheme is usually presented as a full oxidation cycle (Figure 3.2). According to the mechanisms, anodes with low oxygen evolution overpotential such as amorphous carbon, graphite, iridium dioxide, ruthenium dioxide or platinum have an active behaviour, allowing mainly partial oxidation of organics. In contrast, anodes with high oxygen evolution overpotential such as antimony-doped tin oxide, lead dioxide or BDD have a nonactive behaviour and favour complete oxidation of the organics to carbon dioxide.

Based on this approach, different anode materials are listed in Table 3.1 according to their EOTR oxidation power.

The concept of overpotential is a term referring to the potential difference needed for a specific electrolytic reaction experimentally to occur at a given anode material compared to the one thermodynamically predicted by the standard reduction potential. The difference in overpotential for oxygen evolution shown for Ti/Pt and Si/BDD in Table 3.1 is found by CV measurements in an inert electrolyte, where 0.3 and 1.3 V, respectively, are the potential differences observed from the standard

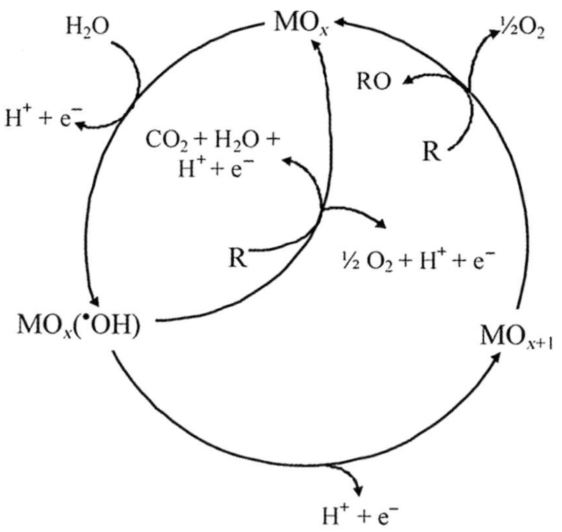

FIGURE 3.2 Scheme of the EO of organics on 'active' and 'nonactive' anodes by the EOTR. *Reprinted from Ref. Comninellis (1994), with permission from Elsevier.*

reduction potential of molecular oxygen ($E^0 = 1.23$ V vs SHE) to the applied potential needed for the anodic current peaks to raise (Figure 3.3). Another term in electrochemistry useful to be aware of is the electrochemical potential window, which denotes the potential difference observed for a specific electrode material between the cathodic hydrogen evolution peak and the anodic oxygen evolution peak in CV.

In literature, the active anodes are typically denoted low oxidation power anode due to their preference to partial oxidation caused by the

TABLE 3.1 Oxidation Power in Acidic Media of Various Anode Materials Used in EO

Electrode Material	Oxidation Potential (V)	Overpotential of O_2 Evolution (V)	Adsorption Enthalpy of $MO_x(^•OH)$	Oxidation Power of the Anode
RuO_2-TiO_2	1.4−1.7	0.18	Chemisorption of OH radical	
$IrO_2-Ta_2O_5$	1.5−1.8	0.25		
Ti/Pt	1.7−1.9	0.3		
Ti/PbO_2	1.8−2.0	0.5	↑	↓
$Ti/SnO_2-Sb_2O_5$	1.9−2.2	0.7		
p-Si/BDD	2.2−2.6	1.3	Physisorption of OH radical	

Prepared after Ref. Comninellis et al. (2008).

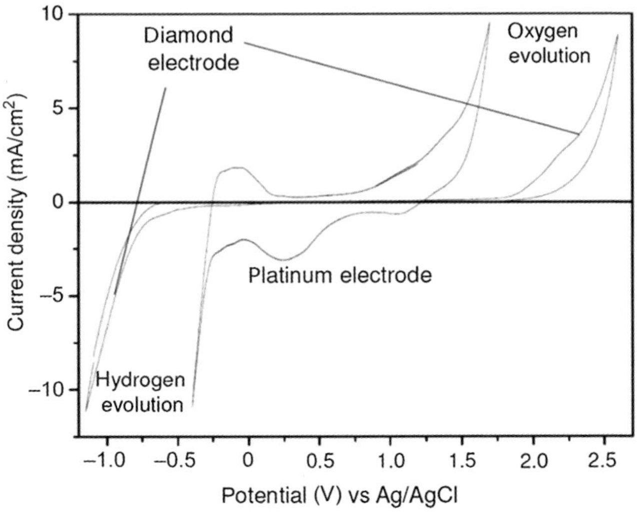

FIGURE 3.3 Cyclic voltammograms of a platinum- and a boron-doped diamond (BDD) electrode in 0.2 M H_2SO_4 in the region between hydrogen and oxygen evolution. *Reprinted from Ref. Kraft (2007), with permission from the author.*

strong electrode—hydroxyl radical interaction. A typically example is the IrO_2 anode, which was used for the documentation of the formation of the higher active oxidation state IrO_3 (Fierro et al., 2007, 2008; Foti et al., 1999). The most important type of the high oxidation power nonactive anodes are the reasonably newly developed BDD anodes, which since the millennium has received most attention in published research on this subject (Anglada et al., 2009, Boye et al., 2006; Canizares et al., 2003, 2004a, 2004b, 2008, 2009a; Chatzisymeon et al., 2009; Duo et al., 2004; Faouzi et al., 2007; Ferro and De Battisti, 2002; Holt et al., 2003; Kapalka et al., 2008a, 2008b, 2008c, 2008d; Kapalka et al., 2007; Kraft, 2007; Marselli et al., 2003; Martinez-Huitle et al., 2008; Mascia et al., 2010; Palmas et al., 2007; Panizza and Cerisola, 2005; Panizza et al., 2001, 2008). The evidence of hydroxyl radical formation on the BDD anodes during anodic polarisation have been found by means of spin trapping during electrolysis of 5.5-dimethyl-1-pyrroline-*N*-oxide analysed by electron spin resonance spectroscopy (Marselli et al., 2003).

An example of the difference in anodic oxidation power on the two classes of anodes is seen in Figure 3.4. Triethyl phosphate is a degradation product of organophosphoric pesticides as parathion and is hard to oxidise further. The superior oxidation power of the nonactive BDD anode due to the actions of the physisorbed hydroxyl radicals is clear compared to the active Ti/Pt-Ir-type anode utilising MO_{x+1} (Figure 3.4(A)). TOC analysis confirms that full mineralisation occurs during the BDD

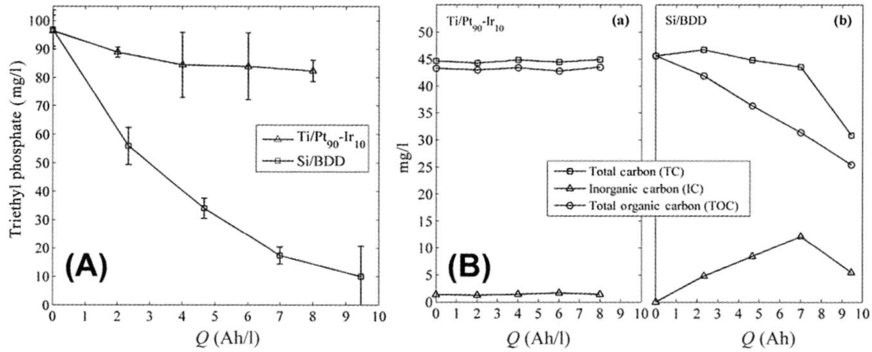

FIGURE 3.4 (A) Evolution in concentration of triethyl phosphate (TEP) during electrolysis at 100 mA/cm^2 in 0.050 M NaCl. (B) Evolution in elemental carbon speciation during electrolysis of TEP at 100 mA/cm^2 in 0.050 M NaCl by (a) Ti/Pt$_{90}$-Ir$_{10}$ and (b) Si/BDD.

electrolysis of triethyl phosphate (TEP) and that TOC remains constant at Ti/Pt-Ir (Figure 3.4(B)).

The nonactive anodes are due to their capabilities of full mineralisation often considered ideal for wastewater treatment or other polluted water bodies, and the development of BDD anodes has definitely placed EO as a true AOP. However, despite the advantages of BDD, their high costs, mechanical resistance issues and difficulties in finding an appropriate substrate on which to deposit the thin diamond layer are major drawbacks. According to the literature, stable diamond film can currently only be deposited on silicon, tantalum, niobium and tungsten, but these material are not considered suitable for large-scale use, due to the poor conductivity and brittleness of silicon and the high costs of the latter rare earth metals (Panizza, 2010). Deposition on the most preferred substrate in electrode manufacturing, titanium, results so far in BDD anodes with cracks not suitable for industrial use (Panizza, 2010).

In addition, full mineralisation may not always be the preferable treatment strategy, since energy may be wasted on oxidising harmless short-length oxidation products as carboxylic acids, which on the contrary in some pretreatment application scenarios may serve as necessary carbon substrates in a subsequent biological treatment. Also, in practise the categories of electrode classification are not as sharply defined as most anodes will exhibit a mixed behaviour, since both reaction paths may participate in parallel in organic oxidation and oxygen evolution reactions.

The EOTR requires the transport of pollutants directly to or very near to the surface of the anode. However, another feature which influences the optimal choice of anode material is the electrochemical degradation pathway termed indirect oxidation, where reversible or irreversible oxidation of ions or other constituents in the water matrix produces oxidants capable of bulk oxidation of the organics.

3.2.4. Indirect Oxidation Pathways

The idea of indirect oxidation is to prevent electrode fouling, avoiding direct electron exchange between the organics and the anode surface and utilise the electrocatalytic potential of other anode materials. Therefore, the pollutants are oxidised through the mediation of some electrochemically generated redox reagents, which act as intermediaries for transfer of electrons between the electrode and the organics (Panizza and Cerisola, 2009). However, intended or not, the indirect oxidation processes play a role whenever a real polluted water matrix is treated through electrolysis, and if of importance for the degradation of the target pollutant, it can be emphasised and enhanced through addition of extra electrolytes.

Oxidation mediators can be metallic redox couples such as Ag(II/I), Ce(IV/III), Co(III/II), Fe(III/II) and Mn(III/II), or strong oxidising chemicals such as active chlorine species, ozone, hydrogen peroxide, persulphate, percarbonate and perphosphate.

3.2.4.1. *Mediated Oxidation by Metallic Redox Couples*

When metallic redox couples are applied as redox reagents, the process is typically termed mediated electrochemical oxidation (MEO). Metal ions in acidic solutions are anodically oxidised from their stable oxidation state (M^{z+}) to the higher reactive oxidation state ($M^{(z+1)+}$) in which they attack the organic feed, degrading it into carbon dioxide, inorganic salts and water:

$$M^{z+} \rightarrow M^{(z+1)+} + e^- \quad (3.16)$$

$$M^{(z+1)+} + R \rightarrow M^{z+} + CO_2 + H_2O \quad (3.17)$$

Reaction (3.17) returns the couple back to a stable state (M^{z+}) and it is then recycled through the cell in order continuously to regenerate further reaction. MEO through metallic redox reagents is mainly applied for pollution abatement in divided cells, where the anode and cathode are separated by a porous separator or an ion-exchange membrane in order to prevent reduction of the oxidative agent at the cathode (Panizza and Cerisola, 2009). The MEO process is preferably used to treat solid waste or highly concentrated solutions with an organic content higher than 20%, in order to avoid or limit the need for implementation of a solution–solution separation step to retain the key metallic ions from the discharged solution. For total organic oxidation, a redox couple with a high oxidation potential must be chosen as, for instance, the Ag(I/II) ($E^0 = 1.98$ V). When this redox pair is applied, Ag(II) ions are generated by anodic oxidation of Ag(I) ions in concentrated nitric acid in the anolyte part of the divided cell, forming a dark brown inorganic complex with nitrate ions ($AgNO_3^+$), which slows down the Ag(II) reduction by water (Juttner et al., 2000):

$$Ag^+ + NO_3^- \rightarrow AgNO_3^+ + e^- \quad (3.18)$$

The corresponding cathodic reaction in the other half cell is the reduction of nitric acid or NO, which can be further oxidised to nitric acid by bubbling of air through the catholyte compartment. The Ag(II) complex will mineralise the organic waste, which is fed to the anolyte compartment:

$$AgNO_3^+ + R + H_2O \rightarrow Ag^+ + CO_2 + H^+ + NO_3^- \quad (3.19)$$

During the oxidation, the anolyte remains clear since the steady state bulk oxidation of $AgNO_3^+$ is very low. Then, when the solution turns dark brown, it is evidence of build-up of $AgNO_3^+$ in the anolyte and full removal of the organics. Other redox couples which have been demonstrated to provide full mineralisation of various organic pollutants are Co(II/III) and Ce(III/IV) (Panizza and Cerisola, 2009).

Redox couples more abundant in natural waters as Fe(II/III) and Mn(II/III) are not suited for total oxidation of organics, because of their low oxidation potential ($E^0 = 0.77$ V and $E^0 = 1.56$ V, respectively), but they can be used for selective partial oxidation of organic compounds. However, when treating polluted water rich in iron and manganese at potentials above oxygen evolution, the oxidised ionic iron and manganese tend to precipitate as insoluble oxides after reaction with molecular oxygen produced by the anodic oxygen evolution side reaction. These reactions may induce co-precipitation of metallic pollutants, and this process known as electro-coagulation or electro-precipitation finds industrial use for removal of heavy metals, where dissolution of a solid iron (or alumina) anode secures a steady feed of Fe(II) (Chen, 2004). However, the electro-coagulation process can as showed also be a side process of the anodic oxidation.

3.2.4.2. *Mediated Oxidation by In Situ-Generated Chemical Oxidants*

Wastewater effluents containing low concentrations of organic pollutants (organic content less than 20%) can be treated by indirect electrolysis generating in situ chemical reactants and converting them into less harmful products (Panizza and Cerisola, 2009). The oxidising chemicals can be electro-generated either by anodic oxidation such as formation of active chlorine, ozone and persulphate, or by cathodic reduction such as

TABLE 3.2 Thermodynamic Formation Potentials of Several Oxidants

Oxidant		Standard Reduction Potential, E^0 (V)
Hydroxyl radical	($H_2O/OH^{\bullet}$)	2.80
Ozone	(O_2/O_3)	2.07
Peroxodisulphate	($SO_4^{2-}/S_2O_8^{2-}$)	2.01
Hydrogen peroxide	(H_2O/H_2O_2)	1.77
Permanganate ion	(MnO_2/MnO_4^-)	1.67
Hypochlorous acid	($Cl^-/HOCl$)	1.48
Chlorine	(Cl^-/Cl_2)	1.36
Dichromate	($Cr^{3+}/Cr_2O_7^{2-}$)	1.23
Oxygen	(H_2O/O_2)	1.23
Hypochlorite	(Cl^-/ClO^-)	0.81

Prepared after Ref. Tröster et al. (2002).

formation of hydrogen peroxide. A list with oxidants sorted after their standard reduction potential is shown in Table 3.2. The most traditional and intensively studied indirect EO agent is active chlorine in a process often referred to as electro-chlorination.

3.2.4.2.1. ELECTRO-CHLORINATION OF ORGANIC POLLUTANTS

Due to the natural abundance of chloride in most polluted waters, chlorine-mediated oxidation is the most widely employed in situ generated chemical oxidant for wastewater treatment. The use of chlorine-mediated oxidation or electro-chlorination has been reported for oxidation of a wide range of model pollutants and has in particular been shown to be suitable for the treatment of waste waters with high natural concentrations of sodium chloride. However, despite the wide use of active chlorine species as oxidation mediators, the exact natures of the electrochemical and chemical reactions that take place during EO in chloride media are complex and not entirely known (Panizza and Cerisola, 2009). It has been shown that indirect transfer of oxygen to organic molecules can be obtained both on the anode surface through adsorbed oxy-chloro species or in the bulk of the solution through chlorine, hypochlorous acid or hypochlorite depending on pH.

$$2Cl^- \rightarrow Cl_2 + 2e^- \tag{3.20}$$

$$Cl_2 + H_2O \rightleftharpoons HOCl + H^+ + Cl^- \tag{3.21}$$

$$HOCl \rightleftharpoons OCl^- + H^+ \tag{3.22}$$

An explanation of the mediating role of chloride ions has been proposed by the group of Bonfatti et al. (2000a,b). They extended the scheme of the EOTR to the cases where oxygen transfer is carried out by adsorbed oxychloro species, which are considered as intermediates of the chlorine evolution reaction instead of hydroxyl radicals (Figure 3.5).

The presence of chloride ions was proposed to inhibit the oxygen evolution reaction causing an increase in anode potential and, therefore, a higher reactivity of adsorbed oxidised chloride species. They observed that the oxidation of organics in the presence of chloride mainly depends on chloride concentration, solution temperature, and pH, and that it was basically independent of the electrode. However, despite the comprehensive work of De Battisti, some differences in the efficiency of the electro-chlorination process on active and nonactive anodes in the presence of organics have been noticed and will be presented in the next section.

3.2.4.2.2. ANODE MATERIALS AND OPERATIVE PARAMETERS OF ELECTRO-CHLORINATION

In the literature, the most used electrode materials for in situ generation of active chlorine are based on platinum or on a mixture of metal oxides (e.g. RuO_2, TiO_2 and IrO_2) that have a good electrocatalytic properties for chlorine evolution as well as long-term mechanical and chemical stability. The ruthenium, titanium and iridium oxide anodes belong to a class of catalytic electrodes known as dimensional stable anodes (DSAs). The metal oxides are often mixed in binary or tertiary systems and coated on metal, mainly titanium, substrate. The DSAs have since the 1960s been used in large scale in the chlor-alkali industry, due to high stability under

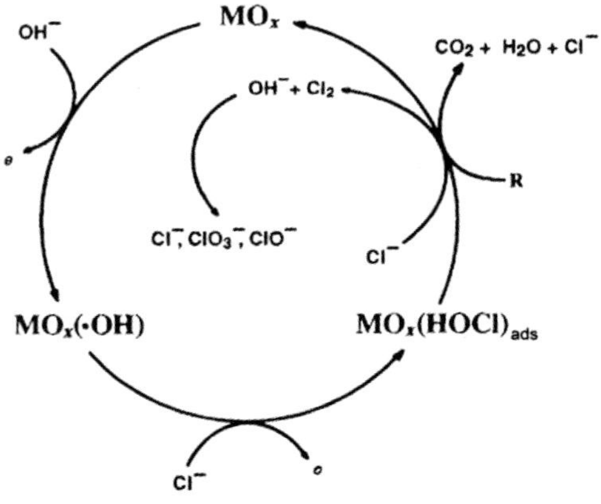

FIGURE 3.5 Scheme of chloride-mediated electrolysis proposed by Bonfatti et al. *Reprinted from Ref. Bonfatti et al. (2000b) with permission from the Electrochemical Society.*

high current density loading and high current efficiency of the chlorine gas evolution caused by good catalytic properties for this reaction (Bergmann and Koparal, 2005). The history and success of DSAs are thoroughly reviewed by Trasatti (2000). Other electrodes can also be used such as graphite and nonactive anodes as PbO_2 and BDD. However, in the presence of higher concentrations of organic pollutants in chloride media, the otherwise superior performance of the nonactive anodes for EOTR is seriously threatened by the more efficient chlorine evolution anodes.

In electrochemical treatment of any water body at potentials above oxygen evolution, the chloride electrolysis process generating the adsorbed or bulk chlorine species amendable for indirect chemical oxidation coexist with the direct EOTR process giving rise to a very complex chemical reaction system, which does not easily allow predicting the role of operative parameters on the performances of the process. This prompts a difficult optimisation of the process and to discrepancies between literature data. One of the most recent and ambitious papers treating this subject is Scialdone et al. (2009), which theoretically and experimentally investigated the role of operative parameters of EO of aqueous organics in the absence and presence of sodium chloride. Applying oxalic acid as model compound on BDD (nonactive anode) and $IrO_2{-}Ta_2O_5$ (DSA type, active anode), they found that in the presence of a 'suitable amount' of chlorides, the anodic oxidation of oxalic acid was partially replaced by a homogeneous indirect oxidation process. The interesting thing was that a very different influence of the nature of the anodic material, the flow rate, and the current density on the performance of the process arose in the absence and the presence of chlorides so that optimisation of the two processes required very different operative conditions. In the absence of chlorides, highest current efficiency was obtained at BDD as expected, when most part of the process was under charge transfer controlled kinetics (i.e. when low current densities and high flow rates were imposed favouring the EOTR). On the other hand, in the presence of chlorides, higher current efficiencies were generally obtained at the DSA anode, when high current densities and low flow rates were imposed (i.e. when a high concentration of chemical oxidants was obtained as a result of the chloride electrolysis). As a result of the different effects of operative parameters on the processes, the addition of chlorides to the electrolytes caused an increase or decrease in the abatement of organics depending on the adopted operative conditions and applied anode materials. In particular, in the case of DSA anode the addition of chlorides gave rise to a dramatical increase in the current efficiency (Scialdone et al., 2009).

The electro-chlorination process can lead to either partial oxidation or full mineralisation of the organics depending on the oxidation abilities of the generated active chlorine species towards the target organics.

However, the main drawback and concern of the electro-chlorination process is a likely formation of chlorinated organic compounds during electrolysis, resulting in increased wastewater toxicity, as it is known from common chlorination chemistry. Indeed, organo-chlorinated compounds are detected during electro-chlorination, but in most cases they are degraded during further electrolyses. It has also been demonstrated that selecting appropriate experimental conditions, glucose as model compound can be mineralised by chlorine-mediated electrolysis without formation of chlorinated organics (Bonfatti et al., 2000a). The issues and concerns of by-product formation during the application of EO are further discussed in Section 3.4.

3.2.4.3. *Anodic Generation of Other Strong Oxidants*

Ozone is another strong chemical oxidant that can be electrochemically produced by anodic oxidation and employed for water disinfection and treatment. Because of its high oxidation potential ($E^0 = 2.1$ V) there is an increasing use of ozone in wastewater treatment, and one of the advantages of ozone oxidation is that it does not produce secondary pollutants, due to its decomposition into molecular oxygen.

$$3H_2O \rightarrow O_3 + 6e^- + 6H^+ \tag{3.23}$$

$$O_2 + H_2O \rightarrow O_3 + 2H^+ + 2e^- \tag{3.24}$$

However, ozone is known to be capable of formation of disinfection by-products (DBPs) as bromate (BrO_3^-), when used in treatment of water containing bromide ions (von Gunten, 2007).

Other strong oxidants, which can be used as mediators in the indirect EO of organics, are persulphate ($S_2O_8^{2-}$), percarbonate ($C_2O_6^{2-}$) and perphosphate ($P_2O_8^{4-}$) (Canizares et al., 2009b).

$$2SO_4^{2-} \rightarrow S_2O_8^{2-} + 2e^- \tag{3.25}$$

$$2CO_3^{2-} \rightarrow C_2O_6^{2-} + 2e^- \tag{3.26}$$

$$2PO_4^{3-} \rightarrow P_2O_8^{4-} + 2e^- \tag{3.27}$$

These compounds are produced by anodic oxidation of the corresponding anions in the solution, but they are only generated efficiently by nonactive anodes such as the BDD and PbO_2 electrodes, due to the need of high oxygen overpotentials in order to reach the formation potentials (Panizza and Cerisola, 2009, Serrano et al., 2002). In addition, the formation is usually not only a direct electrochemical process, but also a sequence of electrochemical and chemical stages (Canizares et al., 2009b).

3.2.4.4. Cathodic Generation of Hydrogen Peroxide

Hydrogen peroxide can be continuously supplied to an acidic contaminated aqueous solution by a two-electron reduction of oxygen gas, directly injected as pure oxygen or bubbled air, at the cathode (Brillas et al., 2009).

$$O_2 + 2H^+ + 2e^- \rightarrow H_2O_2 \tag{3.28}$$

Preferably, a divided cell with a cationic membrane is applied in order to reduce the loss of hydrogen peroxide due to anodic oxidation to molecular oxygen through hydroperoxide radicals ($HO_2^\bullet$) as intermediates. Other major side reactions decreasing the current efficiency are further electrochemical reduction of hydrogen peroxide at the cathode and, to a lesser extent, disproportion in the bulk to molecular oxygen and water. The most widely applied cathode materials for the hydrogen peroxide electro-generation are based on carbon in various configurations (carbon felt, reticulated vitreous carbon, carbon sponges, carbon nanotubes or graphite) (Brillas et al., 2009).

Hydrogen peroxide is despite its widespread use for odour control, bleaching purposes, etc. a weak oxidant, when applied solely. In relation to inorganic and organic oxidation it is only reported effective for oxidising reduced sulphur compounds, cyanides, aldehydes, formic acid and some nitro-organic and sulpho-organic compounds (Harrington and Pletcher, 1999). Its oxidation power can be notably improved by combination with ozone, ultraviolet radiation, and transition-metal catalysts such as iron ions breaking the peroxide bond and producing hydroxyl radicals. The latter technology is Fenton's chemistry, which is the most widely spread technology adapted to indirect electrochemical hydrogen peroxide treatment and termed the electro-Fenton process.

The full mechanisms of the Fenton process are extremely complex, but are very thoroughly presented in a clear manner by Pignatello et al. (2006), and will not be given in full details here since it is already discussed in Chapter 2. However, it is generally accepted that the process is initiated by the formation of a hydroxyl radical as in accordance with the classic Fenton's reaction.

$$Fe^{2+} + H_2O_2 \rightarrow Fe^{3+} + {}^\bullet OH + OH^- \tag{3.29}$$

The hydroxyl radical then participates not only in a number of radical chain reactions, but is most importantly also involved in oxidation of organics. In the electro-Fenton process, the continuously in situ produced H_2O_2 is mixed with an iron catalyst in the solution subjected to treatment in order to produce the hydroxyl radicals in the bulk via the Fenton's process. In addition to the obvious advantages compared to the traditional chemical Fenton technology (no transport, storage and handling of

H_2O_2), the overall organic removal rates are usually higher because of the continuous regeneration of Fe^{2+} at the cathode.

$$Fe^{3+} + e^- \rightarrow Fe^{2+} \tag{3.30}$$

Electro-Fenton is the foundation of a range of related processes developed over the past decade. The combination with photochemistry in photoelectro-Fenton in order to promote the hydroxyl radical production and even solar photoelectro-Fenton has been studied extensively by the group of Brillas using undivided cells combining the reductive hydrogen peroxide processes with the EOTR at the anode and the ultraviolet and visible irradiation (Brillas et al., 2009).

3.2.5. Figures of Merit of EO

The performance indicators, which have been used for assessing the efficiencies of the EO processes, are many. The indicators have been established in order to optimise operating parameters and compare electrochemical cells with different geometries and designs. The most often used and the ones evaluated in the present thesis are defined below.

3.2.5.1. Current Efficiency

Current efficiency is generally defined as the ratio of the charge used for the oxidation of each compound to the total charge passed during electrolysis, and hence states if the electrons in the electrical circuit all come from oxidation of the target organics or if side reaction are dominant. In the literature, different expressions of current efficiency have been proposed: instantaneous current efficiency (ICE), electrochemical oxidation index (EOI) and general current efficiency (GCE) (Panizza and Cerisola, 2009). Two methods for the determination of ICE and GCE have been proposed, the oxygen flow rate method and the COD method. Determination of COD is a common parameter in wastewater treatment monitoring, measuring the amount of chemically added oxygen needed in order to oxidise all compounds in a sample amendable for oxidation. This parameter then provides the number of electrons needed to be removed from the solution in order to obtain full mineralisation (4 mol e^- per mole O_2).

$$O_2 + 4H^+ + 4e^- \rightarrow 2H_2O \tag{3.31}$$

The COD method is the one applied in the current research and ICE is calculated using the following relationship:

$$\mathrm{ICE} = \frac{(\mathrm{COD}_t - \mathrm{COD}_{t+\Delta t})}{8I\Delta t} FV \tag{3.32}$$

$(COD)_t$ and $(COD)_{t+\Delta t}$ are the CODs at times t and $t+\Delta t$ (g/l), I is the current intensity (A), F is Faraday's constant (96,485 C/mol), V is the electrolyte volume (l) and 8 is the oxygen equivalent mass (32 g O_2/mol/4e$^-$).

In the same way, GCE represents an average value of the current efficiency between initial time $t=0$ and t.

$$\mathrm{ICE} = \frac{(\mathrm{COD}_0 - \mathrm{COD}_t)}{8It} FV \tag{3.33}$$

The EOI is the average value of the current efficiency during overall oxidation and.is determined based on ICE.

$$\mathrm{EOI} = \frac{\int_0^{\tau} \mathrm{ICE}\ \mathrm{d}t}{\tau} \tag{3.34}$$

The parameter τ represents the time at which ICE is almost zero. If considered during a period of sufficiently short time, particular in the beginning of electrolysis, the EOI can to some extend be considered specific for the compound investigated and supply information of the reactivity of the compound.

3.2.5.2. *Partial Oxidation*

Full mineralisation is not always the aim, and the extent of partial oxidation can be a useful parameter to follow. It is calculated from COD and TOC measurements using the following relationship:

$$\varepsilon = \frac{\mathrm{COD}_0 \cdot \left(\frac{\mathrm{TOC}}{\mathrm{TOC}_0}\right) - \mathrm{COD}}{\mathrm{COD}_0 - \mathrm{COD}} \tag{3.35}$$

3.2.5.3. *Specific Energy Consumption*

An important cost optimisation parameter is the determination of the amount of energy needed in order to reach removal goals. The energy consumption, E_{sp}, of the EO process is typically expressed as the energy consumed in kWh for removal of 1 kg of COD. It is based on the average cell voltage, U(V), applied in order to reach the desired current intensity.

$$E_{sp} = \frac{(tUI/V)}{\Delta\mathrm{COD}} \tag{3.36}$$

t (h) is the time of electrolysis and ΔCOD is the difference in initial and final COD (g/l).

3.2.6. Kinetic Models

Several attempts of modelling the kinetics in EO have been published within the last decade. The approaches have been different, some using purely numerical modelling (Polcaro et al., 2002, 2003, 2007, 2009), some using statistical work tools such as design of experiments for optimisation (Chatzisymeon et al., 2009b), but the so far generally most accepted kinetic models are grounded on theoretical electrochemical considerations. Polcaro et al. (Polcaro et al., 2002, 2007, 2009) have developed mathematical models based on the reaction between organic substrates adsorbed on inert electrode and hydroxyl radicals and on the indirect oxidation in the bulk phase by electrogenerated oxidising agents. Canizares et al. have proposed different models, which divide the electrochemical reactors into several zones: those close to each of the electrodes and the bulk zone. In each zone, the concentration is assumed to be constant with the position (Canizares et al., 2004a, 2004b). Comninellis and co-workers developed different theoretical models based on the EOTR mechanism for nonactive BDD (Kapalka et al., 2008; Panizza et al., 2008; Comninellis, 1994) and active metal oxide anodes (Simond et al., 1997), while Scialdone et al. (2009); Scialdone (2009) has proposed expansions to these models for the direct oxidation of oxalic acid on BDD and the indirect electro-chlorination of organics on DSAs, from which the conclusions concerning the role of operating parameters in the presence of chloride was reviewed in Section 3.2.4.2.1.

Despite the high complexity of the electrochemical process and the multiple pathways to obtain organic oxidation, the basic approaches in the models are similar. The most straightforward and widely developed is the EOTR model considering the kinetics in the mineralisation reaction on BDD, but the progressing procedure applies generally. The main steps will be provided in this section in order to demonstrate the different domains under which the EO can be operated, ending with a short bullet point style guide proposed by Comninellis and co-workers on how to optimise a mineralisation process of an organic pollutant based on the EOTR using BDD anodes. The full development of the model is found in Kapalka et al. (2008) and Panizza et al. (2008).

3.2.6.1. Electrochemical Mineralisation of Organics on BDD

The starting point is to consider the current efficiency of the organic mineralisation, which is assumed to take into account only two parallel reactions; the main EOTR reaction (3.37) and the side reaction of oxygen evolution (3.38), where hydroxyl radicals in both cases act as intermediates.

$$x\mathrm{BDD}\,(^{\bullet}\mathrm{OH}) + \mathrm{R} \rightarrow x\mathrm{BDD} + \text{mineralization products} + \mathrm{yH}^{+} + \mathrm{ye}^{-} \tag{3.37}$$

$$\text{BDD}\ (^{\bullet}\text{OH}) \rightarrow \text{BDD} + \tfrac{1}{2}\text{O}_2 + \text{H}^+ + \text{e}^- \tag{3.38}$$

This initial assumption omits an additional positive indirect oxidation contribution to the mineralisation from chemical oxidants generated from conceivable electroactive ions present in solution in possible much higher concentration than the pollutant. This is one of the limitations of the model.

What is additionally assumed is that (1) adsorption of the organic compounds at the electrode surface is negligible; (2) all organics have the same diffusion coefficient, D, and (3) the global rate of the electrochemical mineralisation of organics is a fast reaction, and it is controlled by mass transport of organics to the anode surface, which implies that the rate of the mineralisation is independent of the chemical nature of the organic compound present in the electrolyte.

Under these conditions, the limiting current density, i_{lim} (A/m^2) for the electrochemical mineralisation of an organic compound (or a mixture of organics) under given hydrodynamic conditions can be written as

$$i_{\text{lim}} = nFk_{\text{m}}C_{\text{org}} \tag{3.39}$$

n is the number of electrons involved in the mineralisation, F is Faraday's constant, k_{m} is the mass transport coefficient (m/s) and C_{org} is the concentration of organics in solution (mol/m^3). When comparing the electrochemical mineralisation of a generic compound with the chemical mineralisation using molecular oxygen, it is possible to obtain the relation between the organics concentration, C_{org}, and COD (mol O_2/m^3). In this way, the limiting current density of the electrochemical mineralisation of organics at a given time t during electrolysis can be related with the COD of the solution:

$$i_{\text{lim}}(t) = 4Fk_{\text{m}}\text{COD}(t) \tag{3.40}$$

where 4 is the number of exchanged electrons per mole of O_2 as explained in Section 3.2.2.

Working under galvanostatic conditions, depending on the applied current density i (A/m^2), different operating regimes have been identified by Kapalka et al., (2008) that are showed schematically in Figure 3.6(A) and mathematically in Table 3.3.

The development of the expressions are found in Kapalka et al., (2008) and are based on a defined characteristic parameter α of the electrolysis processes, which is the ratio between the applied operating current, i_{appl}, and the limiting current, i_{lim}, calculated after (3.40):

1. $i_{\text{appl}} < i_{\text{lim}}$: The electrolysis is under current control, organic intermediates are formed during the oxidation, the ICE is 100% and COD decreases linearly with time. This behaviour persists until a critical time, t_{cr} (3.41), or to a critical conversion of organics, X_{cr} (3.42),

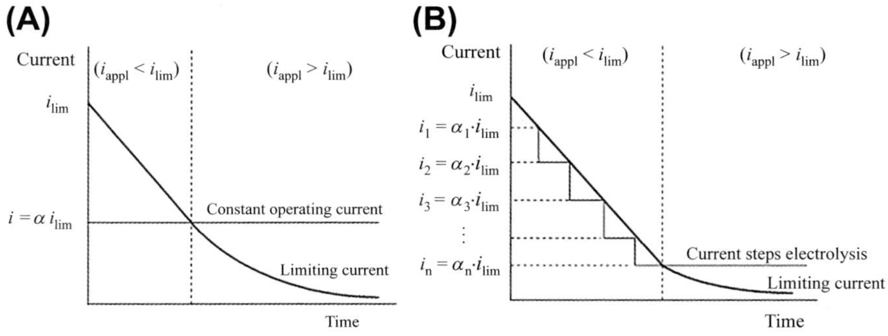

FIGURE 3.6 (A) Schematic current–time curves showing operating regions during constant operating current, *i*. (B) A similar curve showing operating regions during current step electrolysis. *Reprinted from Ref. Panizza et al. (2008) with permission from Elsevier.*

corresponding to the time, or the conversion, at which the limiting current density is equal to the applied current density where $\alpha = 1$:

$$t_{cr} = \frac{1-\alpha}{\alpha} \cdot \frac{V_r}{Ak_m}, \quad \alpha = \frac{i}{i^0_{lim}} \tag{3.41}$$

$$X_{cr} = 1 - \alpha \tag{3.42}$$

A is the surface area of the anode (m^2) and V_r is the reservoir volume (m^3).

2. $i_{appl} > i_{lim}$: The electrolysis is under mass transport control, organic compounds are completely mineralised to CO_2 and secondary reactions (such as oxygen evolution) commence, resulting in a decreasing current efficiency. Under these conditions, ICE is below 100% and COD removal follows an exponential trend.

TABLE 3.3 Equations That Describe ICE, COD, τ and E_{sp} Evolution during Oxidation of Organics at BDD Anode

	Under Current Limited Control, $i_{appl} < i_{lim}$	**Under Mass Transport Control, $i_{appl} > i_{lim}$**
ICE	$ICE = 1$	$ICE = \exp\left(-\frac{Ak_m}{V_r}t + \frac{1-\alpha}{\alpha}\right)$
COD (mol O_2/m^3)	$COD(t) = COD_0\left(1 - \frac{\alpha A k_m}{V_r}t\right)$	$COD(t) = \alpha COD_0 \exp\left(-\frac{Ak_m}{V_r}t + \frac{1-\alpha}{\alpha}\right)$
τ (s)	$\tau_{cr} = \frac{XV_r}{\alpha A k_m}$	$\tau_{cr} = -\frac{V_r}{Ak_m}\left[\ln\left(\frac{1-X}{\alpha}\right) - \frac{1-\alpha}{\alpha}\right]$
E_{sp} (kWh kg/COD)	$E_{sp} = \frac{1}{3600}\frac{F}{8}(V_d + R_c A \alpha i^0_{lim})$	$E_{sp} = \frac{1}{3600}\frac{F}{8}(V_d + R_c A \alpha i^0_{lim}) \times \frac{1-\alpha[1+\ln(1-X/\alpha)]}{X}$

Ref. Panizza et al. (2008).

When considering the mass balances over the electrochemical cell and the reservoir in a batch recirculation setup, models of temporal evolution in COD and ICE during electrolysis have been developed by Kapalka et al., (2008) for both the presented operating regimes (Table 3.3). From these models, equations for the prediction of needed electrolysis time, τ, and specific energy consumption, E_{sp}, to reach a target COD removal has been derived as well.

The latest trend in the published literature is to apply multiple current step electrolysis in order to optimise the performance of the batch recirculation system for electrochemical organic mineralisation on BDD, and thereby maintain a nearly 100% ICE as long as possible during electrolysis (Figure 3.6(B)). The applied current density is in steps adjusted to always be near, but below, the instantaneous limiting current, and still in the potential region of water discharge to avoid fouling (Kapalka et al., 2008). Panizza et al. have extended the models in Table 3.3 to apply for predictions on the evolution in COD, current efficiency, electrolysis time and specific energy consumption by this modulated approach. They have shown that if the number of current steps is sufficiently high (corresponding to working in a semicontinuous control mode), the performance of the reactor is comparable with that achievable in an ideal diffusion controlled process without mass transfer limitations with an efficiency of 100% (Panizza et al., 2008).

3.2.6.2. Approach to Optimisation of Electrochemical Mineralisation on BDD

The energy consumption for the electrochemical mineralisation of organics decreases strongly with increasing average current density (expressed as EOI, see (3.34)) and reaches a minimum value at EOI = 1. In order to work under these favourable conditions, electrolysis has to be carried out under programmed current, in which the applied current density during electrolysis is adjusted to the limiting value. The following steps have been proposed for optimal treatment of waste water using EOTR on BDD based on the above model (Kapalka et al., 2008):

1. Measure the initial COD of the waste water
2. Estimate k_m of the electrolytic cell under fixed hydrodynamic conditions. This can be achieved using a given concentration of ferrocyanide, $Fe(CN)_6^{4-}$ (50 mM), in a supporting electrolyte (1 M NaCl) and measuring the limiting current, I_{lim}, (A) for the anodic oxidation of $Fe(CN)_6^{4-}$. The mass transfer coefficient can then be calculated using the relation:

$$k_m = \frac{I_{lim}}{FA\left[Fe(CN)_6^{4-}\right]} \tag{3.43}$$

3. Estimate the initial limiting current density, $i_{\lim}$, by Eqn (3.40)
4. Calculate the time constant of the electrolytic cell, τ_c, using the relation

$$\tau_c = \frac{V_r}{A k_m} \tag{3.44}$$

5. The temporal evolution of the limiting current during electrolysis is given by

$$i_{\lim} = i^0_{\lim} \exp\left(-\frac{t}{\tau_c}\right) \tag{3.45}$$

Start electrolysis by application of a current density corresponding to the limiting value calculated in (3).

6. Adjust the current density during electrolysis to the time-dependent limiting value according to (3.45)
7. The electrolysis time, τ, needed in order to achieve the target final COD is found using the relation

$$\tau = -\tau_c \log \frac{COD_{final}}{COD_0} \tag{3.46}$$

The presented model and guidelines apply for the EOTR of organics, which by BDD anodes can lead to full mineralisation, and has been shown to provide good agreement compared to experimental model solution results. However, the model has its limitation in the sense that it assumes that all organic oxidation proceeds at or very near the anode surface, and that all side reactions are purely energy consuming and of no use with regards to organic oxidation, i.e. no positively contributing indirect homogeneous chemical oxidation by generated oxidants. Real polluted water will in almost any case contain electroactive ions in some concentrations, i.e. chloride, sulphate, etc. When considering micropollutants as the target organics for the EO process, the concentrations of these electroactive ions may be significantly higher, not having the same mass transfer limitations as the organics.

In treating these waters, the obtained removal of COD can be faster than predicted by the model, if the target organics are amendable to chemical oxidation, since it may be possible to operate at a higher current density than predicted and still maintain a high ICE. Further modelling and discussion of the role of indirect oxidation by active chlorine species by different anode materials and optimisation of treatment processes, where the indirect oxidation pathway is of importance, has been treated by Scialdone et al. (2009). A simple way to implement the indirect oxidation is to add a chemical reaction contribution to the model, where k is a chemical oxidation rate constant (1/s):

$$r'_{\text{chem}} = k\text{COD}_t \quad (3.47)$$

This term should only be considered in case that ICE < 1, as the maximum decrease in the anodic oxidation of COD is limited by the value of the current intensity applied.

This introduction to the fundamental and theoretical principles in EO provides the foundations for the rest of the chapter, with the next section presenting the versatility of EO regarding degradation of different kinds and classes of pollutants.

3.3. THE VERSATILITY OF EO DEGRADATION

Especially since the millennium, a vast amount of papers have dealt with EO and its potential efficiency and applicability for the treatment of many types of primarily industrial effluents of different kinds and as a technique for the abatement of many different aqueous organic pollutants considered of concern for the environment and human beings. Comprehensive reviews on EO treatment of waste water are published by Martinez-Huitle and Ferro (2006) and on decontamination of dye-containing waste water by Martinez-Huitle and Brillas (2009). The intention of this section is not to extend beyond these reviews, but to present some selected studies of treatment of both real waste water and specific compounds in order to demonstrate the versatility of the EO process and its many applications.

3.3.1. Examples of EO for Wastewater Treatment

3.3.1.1. Olive Oil Mill Waste Water

Olive mill waste water (OMWW) constitutes a serious environmental problem in the Mediterranean region, due to features as seasonal and localised production, low flow rates and high organic load up to 220 g/l of COD mainly consisting of phenols, lipids, sugars, and tannins (Gotsi et al., 2005). Its treatment by electrolysis has been the topic of several papers applying EO in different configurations concerning cell designs and active (DSA, Ti/Pt) vs nonactive anodes (BDD) (Papastefanakis et al., 2010; Chatzisymeon et al., 2009a,b; Un et al., 2008; Giannis et al., 2007; Panizza and Cerisola, 2006; Gotsi et al., 2005; Israilides et al., 1997). The most recent paper by Papastefanakis et al. (2010) studied the removal of COD, colour, etc. using a DSA-type Ti/RuO_2 anode, and achieved acceptable removals of phenols (84%), colour (86%), COD (52%) and TOC

(38%) after oxidation at 28 Ah/l (50 mA/cm^2) and addition of NaCl. Both partial and full mineralisation occurred, but important to mention is that ecotoxicity was eliminated as well despite the residual COD. The treatment using DSAs was very much dependent on indirect oxidation routes, and the overall performance could be dramatically affected by the presence of various salts. Using almost one-third of the current density (15 mA/cm^2), the passing of the same amount of charge (28 Ah/l) through the OMWW resulted in an increase in COD removal from less than 7% in a salt-free test volume to 54% after addition of 20 mM NaCl.

3.3.1.2. *Landfill Leachate*

Landfill leachate is another type of polluted water, which treatment by EO has been studied in several papers (Anglada et al., 2009, 2010; Bashir et al., 2009; Rao et al., 2009; Chiang et al., 1995). Leachate is produced when rain water percolate through dumps of municipal waste disposed at landfills, and is characterised by a very complex composition containing a large number of compounds, many of them recalcitrant to biological conversion (Anglada et al., 2009). In leachate, in particular the ratio between biological and chemical oxygen demands (BOD/COD) and the concentration of ammonium, and heavy metals are parameters of interest. Heavy metals may be reduced at the cathode in an undivided cell or removed by electro-coagulation, but COD and ammonium are the parameters mainly targeted by EO. One of the first series of papers on this subject was published by Chiang et al., 1995 where complete removal of 2.6 g/l NH_4^+ and 92% reduction of COD were obtained after 4 h by addition of 0.75% NaCl and using tertiary Sn–Pd–Ru DSA. Recently, a pilot scale study using 1.05 m^2 BDD showed that efficient treatment by this anode material was possible too, and NH_4^+ removal was due to indirect chlorine oxidation and COD removal was mainly due to surface or near-surface EOTR (Anglada et al., 2009). The general assessment was that EO is a very suitable technique for treatment of the biorefractory part of leachate compared to other AOPs (Anglada et al., 2009), especially in combinations with pretreatment methods as membrane filtration in order to decrease volumes and concentrate pollutants.

3.3.1.3. *Dyehouse Effluents*

Treatment of dyehouse effluents or other waste waters containing synthetic organic dyes is another area, which has received lots of attention using EO (Martinez-Huitle and Brillas, 2009; Zaviska et al., 2009; Chatzisymeon et al., 2006; Fang and Al-Abed, 2007; Faouzi et al., 2007; Panizza and Cerisola, 2007, 2009; Saez et al., 2007). Discharge of these coloured compounds in the environment causes considerable nonaesthetic pollution and serious health risk factors (Martinez-Huitle and Brillas, 2009). Both the EOTR by BDD, indirect active chlorine oxidation and the electro-Fenton processes have proved to be efficient routes for bleaching and degradation of dyes. Chatzisymeon et al. (2006) have published a study on treatment of dyehouse effluent using a Pt-Ir-Ta-coated titanium anode. They found that EO with the applied anode was capable of oxidising the chromophore groups of dyes achieving decolourisation of synthetic waste water already after 10−15 min with low energy consumption. However, energy consumption increased significantly treating the real dyehouse effluent. The extent of COD removal varied between 30% and 90% after 180 min, depending on the operating conditions, but the treatment performance was generally found to be improved by increased salinity, current intensity and decreased pH. An alarming observation was that ecotoxicity of the dye effluent increased in the final samples from a low initial level, despite the colour and COD removal. A proposed and feasible explanation was the formation of persistent toxic by-products (Chatzisymeon et al., 2006).

3.3.2. Examples of EO for Degradation of Persistent Organic Pollutants

Persistent organic pollutants (POPs) constitute a serious problem and threat to the safety and health of humans and other mammals. Numerous POPs have been the subject of study by EO, and later in this chapter case studies with treatment of pesticides and polycyclic aromatic hydrocarbons (PAHs) are presented. Some selected studies with different classes of other recalcitrant POPs are highlighted here.

3.3.2.1. Chloromethylphenoxy Herbicides

The presence of herbicides in natural waters comes from their widespread use for agricultural and nonagricultural purposes generating continuously rinsate, waste product and contaminated soil. The degradation of 4-chloro-2-methylphenoxyacetic acid and two other related herbicides was studied by Boye et al. (2006) in an inert 1 M $HClO_4$ electrolyte using BDD. In short terms of this comprehensive study, they showed that all herbicides were degraded and totally mineralized following first-order kinetics in the batch recirculation system under mass transfer control − as predicted by the theoretical model (Section 3.2.3). The intermediate aromatic products and generated carboxylic acids were identified and a general pathway for the electrochemical mineralisation of herbicides was proposed.

3.3.2.2. Acetaminophen (Paracetamol)

Pharmaceutical compounds are increasingly often found as contaminants in sewage effluents, surface water, and ground water (Waterston et al., 2006). An example is acetaminophen (*N*-(*p*-hydroxyphenyl)-acetamid), which has been found in sewage at concentrations up to 6 μg/l and is in this context considered as a micropollutant. The risk of discharge of pharmaceuticals into nature mainly concerns development of resistant strains of microorganisms, but other unknown implications may develop in the future. EO of acetaminophen was studied by Waterston et al. (2006) using different anode materials, and nonactive types of anodes (BDD and Ti/SnO_2) led to full mineralisation, whereas Ti/IrO_2 generated benzoquinone as the exclusive product except at very long electrolysis times (again according to the theory suggesting different classes of anodes). Current efficiencies for mineralisation up to 26% were achieved without optimisation of the cell design.

3.3.2.3. *2,4,6-Trinotritoluene*

A problematic pollutant mostly related to army depots and artillery sites in northern America, is the family of nitro-aromatic compounds such as trinitrotoluene (2,4,6-trinotritoluene (TNT)) and dinitrotoluene (2,4-DNT, 2,6-DNT), due to its large manufacture and use as explosives (Ayoub et al., 2010). The electrochemical aqueous removal of this pollutant has been studied by Rodgers and Bunce (2001), not through oxidation, but by direct reduction at a Ni plate cathode in a divided cell. The reactions are a four-electron reduction to phenylhydroxylamine at $pH > 4$, and a six-electron reduction to aniline at lower pH, involving both electrochemical and chemical steps. The experiments gave high chemical yield of the products, and the treatment process was further extended by reoxidation of the reduced products at the Ti/IrO_2 anode to induce oligomerisation of the organic products and ease the removal of the products. Electro-Fenton has also with success been showed to remove TNT from solution through mineralisation in the cathodic chamber in a combined direct reduction and indirect Fenton oxidation process (Chen and Liang, 2009).

3.3.2.4. *Perchloroethylene*

A very common family of pollutants is chlorinated compounds as perchloroethylene (PCE), trichloroethylene, etc. and as an example chlorinated solvents constitute the main source for contaminated sites and ground water pollution in Denmark (Miljøstyrelsen, 2007). The EO of chlorinated solvents in aqueous solution has so far been little explored with a few papers focussing mainly on the cathodic reduction reaction, where the conversion of the organics occur in the potential region near hydrogen evolution reaction decreasing current efficiency (Sonoyama and Sakata, 1999). However, a recent paper by Saez et al. (2009) has compared cathodic, anodic and combined degradation of PCE in water and studied the influence of several operating parameters. They found that an undivided configuration provided the best results under the experimental conditions investigated. The main degradation of PCE still occurred at the cathode, but the contributing anodic oxidation made the final level of intermediate products much lower than in the case of the pure cathodic reduction. These results have just currently been supported by Scialdone et al. (2010) using BDD as anode and silver as cathode for abatement of PCE and DCE, and confirmed the superior performance of the undivided cell.

The number of published studies on applied use of EO for degradation of troublesome organic pollutants is as briefly demonstrated comprehensive and still growing. Most of the papers demonstrate that the possibilities are promising and larger scale use in the industry should be achievable, when the technology has matured. The coming chapters in this thesis will present the most important results and implications found in our research of using EO in the area of environmental engineering, introduced with case story studies on real polluted water matrices followed by detailed model solution studies investigating fundamental coherences and influences of some of the many parameters involved.

3.4. THE CHALLENGE OF MINIMISING THE BY-PRODUCT FORMATION

AOPs are as emphasised in this chapter able to mineralise organic substances to relatively harmless ions, carbon dioxide and other nontoxic compounds. Therefore, the overwhelming majority of studies on use of AOPs emphasise the benefits of the technology, i.e. detoxification of waste water, decolourisation and disinfection. However, in many cases more careful consideration is necessary. It is known that when any AOP is used for treatment of real water with organic pollutants, organic by-products may and can be formed, which can be more dangerous than the initial compounds as referred to in earlier chapters. Likewise, inorganic by-products as chlorite, chlorate, perchlorate and bromate can be formed during the EO treatment if chloride and/or bromide are present, and common DBPs as trihalomethanes (THMs) and chloramines may also be of concern. Especially, inorganic by-product formation has so far often been neglected by equipment producers and users, but publications from different authors demonstrate considerable risk.

In this section, some selected studies on the extent of formation of inorganic by-products during electrochemical disinfection and treatment are reviewed in order to outline the main issues in this area.

3.4.1. Inorganic By-product Formation during EO

The main contributor underlining the need for awareness of the formation of inorganic by-products during use of EO is M.E.H. Bergmann from Anhalt University of applied Sciences, Germany. He and co-authors have in a number of papers demonstrated that especially the formation of perchlorate needs to be considered and included as an evaluation criterion, when new environmentally friendly AOPs are developed (Bergmann et al., 2009a,b). Especially, electrolysis by BDD of dilute chloride solutions promoted the formation of chlorate and perchlorate, due to the radical oxidation mechanisms. Perchlorate was also found by mixed oxide and Pt anodes, but in three orders of magnitude lower concentrations (Bergmann et al., 2009a), observations which were confirmed by Polcaro et al. (2008). Chloride in the electrolyte was of course needed, but chlorate and perchlorate could be formed by electrolysis of hypochlorite and chlorite ions as well, and the formation mechanisms were shown to be very complex (Bergmann and Rollin, 2007).

Another inorganic by-product, which can be formed during electrochemical disinfection using the natural chloride content of water, is chlorine dioxide gas. This molecule is in itself a highly efficient disinfectant, but Bergmann and Koparal (2005) showed that it can be formed under several conditions and could be measured in the lower milligrams per litre range. Once again they emphasise that technical cells should be designed and controlled better in order to ensure health safety. Despite efficient disinfection, it is not an applicable solution if inorganic by-products are formed, and for that reason BDD is not, compared to DSA and platinum anodes, suited for disinfection purposes.

3.4.2. Halogenation of PAHs during EO

The risk of formation of halogenated organic by-products is as clearly stated in this thesis a matter of concern for the further application of EO as a technique for pollution abatement. The many reactive species generated in the electrolysis process may induce unwanted side reactions besides the target oxidation reactions, and when electroactive halide ions are present in sufficient concentration the oxidised state may participate in substitution reactions with carbon bonded hydrogen on the organics leaving the by-products more toxic and recalcitrant to further oxidation. Halogens are from organic synthesis chemistry known to participate in especially electrophilic aromatic substitution reactions, and when already substituted on the organic, halogens are due to their electronegative nature weakly deactivating groups, which means that primarily mono halogenated organics can be expected to react slower to further oxidation than the parent compound. An example of halogenation of the parent organic is shown in Figure 3.7, where primarily brominated naphthalene was found during treatment of PAH-polluted model sea water.

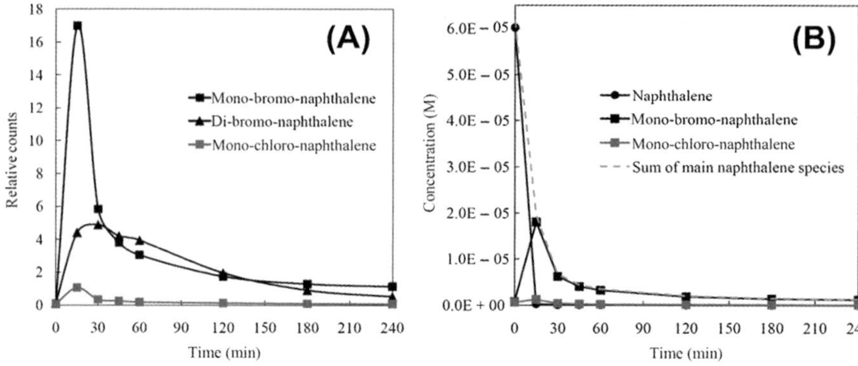

FIGURE 3.7 (A) The evolution in brominated and chlorinated by-products measured as relative counts during EO in mixed 0.0010 M KBr and 0.10 M NaCl electrolyte at 25 mA/cm^2 and pH 6. (B) Concentration comparison of parent naphthalene and monohalogenated naphthalene species. *Reprinted from Ref. Muff and Søgaard (2011) with permission from Elsevier.*

In this example, the halogenated by-products were removed at prolonged electrolysis times.

In this way, halides often play a double role as the good and the bad guys of environmental electrolysis. The rate of electrochemical mineralisation of oxalic acid was significantly increased at a platinum anode due to the introduction of halides following the order $F^- \geq Br^- > Cl^-$ (Martinez-Huitle et al., 2005) and many other examples of this enhancement exist as presented in the prior sections. Since fluoride in considerable concentrations is not found frequently in natural waters, bromide and chloride are the most interesting halides in relation to practical applications. However, several studies have documented the presence of residual refractory organic halogens (measured as adsorbable organic halogens) after electrolysis in especially chloride media (Rajkumar et al., 2005; Neto and De Andrade, 2009; Comninellis and Nerini, 1995; Boudreau et al., 2010; Rajkumar and Palanivelu, 2004), and it must be emphasised that successful remediation of a waste is not achieved once the original contaminant has disappeared, if the reaction by-products are more toxic or persistent than the original contaminant.

3.5. EXAMPLES OF APPLICATION OF EO

The possible application of EO is present whenever facing the need of treatment of water polluted with nonbiodegradable organics. Many types of industrial waste water are as presented in Section 3.3 obvious subjects to EO treatment. In this section some new examples of application of this technology in the areas of polluted ground water remediation and polluted sea water will be presented. Finally, examples of the applications within water disinfection will be discussed.

3.5.1. Case Study 1 – Ground Water Remediation

As thoroughly described in Chapter 2, in situ chemical oxidation (ISCO) using different chemical oxidants is a powerful technique for degradation of organic pollutants below the water table at a contaminated site. Recirculation of the oxidant solution is frequently applied. However, EO may be an efficient innovative approach to address the polluted soil. A strong oxidant solution can electrochemically be generated from the ions present in the hot spot water concurrent with organic oxidation by EOTR and reinfiltration back in situ of the oxidised solution utilising the chemical oxidation potential for remediation of the polluted soil matrix.

Initial steps to this approach have been tested on water from a very complex contaminated site containing a range of organophosphoric pesticides and natural degradation products. The full study is published here (Muff et al., 2009).

3.5.1.1. Results and Implications

COD was used as overall parameter for evaluation of the oxidation performance. A Ti/Pt_{90}-Ir_{10} anode was used in the batch recirculation experiments, which showed that COD was readily removed by EO following first-order kinetics. The natural salinity of the water was high, 0.7 w/w%, and it was expected that indirect active chlorine oxidation was an important pathway for the COD removal. Galvanostatically, addition of sodium chloride up to 2.0 w/w% did slightly enhance the removal rate, while a further increase up to 4.0 w/w% showed a reverse effect and a significant deviation from the first-order kinetics. A proposed explanation was that the increased electrolyte concentration and 'cloud' of negatively charged ions in the diffusive part of the double layer inhibit the transport of oxidation products from the chloride electrolysis out in the bulk phase amendable for oxidation of organics. Operated potentiostatically, the increased sodium chloride concentrations resulted in increased current densities and hence increased COD removal rates, and a linear correlation was demonstrated between the first-order rate constant and the resulting current density (Figure 3.8(A)).

The adverse effect of COD removal induced by chloride disappeared, since the overall rate of reactions (current density) was allowed to increase resulting in a higher active chlorine concentration and hence higher COD removal. This showed the importance of the indirect oxidation pathway. When the evolution of COD was evaluated against the specific amount of charge passed through the solution at different current

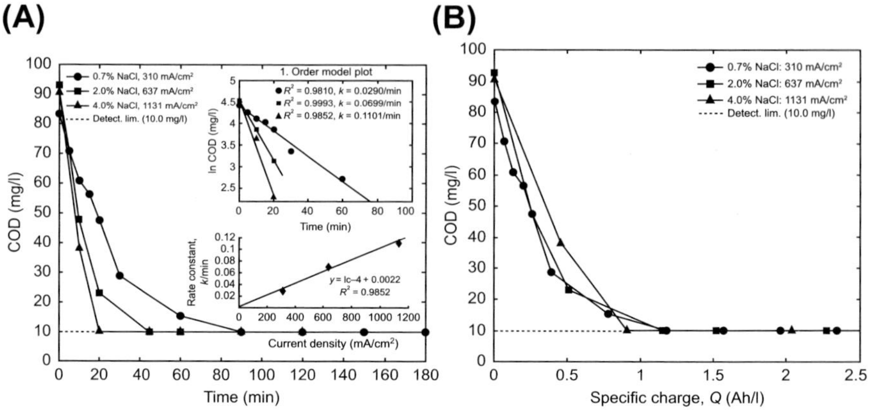

FIGURE 3.8 (A) Evolution in COD during potentiostatic electrolysis by increased salinities and current densities. *(Muff et al. (2009) reprinted with permission from Elsevier.)* (B) The same COD data evaluated against specific charge passed through the solution.

densities (Figure 3.8(B)), the data were almost convergent, typically seen for indirect oxidation-controlled COD removal processes. However, the efficiency at 4.0 w/w% was slightly decreased, probably due to build up of active chlorine faster than removal by organic oxidation.

Analysis of the specific organophosphoric pesticides showed that the parent compounds, parathion, methyl-parathion and malathion, and all the degradation products were degraded during the EO process (Figure 3.9).

Only a single tri ester, *O,O,O*-triethylphosphoric acid (TEP), showed a recalcitrant behaviour and was found in residual concentrations. This

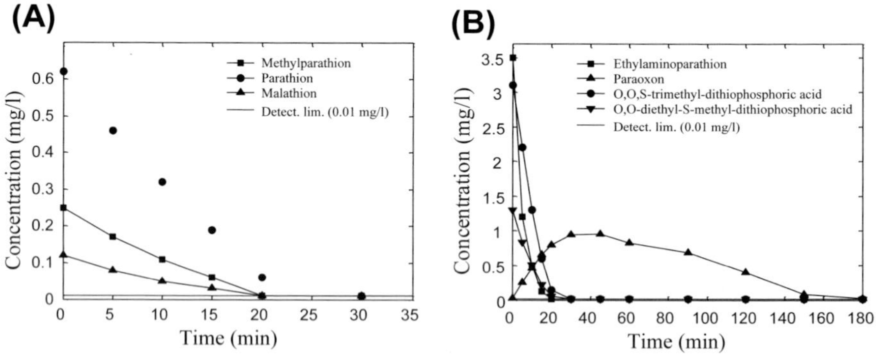

FIGURE 3.9 (A) Degradation of the parent pesticides during electrolysis at natural 0.7 w/w% salinity and 310 mA/cm^2 current density. (B) Evolution in natural occurring degradation products. *From Ref. Muff et al. (2009) reprinted with permission from Elsevier.*

compound was used as example in Figure 3.4. Related to the specific case, TEP does not pose a significant risk to the environment in the residual concentration level found in the natural drainage water after electrolysis.

Initial steps were taken to investigate the soil remediation possibilities. An electrochemical oxidant solution primarily comprising active chlorine species was generated and intermixed with the raw polluted ground water. It was found that half and half mixing of the polluted drainage water and electrochemically generated oxidant removed all the organophosphoric compounds present except TEP (Figure 3.10). Similar results were obtained using chemical hypochlorination in comparable active chlorine concentration, but with a higher level of residual TEP.

This electrochemical recirculation and reinfiltration twist to traditional ISCO is considered a very interesting approach in the use of EO within soil and ground water remediation. The high chloride concentration in the case study presented made the Ti/Pt-Ir anode suitable due to the good properties for chlorine evolution by this material. At different sites where other ions as sulphate, phosphate or (hydrogen) carbonate are present in high concentrations, BDD is a very interesting possibility for the on-site

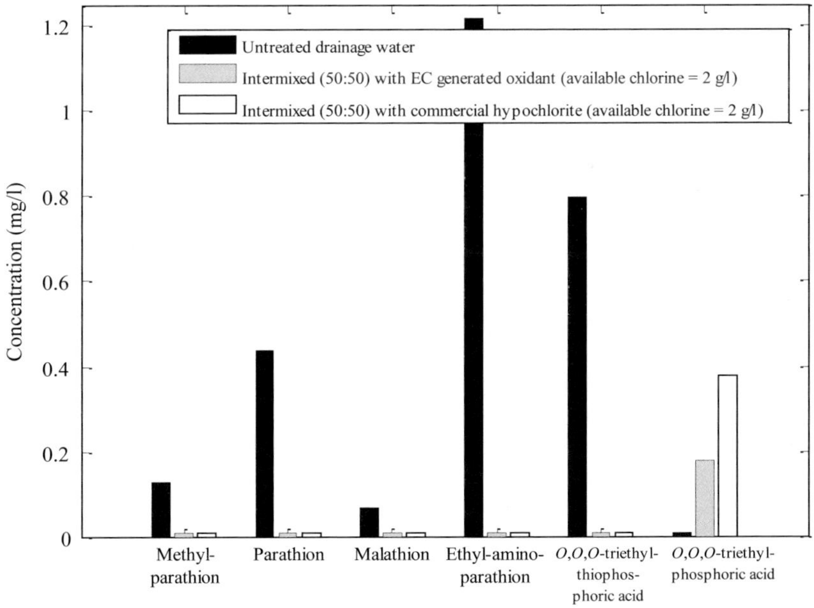

FIGURE 3.10 Initial and final concentrations of the organic contaminants at the site after 1:1 ratio intermixing of EO-generated oxidant and the raw drainage water. *From Ref. Muff et al. (2009) reprinted with permission from Elsevier.* (For colour version of this figure, the reader is referred to the online version of this book.)

generation of strong oxidants. In addition, chlorine is infamous for the risk of by-product formation and this behaviour needs to be studied in more detail before this oxidant solution can be recommended.

3.5.2. Case Study 2 – Runoff Water from a Sediment Removal Process

In this case PAHs are considered as the contaminant of concern. PAHs are an important family of POP-related compounds that due to their benzene analogue structures having two or more fused rings in diverse alignments can be mutagenic, teratogenic and carcinogenic for humans, superior animals and aquatic organisms. They are by-products of incomplete combustion or hydrolysis of organic materials and are widely distributed in the environment from both natural and industrial sources such as forest fires, burning of fossil fuels in diesel engines, wood stoves and power plants, tar and creosote production, etc. PAHs are hydrophobic compounds and their persistence in the environment is mainly due to their low water solubility and a tendency strongly to bind with clay minerals and organic matter present in soil and sediment (Panizza and Cerisola, 2006). Once PAHs enter into the water systems, it is difficult to remediate these by conventional methods, as PAHs are essentially recalcitrant, persistent and nonreactive in water due to their highly stable physicochemical characteristics (Alcantara et al., 2008).

3.5.2.1. Background

Sediments act as a sink for hydrophobic, recalcitrant and hazardous compounds, and organic pollution of sediment is a worldwide problem. The pollution may be complex, but the most important organic compounds as stated in the literature are PAHs and organo-tin compounds such as tributyltin (TBT), di-, and monobutyltin (Perelo, 2010). As long as resting on the seabed, contact with humans and aquatic organisms (except organisms living in the sediment) are not an alarming problem due to the often low solubility of these compounds. However, when harbours are dredged in order to maintain sail channels, the contaminants hidden in the dark are mixed and suspended with the sea water.

In Denmark, the most frequently used method to remove slightly contaminated sediment is to pump it ashore to artificial inland lakes or upland sites just next to the shore, in a water-consuming process using five parts water per one part sediment in order to make the media pumpable (Figure 3.11). The water is then discharged back into the sea, but has now changed status to process water and is subjected to strict discharge limits (PAH: 0.010 μg/l). Traditionally, sand filtration has been the preferred solution, since removal of the suspended part of the pollutants adsorbed to sediment particles has been sufficient. But now,

FIGURE 3.11 **Marking of the upland site next to the natural channel into Odense Harbour (Denmark) that serves as deposit for removed sediment.** *Google.* (For colour version of this figure, the reader is referred to the online version of this book.)

increasingly strict demands to the removal have made it necessary to come up with new approaches to attach the pollutants in the water phase. The task is to remove already low concentrations of organic PAHs to even lower concentrations at an acceptable expense, a task often faced when working with treatment of micropollutants, where mass transfer limitations are hard to avoid due to the small concentrations.

3.5.2.2. Results and Implications

Due to the high salinity and hence high conductivity typically found in the sea water media, EO was straightforward to investigate as a possible treatment solution. The initial test runs using the Ti/Pt-Ir anode material showed that the compounds in the main family of pollutants present, the PAHs, were all degraded after electrolysis using 13.2 kWh/m^3 of energy (Figure 3.12). In another treatment approach, intermixing of electrochemically generated oxidant and the harbour water was again used to investigate the possibility of passing only a minor part of the discharge stream through the electrochemical cell and utilising the remaining residual oxidant to degrade the pollutants in the rest of the stream through indirect oxidation in an electro-chlorination-like process. An intermixing ratio of 1 to 3 (oxidant to sea water) removed all PAHs in the final sample using about 19 kWh/m^3 treated sea water (Figure 3.12), but both processes are subject to further optimisation. The organo-tin compounds were not found in the sea water sample obtained from Odense Harbour and was thus not evaluated.

These results induced a much more detailed investigation of the electrochemical degradation of PAHs in aqueous solution. Three PAHs were selected for model studies; naphthalene – the most soluble (two rings)

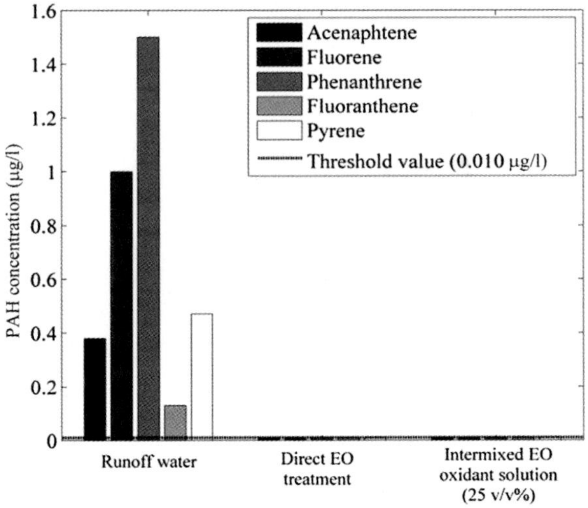

FIGURE 3.12 **Initial and final concentrations of PAHs compounds from tests with runoff process water from Odense Harbour.** Two different treatment approaches applied; direct passing through the cell and an indirect electro-chlorination approach. *Muff, J., Søgaard, E. 2010. Water Science and Technology, 61.8, 2043–2051. Reprinted with permission from IWA publishing.*

and anticipated most reactive species; fluoranthene and pyrene – since they were among the ones present in the harbour sample, both structured of four rings, but in different alignments. The solubility of the PAHs in the model solutions was artificially increased using small volumes of acetonitrile as cosolvent to the demineralised water.

The experiments showed that all three PAHs could be oxidised through a direct oxygen transfer reaction when treated in a 0.10 M Na_2SO_4 electrolyte using the Ti/Pt-Ir anode. Due to the use of fluorescence as detector, it was the initial oxidation step of the PAH and the subsequent loss of the specific emission wavelength, which was measured during the electrolysis. In order to be certain that the loss of fluorescence was due to actual degradation, gas chromatographic analysis was used for confirmation in one of the runs.

The presence of electroactive chloride ions in the electrolyte significantly enhanced the PAH oxidation rates due to indirect oxidation pathways. Naphthalene was influenced the most, probably due to its higher solubility compared to the other PAHs and thus higher affinity to participate in the indirect homogeneous chemical oxidation reaction. An important discovery was that shifts of applied current density showed a mixed behaviour, not as straightforward as with the pesticides previously discussed. It was studied in 0.14 M NaCl electrolytes and when the current density was decreased stepwise from 200 mA/cm^2, a local minimum value of the rate constant was found at 50 mA/cm^2, with a slightly higher reaction rate observed at the lower 25 mA/cm^2. The trend is considered to be caused by a shift in main oxidation routes, where the indirect chlorine oxidation at the low applied cell potentials becomes less important as the chloride electrolysis is challenged more by the EOTR by MO_{x+1}.

This observation indeed demonstrated some of the optimisation challenges that can be faced in EO. A fast kinetics is definitely important, since the rate of oxidation determines the needed retention times in reaction vessels (reservoir in a semibatch recirculation process) and hence the dimensions of these vessels according to fit the needed capacity. Faster kinetics means smaller dimensions and cheaper vessels. However, the efficiency is another just as important parameter as energy waste equals higher costs. COD measurements are the traditional parameter in play when calculating the efficiency of the electrochemical performance. However, when considering micropollutants with (in relation to waste water) very low organic concentrations of below 0.5–2 mg/l in total (corresponds up to ≈6 mg/l of COD), these procedures are not convenient (COD analysis typically has a detection limit of 4 mg/l). A quite simple measure of relative efficiency, the current efficiency constant, k_q, can be used in the evaluation of which settings provided the best efficiency with respect to the individual organic. The constant is defined analogous to the rate constant, but based on concentrations evaluated against the specific charge passed through the solution.

$$\frac{d[\mathrm{PAH}]}{dQ} = -k_q \cdot [\mathrm{PAH}]^n \tag{3.48}$$

When the current efficiency constants of the naphthalene oxidation curves were determined, the reverse trend compared to the kinetics was evident, with much faster oxidation per ampere spent in the process at the lower applied current densities (Figure 3.13). At higher currents, much more energy is wasted on side reactions as primarily oxygen evolution and probably accumulation of residual chlorine. From a plant design perspective, this is of course very important, but the actual removal rate at 25 mA/cm^2 was still less than half of the one obtained at 200 mA/cm^2. This low rate reduces the capacity of a given plant, which might be more expensive than cost of the lower energy efficiency.

The observations from this study clearly demonstrate some of the challenges faced when optimising treatment processes, electrochemically founded or not. However, EO was considered as a promising solution to this process water treatment issue, particularly due to the fact that part flow treatment was possible due to the chemical oxidation

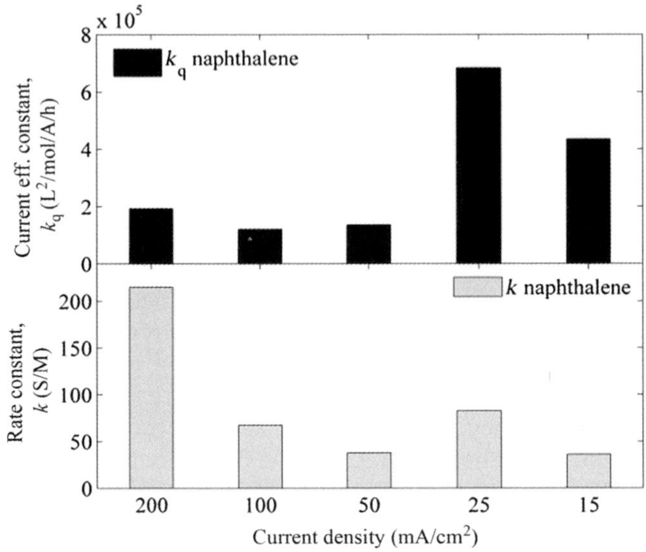

FIGURE 3.13 **Comparison of the second-order reaction rate constants, k, and the current efficiency constants, k_q, for naphthalene in 0.14 M NaCl electrolyte.** *Muff, J., Søgaard, E. 2010. Water Science and Technology, 61.8, 2043–2051. Reprinted with permission from IWA publishing.*

of the PAHs by electro-generated active chlorine species. However, very important questions considering the issues of by-product formation was again raised as the main limitation so far and let to a detailed study on the formation of halogenated PAHs during EO of simulated sea water, a study and discussion that is currently under review. The conclusions of this work are that halogenation of the parent organics are a matter of consideration, when treating PAH-rich sea water by EO. Differences were seen among the PAHs, and naphthalene as the most soluble compound was the only one to be halogenated in detectable amounts. In a single sodium chloride electrolyte, up to 13% of the initial naphthalene was chlorinated at the peak concentration during treatment before it subsequently was removed. However, even small concentrations of added bromide in the electrolyte suppressed the formation of chlorinated by-products for an exchange with primarily mono- and di-brominated naphthalene in peak concentrations up to 30–39% of the initial naphthalene. Most important, all the considered by-products were despite a more recalcitrant behaviour degraded subsequent to formation. This means that considering the removal of the PAHs in question by EO, treatment times need to be prolonged to ensure safe discharge of the treated water, with an acceptable low level of halogenated by-products.

3.5.3. Case Study 3 – Electrochemical Disinfection

Disinfection of water is maybe the most important process within drinking water treatment and generally constitutes the final step in the treatment. Since it represents the last barrier against pathogenic microorganisms, its effectiveness is a crucial point in ensuring public health. Obviously, production of drinking water from surface water requires very high standards of disinfection, and electrochemical techniques have emerged as one of the most feasible alternatives to the common traditional chlorination, which has drawbacks as formation of several potentially toxic by-products.

The most popular electrochemical method is not surprisingly electrochlorination, either through on-site production of disinfectants from brine or direct production of oxidants from the water to be treated by the electrolysis unit. Both processes have been shown to provide significantly better disinfection efficiencies compared to traditional chlorination, which generally are attributed to production of other strong oxidants than active chlorine species in the disinfectant. Especially, the role of reactive oxygen species (ROS), to which category the hydroxyl radicals, atomic oxygen, hydrogen peroxide and ozone belong, has been studied. However, the short lifetime of particularly the radical ROS makes it only possible to underline their role in disinfection using the direct treatment strategy by passing the subject water through the cells. The disinfection efficiency is directly connected to the oxidation power that can be demonstrated through a small experiment. In Figure 3.14, an example

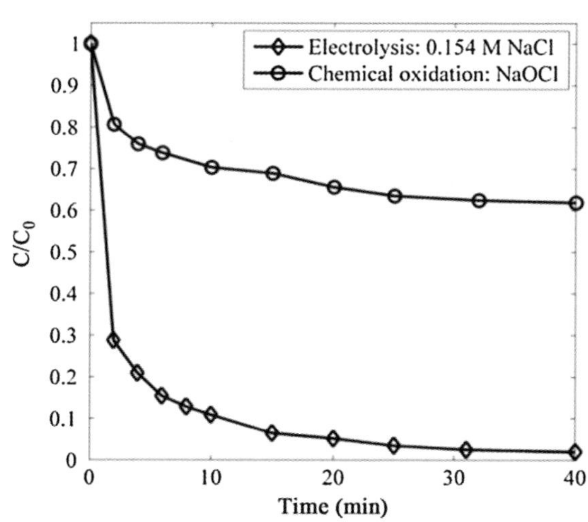

FIGURE 3.14 **Electro-chlorination vs chemical hypochlorination of the organic dye *p*-nitrosodimethylaniline at comparable electrolytic and chemical feeds of active chlorine species.**

showing the superior oxidation power of electro-chlorination compared to chemical hypochlorination is showed considering bleaching of the organic dye *p*-nitrosodimethylaniline.

The electrolytic production of active chlorine is linearly correlated with the applied current intensity, and the bleaching of *p*-nitrosodimethylaniline at 32 mA/cm^2 (1.9 A) was compared to the chemical bleaching by hypochlorite fed at the same rate in the same molar concentration (1.18 ml/min of 14 g/l NaOCl) to the reservoir by a high-performance liquid chromatography pump.

The superiority of the electro-chlorination is evident. The chemical bleaching is despite a fast initial removal of 30% of the colour not capable of any significant further reduction of the solution absorbance within the 40 min it took for the electro-chlorination process almost completely to bleach the solution. These experiments clearly demonstrate that other reactive species than the ordinary active chlorine are generated in the electro-chlorination process, increasing the efficiency under conditions comparable with chemical chlorination.

In a chloride-free electrolyte using BDD, hydroxyl radicals were found by Jeong et al. (2006) to be the most important reactive species for inactivation of *Escherichia coli*. The same authors examined the role of electrode material on the generation of oxidants to be used for disinfection, and found that, in accordance with the theory, hydroxyl radicals were mainly produced by BDD, where they play a key role in generation of ozone (Jeong et al., 2009) in addition to active chlorine.

Numerous electrochemical systems and electrode materials have been tested against a variety of microorganisms, and their effectiveness to the abatement of bacteria, viruses and protozoa is largely dependent on the electrochemical reactor, anode material, electrolyte composition, and electrolysis conditions. Some of the studies within electrochemical drinking water disinfection have recently been reviewed by Martinez-Huitle and Brillas (2008) and Kraft (2008). Table 3.4 shows some other selected examples of logarithmic reductions of relevant microorganisms by different anode materials.

3.5.3.1. Electrochemical Disinfection of **E. coli**

An example of electrochemical disinfection is inactivation of *E. coli*. In saline water this was investigated using the Ti/Pt-Ir anode and compared with common chemical chlorination using the Selleck–Collins model for disinfection efficiency of coliform bacteria in domestic wastewater effluent by chlorination (Selleck et al., 1978).

$$\frac{N}{N_0} = (1 + 0.23 \cdot C \cdot t)^{-3} \tag{3.49}$$

TABLE 3.4 Results of Selected Electrochemical Disinfection Studies

Microorganism	Anode	Current Density (mA/cm^2)	NaCl (%)	Reduction	Reference
E. coli	Pt/Nb	4	0.058	Log-7	Kerwick et al. (2005)
E. coli	Ti/RuO_2 Ti/TiO_2	5	0.058	Log-3	Li et al. (2004)
E. coli	Ti/RuO_2 Ti/TiO_2	8	0.008	Log-3	Li et al. (2002)
E. coli	Graphite	7.5	0.85	Log-3	Stoner et al. (1982)
Legionella[1]	Ti/RuO_2	–	0.1	Log-3	Feng et al. (2004)
Legionella	B-doped diamond	150	0.7	Log-1	Pupunat and Rychen (2002)
Pseudomonas aeruginosa	Ti/RuO_2	20	–	Log-6	Drogui et al. (2001)
MS2	Pt/Nb	24	–	Log-6	Kerwick et al. (2005)
Cryptosporidium p.[2]	–	–	0.001	Log-3	Casteel et al. (2000)

The presented results are only intended to serve as examples, and do not provide a full overview of the research done in this area.
[1]*High-voltage electrical field study.*
[2]*Indirect treatment.*

N_0 is the initial *E. coli* concentration, N is the concentration after chlorination, C (mg/l) is the initial active chlorine concentration and t (min) is the contact time.

The 0.154 M NaCl and *E. coli* electrolyte solution was passed a single time through the cell at different applied current densities. When electrolysed, the sample was quenched with sodium thiosulphate ($Na_2S_2O_3$) to remove residual oxidants. The contact time from the point of entrance to the electrochemical cell to quenching in the sample beaker was 2.3 s calculated from the total volume in cell and hoses (260 ml) and the flow rate (400 l/h = 0.111 l/s). The number of surviving cells was determined by incubation for 24 h at 37 °C and plate counts.

The surviving number of *E. coli* indicated in Figure 3.15(A) as colony forming units per 100 ml rapidly decreased as a function of the applied current, and no living germs were found at 20 A (332 mA/cm). The

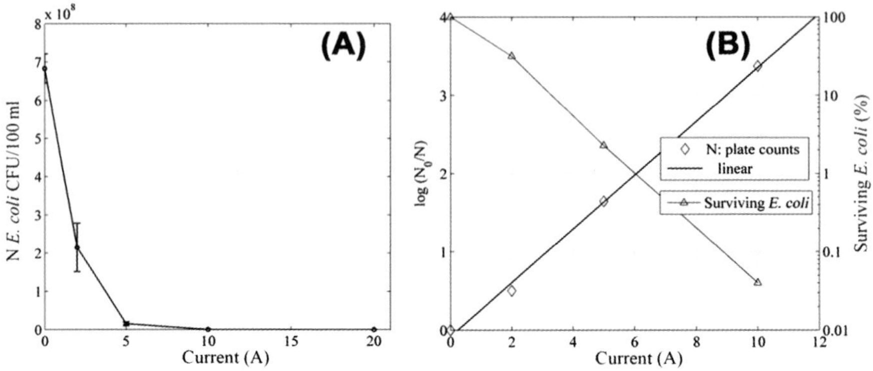

FIGURE 3.15 (A) Electro-chlorination for disinfection of 0.154 M NaCl electrolyte infected with *E. coli* with contact time of 2.3 s. (B) The logarithmic reduction and survival ratio as function of applied current (Muff, 2010).

logarithmic reduction showed a linear dependence on the applied current (Figure 3.15(B)) (Muff, 2010). In order to secure a log-3 reduction (99.9% killing) 9 A was needed at this very short contact time. From the active chlorine production capacity experiments, the concentration obtained in one passage at 9 A can be calculated to 12.7 mg/l (9.4 mg/A/min· 9 A/6.7 l/min). Compared to the pure chemical chlorination, 3.1 min (186 s) of contact time is needed in order to obtain the same disinfection efficiency using purely chemical means of chlorination as calculated using Selleck–Collins. Seen from another perspective, the calculated concentration of active chlorine could by use of the model only cause 24% killing ratio in the 2.3 s and not 99.9%.

Electrochemical disinfection is a very efficient technique, and electro-chlorination is as demonstrated much more efficient than common chemical chlorination. One reason why, is most probably the presence of other oxidising species as the ROS generated in the electrochemical process as has been widely investigated. However, electro-chlorination has been shown also to be more efficient than chlorination even by use of anode materials, which have been shown not to produce significant amounts of oxygen radical species. In this case, chlorine will still be the main disinfectant in play. Since the reaction between chloride and the electrode initially may be of radical nature, a proposed oxidising species generated at the anode surface is the hydrated chlorine radical. They may act as a powerful oxidant in the electro-chlorination process in the cell in the vicinity of the anode, but react further to residual active chlorine species and react as such in the bulk. However, the differences observed on electro-chlorination and chemical chlorination need much more study before the actual responsible mechanisms are revealed.

In addition to ROS production in chlorine-free media, the production and efficiency of peroxodisulphate, peroxodicarbonate and peroxodiphosphate by BDD have been examined as residual disinfectants, but the residual disinfection efficiencies of these oxidants are not as good as active chlorine. However, efficient disinfection of chloride-free water has been obtained using BDD due to the ROS formation (Martinez-Huitle and Brillas, 2008).

3.5.3.2. *Implications of Electrochemical Disinfection*

The practical possibilities of the efficient disinfection provided with the electro-chlorination process are many. Since disinfection typically can do with a single pass, the implementation of the technique is very easy compared to COD removal, where recirculation or some kind of cascade treatment with multiple cells has to be used. Experience has especially been obtained using in-line electro-chlorination for treatment and disinfection of salt water swimming pools, where the enhanced disinfection efficiency from the passing of the cells makes it possible to operate at much lower residual free chlorine concentrations (<0.2 mg/l) that again decreases the concentrations of DBPs such as THMs and chloramines.

The market for drinking water disinfection is in many countries around the world, where surface water constitutes the main source of drinking water, the opportunities are immense. The ease of operation of the electrochemical technique and the fact that it only needs electric power in order to operate may in addition provide great possibilities of exporting the technique to third-world countries, where securing safe drinking water is of highest importance.

A coming market of high relevance too is ballast water disinfection. The introduction of invasive marine species into new environments by ships' ballast water, attached to ships' hulls and via other vectors has been identified as one of the four greatest threats to the world's oceans. In the framework of the International Maritime Organisation the 'International Convention for the Control and Management of Ships Ballast Water & Sediments' was adopted by 74 states in 2004. In short terms, the convention demands that by 2016 the latest, all ballast water from ships has to be disinfected before discharge. The intension is of course to hinder the devastating effect of invasive species on local marine habitats, and electrochemical techniques and electro-chlorination has a great potential in offering a solution for this problem. The high conductivity of sea water (mainly due to high chloride concentration) promoting low cost and the small space requirements of the electrochemical technique are important advantages compared to competitive techniques. Several companies have electrochemically founded solutions in the pipeline of approval, and with more than 100,000 commercial ships in the world, the economic potential is very promising.

3.6. IMPLEMENTATION OF EO IN LARGE SCALE – COMBINED TREATMENT

The implementation of EO for disinfection purposes is in principle rather straightforward using in-line electrolysis, when the conductivity of the water media is sufficient or otherwise through addition of electrolyte as salts or brine solution. For disinfection, typically a single pass of the electrochemical cell is sufficient, when the applied current is high enough to reach the aimed killing ratio (Figure 3.16(A)). For oxidation of organics, recirculation of the waste stream through the electrochemical cell is needed in order to obtain the target degradation levels, and this stipulates some other requirements.

In general two continuous treatment approaches are conceivable; recirculation through the cell from a reservoir in a semi-batch process or cascade treatment with multiple single passes of the waste stream in cells connected in series. So far industrial installations typically use the semi-batch approach in different alignments utilising both direct treatment of the waste stream in the cells and indirect chemical oxidation through addition of an electrochemically produced oxidant (Figure 3.16(B)). Not many scientific studies are published on pilot and/or full-scale treatment of waste water by EO, but the ones that are, to the authors knowledge, all use the semi-batch approach, and this design principle does also seem as the one most easy to implement. A combined approach using semi-batch treatment in two connected stages is also seen (Serikawa et al., 2009).

In many cases, waste water contains both solid and dissolved pollutants, biologically degradable and toxic and refractory organic compounds. In these cases, EO cannot stand alone as a universal solution, but needs to be combined with other treatment principles for the most

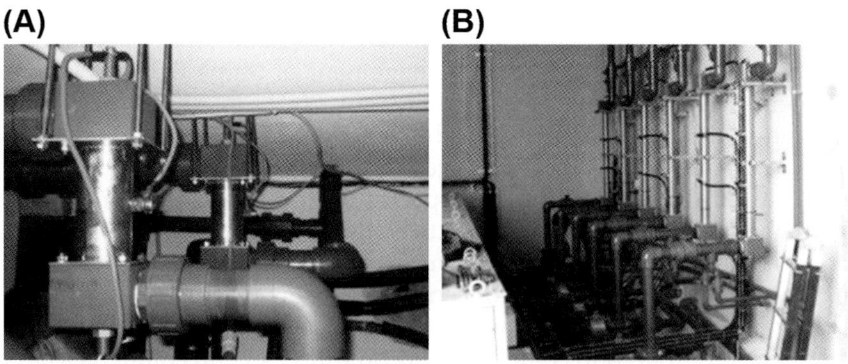

FIGURE 3.16 (A) Two parallel cells for in-line hotel swimming pool treatment. (B) Six parallel cells installed for COD abatement of olive oil mill waste water (OMWW), 500 m^3/d. (For colour version of this figure, the reader is referred to the online version of this book.)

optimal performance. This can be combined with other AOPs in sequential or simultaneous treatment, but most frequently a separation step is needed prior to EO in order to remove the solid particulate waste. In addition, the combination of EO with the cheaper biological treatment is obviously an approach to minimise treatment cost. This combination is frequently stressed as promising and the way forward, since EO pretreatment can enhance the BOD/COD ratio of the waste water and convert toxic and recalcitrant organics to highly oxidised biodegradable carbon source compounds, which are rapidly metabolised by microorganisms. However, EO as pretreatment requires quenching of any residual oxidant that will provide a disaster if entering the biological treatment stage. The concept of integration of AOPs as pretreatment to biodegradation posttreatment has been illustrated by Comninellis et al. (2008), who consider the rate of C–C bond scission by chemical and biological means vs the molecular size of the organics (Figure 3.17).

As seen, AOPs in general provide the fastest degradation rates when the organics have larger size, whereas a breakpoint appears beyond which biological treatment becomes more attractive than chemical oxidation in terms of rate and consequently efficiency of the treatment.

Another combined approach is to use EO for tertiary polishing of waste water before discharge, not only for disinfection purposes, but also for oxidation and removal of pollutants not affected by the preceding treatment stages or to complete the COD removal or other parameters to comply with the discharge limits. An example of this approach is published by Chrysafopoulou et al. (2006), who treated biologically processed waste water in a pilot unit consisting of an electrolytic oxidation step and

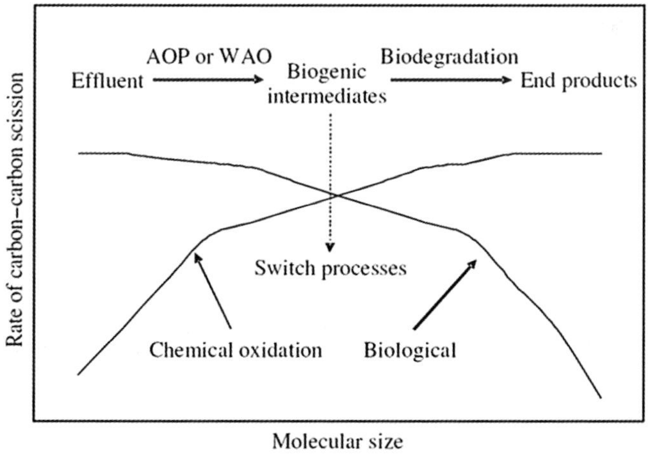

FIGURE 3.17 Concept of integrating AOP pretreatment and biological posttreatment. *Reprinted from Ref. Comninellis et al. (2008) with permission from Wiley Interscience.*

a flotation and coagulation/precipitation step. The results were reported as promising with complete discolouration, disinfection, removal of suspended solids and reduction of COD and BOD. About 1 m^3/h was continuously treated in the setup, with the recirculation flow rate being 7.5 m^3/h, but it was recommended to use higher recirculation rate in future installation. The effluent from the pilot unit complied with the guidelines for agricultural water reuse or aquifer recharge in Greece.

Several other combinations of EO and other treatment techniques have been proposed and studied in smaller scale with some of them shown in Figure 3.18.

Particularly the combination of membrane filtration and EO is promising, since the use of reverse osmosis or other membranes can reduce the capacity requirement to the EO, generating a concentrated stream of not only suspended but also soluble pollutants to be EO treated, which is good to enhance the mass transfer in the cell. This combination has in particular been studied within leachate treatment. However, all the presented treatment schemes in Figure 3.18 have been evaluated with promising results (Anglada et al., 2009), and development of a sustainable process based on the integration of efficient technologies is often stressed as one of the key obstacles that have to be overcome before full-scale implementation of EO can be expected in large scale. Another important upscale issue is to better understand the behaviour of EO described by mathematical models and parameters. As presented we are close by,

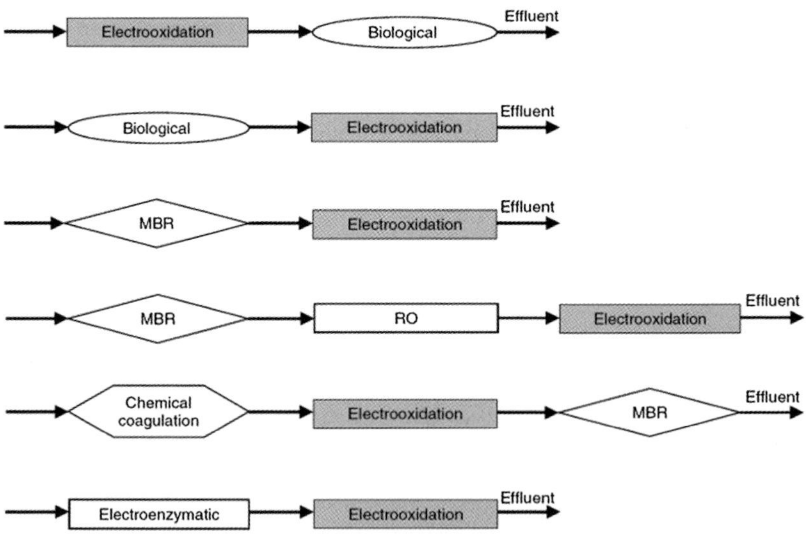

FIGURE 3.18 Flow sheets for different alignments of combined EO treatment of waste water. *Reprinted from Ref. Anglada et al. (2009) with permission from Wiley Interscience.*

but the model has been reported to show deviations when dealing with complex mixtures, where the presence of electro-generated oxidants also contributes to the oxidation. Other models that describe the combined contribution of both mechanisms have been developed, but still lack thorough validation, and the many competitive reactions on the anode and the different species interacting create a very complex system as clearly stressed in this thesis. Future research areas are definitely scaling up issues of EO from the laboratory studies to pilot and full-scale treatment in combination with separation and other advanced oxidation techniques.

GREEN ELECTROCHEMISTRY – COMBINATIONS WITH RENEWABLE ENERGY SOURCES

Green and sustainable remediation is the principle that now increases in attention. Already within the technical area of remediation of contaminated sites, software tools have been developed, where the total impact of different remediation strategies can be evaluated based on carbon footprint analysis, etc., with the overall aim of minimising the impact on the surrounding ecosystem from our otherwise good intentions of taking care of and restoring the environment. Electrochemistry and EO as a technique for degradation of organics have features that render them promising as part of such strategies. Many of the competing techniques for the abatement of refractory pollutants involve transportation and storage of reactants, which compared to EO induce increased emission of carbon dioxide that is not generated by EO, since the primary reactant is the electron itself produced on-site and addition of chemicals is not necessary if used under the right circumstances. If the needed electrical power can be produced by on-site renewable means from, e.g. wind or solar energy, then we are really on the right track. These arguments are in line with the idea of on-site production of oxidant for ISCO applications as earlier discussed.

Another obvious environmental benefit, which still needs to be explored in coming research, is the fact that the main anodic oxidation always in water electrolysis is corresponded by the cathodic hydrogen evolution, and a promising potential is to couple this hydrogen production with fuel cells, which can use hydrogen to generate electrical power as a co-source for the EO treatment unit. For this combination to be feasible other than in drawings, the efficiency and several other parameters of the fuel cell technology needs to be appropriately developed.

3.7. PERSPECTIVES

Water resources as the rest of the natural environment on Earth are under pressure from the ever increasing human activities in our desire for global wealth. As the analytical capabilities improve, traces of these activities are intensively emerging in the form of harmful organic pollutants recalcitrant to natural degradation. Efficient treatment of waste water and polluted ground water is essential in order to prevent the discharge of these pollutants and to secure public health and our access to good-quality drinking water. In this context, EO has the potential to become a very important technique for treatment of almost any kind of polluted water. It is a powerful technique for degradation of a wide range of toxic and recalcitrant organics and fully competitive with the other AOPs with some significant advantages when used under the right conditions. However, EO is still just on the edge of industrial and commercial breakthrough, and more knowledge on the complex chemistry in the cell and the engineering perspectives in designing and optimising an EO treatment unit is needed before the technique can be considered a mature and ready implementable treatment solution.

References

Alcantara, M.T., Gomez, J., Pazos, M., Sanroman, M.A., 2008. Combined treatment of PAHs contaminated soils using the sequence extraction with surfactant-electrochemical degradation. Chemosphere 70 (8), 1438–1444.

Anglada, A., Ortiz, D., Urtiaga, A.M., Ortiz, I., 2010. Electrochemical oxidation of landfill leachates at pilot scale: evaluation of energy needs. Water Sci. Technol. 61 (9), 2211–2217.

Anglada, A., Urtiaga, A., Ortiz, I., 2009. Contributions of electrochemical oxidation to wastewater treatment: fundamentals and review of applications. J. Chem. Technol. Biotechnol. 84 (12), 1747–1755.

Anglada, A., Urtiaga, A., Ortiz, I., 2009. Pilot scale performance of the electro-oxidation of landfill leachate at boron-doped diamond anodes. Environ. Sci. Technol. 43 (6), 2035–2040.

Ayoub, K., Van Hullebusch, E.D., Cassir, M., Bermond, A., 2010. Application of advanced oxidation processes for TNT removal: a review. J. Hazard. Mater. 178 (1–3), 10–28.

Bashir, M.J.K., Isa, M.H., Kutty, S.R.M., Bin Awang, Z., Aziz, H.A., Mohajeri, S., Farooqi, I.H., 2009. Landfill leachate treatment by electrochemical oxidation. Waste Manag. 29 (9), 2534–2541.

Bergmann, H., Koparal, S., 2005a. The formation of chlorine dioxide in the electrochemical treatment of drinking water for disinfection. Electrochim. Acta 50 (25–26), 5218–5228.

Bergmann, M.E.H., Koparal, A.S., 2005b. Studies on electrochemical disinfectant production using anodes containing RuO2. J. Appl. Electrochem. 35 (12), 1321–1329.

Bergmann, M.E.H., Rollin, J., 2007. Product and by-product formation in laboratory studies on disinfection electrolysis of water using boron-doped diamond anodes. Catal. Today 124 (3–4), 198–203.

Bergmann, M.E.H., Rollin, J., Iourtchouk, T., 2009a. The occurrence of perchlorate during drinking water electrolysis using BDD anodes. Electrochim. Acta 54 (7), 2102–2107.

Bergmann, M.E.H., Rollin, J., Koparal, A.S., 2009b. Chlorate and Perchlorate – New Criterions for Environmentally Friendly Processes in Advanced Oxidation.

Bonfatti, F., De Battisti, A., Ferro, S., Lodi, G., Osti, S., 2000a. Anodic mineralization of organic substrates in chloride-containing aqueous media. Electrochim. Acta 46 (2–3), 305–314.

Bonfatti, F., Ferro, S., Lavezzo, F., Malacarne, M., Lodi, G., De Battisti, A., 2000b. Electrochemical incineration of glucose as a model organic substrate – II. Role of active chlorine mediation. J. Electrochem. Soc. 147 (2), 592–596.

Bonfatti, F., Ferro, S., Lavezzo, F., Malacarne, M., Lodi, G., De Battisti, A., 1999. Electrochemical incineration of glucose as a model organic substrate I. Role of the electrode material. J. Electrochem. Soc. 146 (6), 2175–2179.

Boudreau, J., Bejan, D., Li, S.H., Bunce, N.J., 2010. Competition between electrochemical advanced oxidation and electrochemical hypochlorination of sulfamethoxazole at a boron-doped diamond anode. Ind. Eng. Chem. Res. 49 (6), 2537–2542.

Boye, B., Brillas, E., Marselli, B., Michaud, P.A., Comninellis, C., Farnia, G., Sandona, G., 2006. Electrochemical incineration of chloromethylphenoxy herbicides in acid medium by anodic oxidation with boron-doped diamond electrode. Electrochim. Acta 51 (14), 2872–2880.

Brillas, E., Sires, I., Oturan, M.A., 2009. Electro-fenton process and related electrochemical technologies based on Fenton's reaction chemistry. Chem. Rev. 109 (12), 6570–6631.

Canizares, P., Garcia-Gomez, J., Lobato, J., Rodrigo, M.A., 2004. Modeling of wastewater electro-oxidation processes part I. General description and application to inactive electrodes. Ind. Eng. Chem. Res. 43 (9), 1915–1922.

Canizares, P., Garcia-Gomez, J., Lobato, J., Rodrigo, M.A., 2004. Modeling of wastewater electro-oxidation processes part II. Application to active electrodes. Ind. Eng. Chem. Res. 43 (9), 1923–1931.

Canizares, P., Garcia-Gomez, J., Saez, C., Rodrigo, M.A., 2003. Electrochemical oxidation of several chlorophenols on diamond electrodes - Part I. Reaction mechanism. J. Appl. Electrochem. 33 (10), 917–927.

Canizares, P., Hernandez-Ortega, M., Rodrigo, M.A., Barrera-Diaz, C.E., Roa-Morales, G., Saez, C., 2009. A comparison between conductive-diamond electrochemical oxidation and other advanced oxidation processes for the treatment of synthetic melanoidins. J. Hazard. Mater. 164 (1), 120–125.

Canizares, P., Paz, R., Saez, C., Rodrigo, M.A., 2008. Electrochemical oxidation of alcohols and carboxylic acids with diamond anodes – a comparison with other advanced oxidation processes. Electrochim. Acta 53 (5), 2144–2153.

Canizares, P., Saez, C., Sanchez-Carretero, A., Rodrigo, M.A., 2009. Synthesis of novel oxidants by electrochemical technology. J. Appl. Electrochem. 39 (11), 2143–2149.

Casteel, M.J., Sobsey, M.D., Arrowood, M.J., 2000. Inactivation of *Cryptosporidium parvum* oocysts and other microbes in water and wastewater by electrochemically generated mixed oxidants. Water Sci. Technol. 41 (7), 127–134.

Chatzisymeon, E., Dimou, A., Mantzavinos, D., Katsaounis, A., 2009. Electrochemical oxidation of model compounds and olive mill wastewater over DSA electrodes: 1. The case of Ti/IrO2 anode. J. Hazard. Mater. 167 (1–3), 268–274.

Chatzisymeon, E., Xekoukoulotakis, N.P., Coz, A., Kalogerakis, N., Mantzavinos, D., 2006. Electrochemical treatment of textile dyes and dyehouse effluents. J. Hazard. Mater. 137 (2), 998–1007.

Chatzisymeon, E., Xekoukoulotakis, N.P., Diamadopoulos, E., Katsaounis, A., Mantzavinos, D., 2009. Boron-doped diamond anodic treatment of olive mill wastewaters: statistical analysis, kinetic modeling and biodegradability. Water Res. 43 (16), 3999–4009.

Chen, G.H., 2004. Electrochemical technologies in wastewater treatment. Sep. Purif. Technol. 38 (1), 11–41.

Chen, W.S., Liang, J.S., 2009. Electrochemical destruction of dinitrotoluene isomers and 2,4,6-trinitrotoluene in spent acid from toluene nitration process. J. Hazard. Mater. 161 (2–3), 1017–1023.

Chiang, L.C., Chang, J.E., Wen, T.C., 1995. Electrochemical treatability of refractory pollutants in landfill leachate. Hazard. Waste Hazard. Mater. 12 (1), 71–82.

Chiang, L.C., Chang, J.E., Wen, T.C., 1995. Indirect oxidation effect in electrochemical oxidation treatment of landfill leachate. Water Res. 29 (2), 671–678.

Chrysafopoulou, E., Hourdakis, A., Aghelopoulos, M., Mantzavinos, D., Kalogerakis, N., 2006. A pilot scale electrolytic unit for tertiary treatment of industrial effluents. Int. J. Environ. Technol. Manag. 6, 480–488.

Comninellis, C., 1994. Electrocatalysis in the electrochemical conversion/combustion of organic pollutants for waste water treatment. Electrochim. Acta 39 (11–12), 1857–1862.

Comninellis, C., Debattisti, A., 1996. Electrocatalysis in anodic oxidation of organics with simultaneous oxygen evolution. J. Chim. Phys. Physicochim. Biol. 93 (4), 673–679.

Comninellis, C., Kapalka, A., Malato, S., Parsons, S.A., Poulios, L., Mantzavinos, D., 2008. Advanced oxidation processes for water treatment: advances and trends for R&D. J. Chem. Technol. Biotechnol. 83 (6), 769–776.

Comninellis, C., Nerini, A., 1995. Anodic-oxidation of phenol in the presence of NaCl for waste-water treatment. J. Appl. Electrochem. 25 (1), 23–28.

Cui, Y.H., Li, X.Y., Chen, G.H., 2009. Electrochemical degradation of bisphenol A on different anodes. Water Res. 43 (7), 1968–1976.

Dabrowski, A., Mieluch, J., Sadkowski, A., Wild, J., Zoltowski, P., 1975. Pilot scale studies on purification of phenol waste-waters by electrochemical method. Przem. Chem. 54 (11), 653–655.

Daniell, J.F., 1836. On voltaic combinations. Philos. Trans. R. Soc. Lond. 126, 107–124.

Drogui, P., Elmaleh, S., Rumeau, M., Bernard, C., Rambaud, A., 2001. Oxidising and disinfecting by hydrogen peroxide produced in a two-electrode cell. Water Res. 35 (13), 3235–3241.

Duo, I., Levy-Clement, C., Fujishima, A., Comninellis, C., 2004. Electron transfer kinetics on boron-doped diamond part I: influence of anodic treatment. J. Appl. Electrochem. 34 (9), 935–943.

Fang, Y.X., Al-Abed, S.R., 2007. Modeling the electrolytic dechlorination of trichloroethylene in a granular graphite-packed reactor. Environ. Eng. Sci. 24 (5), 581–594.

Faouzi, A.M., Nasr, B., Abdellatif, G., 2007. Electrochemical degradation of anthraquinone dye Alizarin Red S by anodic oxidation on boron-doped diamond. Dyes Pigm. 73 (1), 86–89.

Faraday, M., 1849. Experimental Researches in Electricity, second ed. Richard and John Edward Taylor, London.

Feng, C.P., Suzuki, K., Zhao, S.Y., Sugiura, N., Shimada, S., Maekawa, T., 2004. Water disinfection by electrochemical treatment. Bioresour. Technol. 94 (1), 21–25.

Feng, J.R., Johnson, D.C., 1990. Electrocatalysis of anodic oxygen-transfer reactions – Fe-doped beta-lead dioxide electrodeposited on noble metals. J. Electrochem. Soc. 137 (2), 507–510.

Ferro, S., De Battisti, A., 2002. Electron transfer reactions at conductive diamond electrodes. Electrochim. Acta 47 (10), 1641–1649.

Fierro, S., Nagel, T., Baltruschat, H., Comninellis, C., 2008. Investigation of formic acid oxidation on Ti/IrO_2 electrodes using isotope labeling and online mass spectrometry. Electrochem. Solid State Lett. 11 (7), E20–E23.

Fierro, S., Nagel, T., Baltruschat, H., Comninellis, C., 2007. Investigation of the oxygen evolution reaction on Ti/IrO_2 electrodes using isotope labelling and on-line mass spectrometry. Electrochem. Commun. 9 (8), 1969–1974.

Foti, G., Gandini, D., Comninellis, C., Perret, A., Haenni, W., 1999. Oxidation of organics by intermediates of water discharge on IrO_2 and synthetic diamond anodes. Electrochem. Solid State Lett. 2 (5), 228–230.

Fryda, M., Matthee, T., Mulcahy, S., Hampel, A., Schafer, L., Troster, I., 2003. Fabrication and application of diachem (R) electrodes. Diam. Relat. Mater. 12 (10–11), 1950–1956.

Galvani, A., 1791. De viribus electricitatis in motu musculari commentarius. De Bononiensi Scientiarum et Artium Instituto atque Academia Commentarii 7, 363–415.

Giannis, A., Kalaitzakis, M., Diamadopoulos, E., 2007. Electrochemical treatment of olive mill wastewater. J. Chem. Technol. Biotechnol. 82 (7), 663–671.

Gotsi, M., Kalogerakis, N., Psillakis, E., Samaras, P., Mantzavinos, D., 2005. Electrochemical oxidation of olive oil mill wastewaters. Water Res. 39 (17), 4177–4187.

Harrington, T., Pletcher, D., 1999. The removal of low levels of organics from aqueous solutions using Fe(II) and hydrogen peroxide formed in situ at gas diffusion electrodes. J. Electrochem. Soc. 146 (8), 2983–2989.

Holt, K.B., Forryan, C., Compton, R.G., Foord, J.S., Marken, F., 2003. Anodic activity of boron-doped diamond electrodes in bleaching processes: effects of ultrasound and surface states. New J. Chem. 27 (4), 698–703.

Israilides, C., Vlyssides, A.G., Mourafeti, V.N., Karvouni, G., 1997. Olive oil wastewater treatment with the use of an electrolysis system. Bioresour. Technol. 61 (2), 163–170.

Jeong, J., Kim, C., Yoon, J., 2009. The effect of electrode material on the generation of oxidants and microbial inactivation in the electrochemical disinfection processes. Water Res. 43 (4), 895–901.

Jeong, J., Kim, J.Y., Yoon, J., 2006. The role of reactive oxygen species in the electrochemical inactivation of microorganisms. Environ. Sci. Technol. 40 (19), 6117–6122.

Juttner, K., Galla, U., Schmieder, H., 2000. Electrochemical approaches to environmental problems in the process industry. Electrochim. Acta 45 (15–16), 2575–2594.

Kapalka, A., Foti, G., Comninellis, C., 2010. Basic principles of the electrochemical mineralization of organic pollutants for wastewater treatment. In: Comninellis, C., Chen, G.H. (Eds.), Electrochemistry for the Environment, first ed. LLC: Springer Science, pp. 1–24.

Kapalka, A., Foti, G., Comninellis, C., 2008. Determination of the Tafel slope for oxygen evolution on boron-doped diamond electrodes. Electrochem. Commun. 10 (4), 607–610.

Kapalka, A., Foti, G., Comninellis, C., 2008. Investigation of the anodic oxidation of acetic acid on boron-doped diamond electrodes. J. Electrochem. Soc. 155 (3), E27–E32.

Kapalka, A., Foti, G., Comninellis, C., 2008. Kinetic modelling of the electrochemical mineralization of organic pollutants for wastewater treatment. J. Appl. Electrochem. 38 (1), 7–16.

Kapalka, A., Foti, G., Comninellis, C., 2007. Investigations of electrochemical oxygen transfer reaction on boron-doped diamond electrodes. Electrochim. Acta 53 (4), 1954–1961.

Kapalka, A., Lanova, B., Baltruschat, H., Foti, G., Comninellis, C., 2008. DEMS study of the acetic acid oxidation on boron-doped diamond electrode. J. Electrochem. Soc. 155 (7), E96–E100.

Kerwick, M.I., Reddy, S.M., Chamberlain, A.H.L., Holt, D.M., 2005. Electrochemical disinfection, an environmentally acceptable method of drinking water disinfection? Electrochim. Acta 50 (25–26), 5270–5277.

Kraft, A., 2008. Electrochemical water disinfection: a short review – electrodes using platinum group metal oxides. Platin. Met. Rev. 52 (3), 177–185.

Kraft, A., 2007. Doped diamond: a compact review on a new, versatile electrode material. Int. J. Electrochem. Sci. 2 (5), 355–385.

Li, X.Y., Diao, H.F., Fan, F.X.J., Gu, J.D., Ding, F., Tong, A.S.F., 2004. Electrochemical wastewater disinfection: identification of its principal germicidal actions. J. Environ. Eng. ASCE 130 (10), 1217–1221.

Li, X.Y., Ding, F., Lo, P.S.Y., Sin, S.H.P., 2002. Electrochemical disinfection of saline wastewater effluent. J. Environ. Eng. ASCE 128 (8), 697–704.

Marselli, B., Garcia-Gomez, J., Michaud, P.A., Rodrigo, M.A., Comninellis, C., 2003. Electrogeneration of hydroxyl radicals on boron-doped diamond electrodes. J. Electrochem. Soc. 150 (3), D79–D83.

Martinez-Huitle, C.A., Brillas, E., 2009. Decontamination of wastewaters containing synthetic organic dyes by electrochemical methods: a generel review. Appl. Catal. B Environ. 87, 105–145.

Martinez-Huitle, C.A., Brillas, E., 2008. Electrochemical alternatives for drinking water disinfection. Angew. Chem. Int. Ed. Engl. 47 (11), 1998–2005.

Martinez-Huitle, C.A., Ferro, S., 2006. Electrochemical oxidation of organic pollutants for the wastewater treatment: direct and indirect processes. Chem. Soc. Rev. 35 (12), 1324–1340.

Martinez-Huitle, C.A., Ferro, S., De Battisti, A., 2005. Electrochemical incineration in the presence of halides. Electrochem. Solid State Lett. 8 (11), D35–D39.

Martinez-Huitle, C.A., Ferro, S., De Battisti, A., 2005. Electrochemical incineration of oxalic acid: reactivity and engineering parameters. J. Appl. Electrochem. 35 (11), 1087–1093.

Martinez-Huitle, C.A., Ferro, S., De Battisti, A., 2004. Electrochemical incineration of oxalic acid – role of electrode material. Electrochim. Acta 49 (22–23), 4027–4034.

Martinez-Huitle, C.A., Ferro, S., Reyna, S., Cerro-Lopez, M., De Battisti, A., Quiroz, M.A., 2008. Electrochemical oxidation of oxalic acid in the presence of halides at boron doped diamond electrode. J. Braz. Chem. Soc. 19 (1), 150–156.

Mascia, M., Vacca, A., Polcaro, A.M., Palmas, S., Ruiz, J.R., Da Pozzo, A., 2010. Electrochemical treatment of phenolic waters in presence of chloride with boron-doped diamond (BDD) anodes: experimental study and mathematical model. J. Hazard. Mater. 174 (1–3), 314–322.

Miljøstyrelsen, 2007. Store Jordforureningssager – Afrapportering Fra Den Tekniske Arbejdsgruppe. Miljøstyrelsen, København.

Muff, J., 2010. Applications of Electrochemical Oxidation for Degradation of Aqueous Organic Pollutants (Ph.D. thesis). Esbjerg Institute of Technology, Aalborg University.

Muff, J., Andersen, C.D., Erichsen, R., Soegaard, E.G., 2009. Electrochemical treatment of drainage water from toxic dump of pesticides and degradation products. Electrochim. Acta 54 (7), 2062–2068.

Muff, J., Søgaard, E.G., 2011. Identification and fate of halogenated PAHs formed during electrochemical treatment of saline aqueous solutions. J. Hazard. Mater. 186 (2–3), 1993–2000.

Nernst, W., 1889. Die elektromotorische Wirksamkeit der Ionen. Z. Phys. Chem. Stöchiometrie und Verwandtschaftslehre 4, 129–181.

Neto, S.A., De Andrade, A.R., 2009. Electrochemical degradation of glyphosate formulations at DSA(A (R)) anodes in chloride medium: an AOX formation study. J. Appl. Electrochem. 39 (10), 1863–1870.

Palmas, S., Polcaro, A.M., Vacca, A., Mascia, M., Ferrara, F., 2007. Characterization of boron doped diamond during oxidation processes: relationship between electronic structure and electrochemical activity. J. Appl. Electrochem. 37 (1), 63–70.

Panizza, M., 2010. Importance of electrode material in the electrochemical treatment of wastewater containing organic pollutants. In: Comninellis, C., Chen, G.H. (Eds.), Electrochemistry for the Environment, first ed. LLC: Springer Science, pp. 25–54.

Panizza, M., Cerisola, G., 2009a. Direct and mediated anodic oxidation of organic pollutants. Chem. Rev. 109 (12), 6541–6569.

Panizza, M., Cerisola, G., 2009b. Electro-Fenton degradation of synthetic dyes. Water Res. 43 (2), 339–344.

Panizza, M., Cerisola, G., 2007. Electrocatalytic materials for the electrochemical oxidation of synthetic dyes. Appl. Catal. B Environ. 75 (1–2), 95–101.

Panizza, M., Cerisola, G., 2005. Application of diamond electrodes to electrochemical processes. Electrochim. Acta 51 (2), 191–199.

Panizza, M., Cerisola, G., 2003. Influence of anode material on the electrochemical oxidation of 2-naphthol – part 1. Cyclic voltammetry and potential step experiments. Electrochim. Acta 48 (23), 3491–3497.

Panizza, M., Kapalka, A., Comninellis, C., 2008. Oxidation of organic pollutants on BDD anodes using modulated current electrolysis. Electrochim. Acta 53 (5), 2289–2295.

Panizza, M., Michaud, P.A., Cerisola, G., Comninellis, C., 2001. Anodic oxidation of 2-naphthol at boron-doped diamond electrodes. J. Electroanal. Chem. 507 (1–2), 206–214.

Panizza, M., Michaud, P.A., Cerisola, G., Comninellis, C., 2001. Electrochemical treatment of wastewaters containing organic pollutants on boron-doped diamond electrodes: prediction of specific energy consumption and required electrode area. Electrochem. Commun. 3 (7), 336–339.

Panizza, M., Cerisola, G., 2006. Olive mill wastewater treatment by anodic oxidation with parallel plate electrodes. Water Res. 40 (6), 1179–1184.

Papastefanakis, N., Mantzavinos, D., Katsaounis, A., 2010. DSA electrochemical treatment of olive mill wastewater on Ti/RuO_2 anode. J. Appl. Electrochem. 40 (4), 729–737.

Perelo, L.W., 2010. Review: in situ and bioremediation of organic pollutants in aquatic sediments. J. Hazard. Mater. 177 (1–3), 81–89.

Pignatello, J.J., Oliveros, E., Mackay, A., 2006. Advanced oxidation processes for organic contaminant destruction based on the Fenton reaction and related chemistry. Crit. Rev. Environ. Sci. Technol. 36 (1), 1–84.

Polcaro, A.M., Mascia, M., Palmas, S., Vacca, A., 2003. Electrochemical oxidation of phenolic and other organic compounds at boron doped diamond electrodes for wastewater treatment: effect of mass transfer. Ann. Chim. 93 (12), 967–976.

Polcaro, A.M., Mascia, M., Palmas, S., Vacca, A., 2002. Kinetic study on the removal of organic pollutants by an electrochemical oxidation process. Ind. Eng. Chem. Res. 41 (12), 2874–2881.

Polcaro, A.M., Vacca, A., Mascia, M., Ferrara, F., 2008. Product and by-product formation in electrolysis of dilute chloride solutions. J. Appl. Electrochem. 38 (7), 979–984.

Polcaro, A.M., Vacca, A., Mascia, M., Palmas, S., Pompei, R., Laconi, S., 2007. Characterization of a stirred tank electrochemical cell for water disinfection processes. Electrochim. Acta 52 (7), 2595–2602.

Polcaro, A.M., Vacca, A., Mascia, M., Palmas, S., Ruiz, J.R., 2009. Electrochemical treatment of waters with BDD anodes: kinetics of the reactions involving chlorides. J. Appl. Electrochem. 39 (11), 2083–2092.

Pupunat, L., Rychen, P., 2002. Inactivation of Legionella with the DiaCell water treatment technology. CSEM Scientific Report.

Rajeshwar, K., Ibanez, J., 1997. Environmental Electrochemistry – Fundamentals and Applications in Pollution Abatement. Academic Press.

Rajkumar, D., Kim, J.G., Palanivelu, K., 2005. Indirect electrochemical oxidation of phenol in the presence of chloride for wastewater treatment. Chem. Eng. Technol. 28 (1), 98–105.

Rajkumar, D., Palanivelu, K., 2004. Electrochemical treatment of industrial wastewater. J. Hazard. Mater. 113 (1–3), 125–131.

Rao, N.N., Rohit, M., Nitin, G., Parameswaran, P.N., Astik, J.K., 2009. Kinetics of electrooxidation of landfill leachate in a three-dimensional carbon bed electrochemical reactor. Chemosphere 76 (9), 1206–1212.

Rodgers, J.D., Bunce, N.J., 2001. Electrochemical treatment of 2,4,6-trinitrotoluene and related compounds. Environ. Sci. Technol. 35 (2), 406–410.

Saez, C., Panizza, M., Rodrigo, M.A., Cerisola, G., 2007. Electrochemical incineration of dyes using a boron-doped diamond anode. J. Chem. Technol. Biotechnol. 82 (6), 575–581.

Saez, V., Vicente, M.D.E., Frias-Ferrer, A.J., Bonete, P., Gonzalez-Garcia, J., 2009. Electrochemical degradation of perchloroethylene in aqueous media: an approach to different strategies. Water Res. 43 (8), 2169–2178.

Scialdone, O., 2009. Electrochemical oxidation of organic pollutants in water at metal oxide electrodes: a simple theoretical model including direct and indirect oxidation processes at the anodic surface. Electrochim. Acta 54 (26), 6140–6147.

Scialdone, O., Galia, A., Gurreri, L., Randazzo, S., 2010. Electrochemical abatement of chloroethanes in water: reduction, oxidation and combined processes. Electrochim. Acta 55 (3), 701–708.

Scialdone, O., Randazzo, S., Galia, A., Silvestri, G., 2009. Electrochemical oxidation of organics in water: role of operative parameters in the absence and in the presence of NaCl. Water Res. 43 (8), 2260–2272.

Selleck, R.E., Saunier, B.M., Collins, H.F., 1978. Kinetics of bacterial deactivation with chlorine. J. Environ. Eng. Div. ASCE 104 (6), 1197–1212.

Serikawa, R.M., Senda, Y., Sasaki, K., 2009. Industrial E-AOP Application Using Boron Doped Diamond Electrodes.

Serrano, K., Michaud, P.A., Comninellis, C., Savall, A., 2002. Electrochemical preparation of peroxodisulfuric acid using boron doped diamond thin film electrodes. Electrochim. Acta 48 (4), 431–436.

Simond, O., Schaller, V., Comninellis, C., 1997. Theoretical model for the anodic oxidation of organics on metal oxide electrodes. Electrochim. Acta 42 (13–14), 2009–2012.

Sonoyama, N., Sakata, T., 1999. Electrochemical continuous decomposition of chloroform and other volatile chlorinated hydrocarbons in water using a column type metal impregnated carbon fiber electrode. Environ. Sci. Technol. 33 (19), 3438–3442.

Stoner, G.E., Cahen, G.L., Sachyani, M., Gileadi, E., 1982. The mechanism of low-frequency a.c. electrochemical disinfection. Bioelectrochem. Bioenerg. 9 (3), 229–243.

Trasatti, S., 2000. Electrocatalysis: understanding the success of DSA (R). Electrochim. Acta 45 (15–16), 2377–2385.

Troster, I., Fryda, M., Herrmann, D., Schafer, L., Hanni, W., Perret, A., Blaschke, M., Kraft, A., Stadelmann, M., 2002. Electrochemical advanced oxidation process for water treatment using DiaChem (R) electrodes. Diam. Relat. Mater. 11 (3–6), 640–645.

Un, U.T., Altay, U., Koparal, A.S., Ogutveren, U.B., 2008. Complete treatment of olive mill wastewaters by electrooxidation. Chem. Eng. J. 139 (3), 445–452.

Volta, A., 1800. On the electricity excited by the mere contact of conducting substances of different kinds. Philos. Trans. R. Soc. Lond. 90, 403–431.

Von Gunten, U., 2007. The basics of oxidants in water treatment. Part B: ozone reactions. Water Sci. Technol. 55 (12), 25–29.

Waterston, K., Wang, J.W.J., Bejan, D., Bunce, N.J., 2006. Electrochemical waste water treatment: electrooxidation of acetaminophen. J. Appl. Electrochem. 36 (2), 227–232.

Zaviska, F., Drogui, P., Blais, J.F., Mercier, G., 2009. In situ active chlorine generation for the treatment of dye-containing effluents. J. Appl. Electrochem. 39 (12), 2397–2408.

CHAPTER

4

Heterogeneous Photocatalysis

Assistant Professor Morten E. Simonsen

Section of Chemical Engineering, Department of Biotechnology, Chemistry and Environmental Engineering, Aalborg University Esbjerg, Niels Bohrs Vej 8, DK 6700 Esbjerg, Denmark

4.1. INTRODUCTION

The initial step in heterogeneous photocatalysis is the generation of an electron-hole pair by light activation of photocatalytic material. The photogenerated electron-hole pair can take part in different redox reactions which lead to the formation of reactive species, i.e. hydroxyl radicals which can mediate the degradation of organic components. In this process mineralisation of organic pollutants to carbon dioxide and mineral acids can be achieved. Literature reviews have shown that a wide range of organic components such as aliphatics, aromatics, dyes and surfactants can be completely mineralised by semiconductor photocatalysis (Mills and LeHunte, 1997; Carp et al., 2004). In the field of photocatalysis most applications are concerned with water and air purification or self-cleaning coatings/materials for outdoor and indoor uses. Considerable amount of work has been done on TiO_2-modified materials for self-cleaning windows, tiles and building facades (Carp et al., 2004; Mills and Lee, 2002). The self-cleaning effect of TiO_2 arises from the fact that dirt is photocatalytically degraded and washed away due to superhydrophilic properties of the surface. Additionally, such materials have antifogging properties arising from the superhydrophilic properties (Fujishima and Zhang, 2006). Photocatalytically modified cement has also been produced. In this case TiO_2 is used both as an additive in the concrete and as a coating (Carp et al., 2004). Self-cleaning paints for both indoor and outdoor use are also areas of increasing focus. However, further development of TiO_2-modified materials for self-cleaning building surfaces is needed in particular in regard to activity and visible light activation.

Chemistry of Advanced Environmental Purification Processes of Water
http://dx.doi.org/10.1016/B978-0-444-53178-0.00004-3

Early work on water purification using photocatalysis mainly focused on the photomineralisation of organics dissolved in aqueous solution usually employing the semiconductor in the form of a powder suspension. The use of TiO_2 in suspension is efficient due to the large surface area of the catalyst and the absence of mass transfer limitations (the species are close to the photocatalyst surface) (Mills and LeHunte, 1997; Mukherjee and Ray, 1999). However, the use of suspensions requires an additional separation step which is both time consuming and expensive (Mills and LeHunte, 1997; Mukherjee and Ray, 1999; Doll and Frimmel, 2004). In order to overcome these obstacles, immobilisation of the TiO_2 catalyst in the photoreactor seems to be an important step of improvement. The use of heterogeneous photocatalysis for water treatment is beginning to be commercialised and different systems is available on the market.

INFORMATION BOX 1

History of Heterogeneous Photocatalysis

1938: First report on photo-bleaching of dyes by help of UV irradiated TiO_2. The bleaching was reported to be caused by active oxygen species generated on the UV irradiated TiO_2 surface. In this report TiO_2 was reported as a photosensitizer. (Goodeve and Kitchener, 1938)

1956: Mashio et al. reported the use of TiO_2 as a photocatalyst. They investigated the oxidation of different organic solvents such as alcohols and hydrocarbons by UV irradiation of dispersed TiO_2 powders. In their investigation they concluded that the activity was higher for anatase than rutile. (Kato and Mashio, 1956)

1972: Fujishimna and Honda discovered the photocatalytic splitting of water on TiO_2 electrodes. (Fujishima and Honda, 1972)

1977: First report on the use of TiO_2 photocatalysis for destruction of pollutants for environmental purification. Frank and Bard investigated the reduction of CN in water by help of a TiO_2 suspension. (Frank and Bard, 1977)

1985: Report of photocatalytic antibacterial effect. Matsunaga et al. reported that photoactivated TiO_2 effectively can kill Lactobacillus acidophilus, Saccharomyces cerevisiae and Escherichia coli. (Matsunaga et al., 1985)

1995: Fujishima and co-workers discover the superhydrophilic properties of TiO_2 after UV irradiation. (Fujishima et al., 2000)

1997: Highly hydrophilic TiO_2 surfaces with antifogging and self-cleaning properties are reported by Wang et al., (1998).

Although TiO_2 photocatalysis was initially applied for water treatment, attention has also been directed towards air treatment. The photocatalytic detoxification of volatile organic compounds is generally more efficient in the gas phase compared to the liquid phase (Carp et al., 2004).

4.2. THE PRINCIPLE AND MECHANISMS IN SEMICONDUCTOR PHOTOCATALYSIS

The initial process in heterogeneous photocatalysis is the generation of an electron-hole pair by excitation of an electron in the semiconductor. Unlike metals which have a continuum of electronic states, semiconductors have an energy void where no energy levels are available to promote recombination of an electron and hole produced by photoactivation (Figure 4.1) (Linsebigler et al., 1995).

The energy void which extends from the top of the filled valence band to the bottom of the conduction band is called the band gap. Once excitation occurs across the band gap there is sufficient time for the created electron-hole pair to be transferred to adsorbed species on the semiconductor surface. The enlarged section of Figure 4.2 illustrates the excitation of an electron from the valance band to the conduction band initiated by light absorption with energy equal to or greater than the band gap of the semiconductor. After excitation the fate of the separated electron and hole can follow different pathways. Electron-hole pair recombination can occur at (A) the surface or (B) in the bulk of the semiconductor particle. Migration of the photogenerated electrons to the surface of the semiconductor can lead to reduction of an electron acceptor A, illustrated by reaction (C). Furthermore, migration of the photogenerated holes to the surface can lead to oxidation of an electron

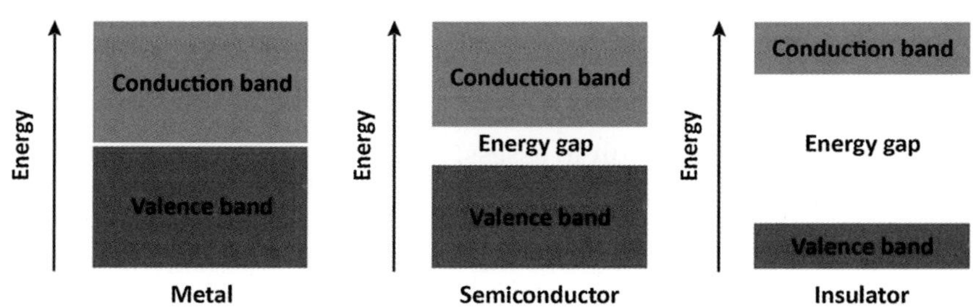

FIGURE 4.1 **Illustration of electronic states in metals, semiconductors and insulators.** (For colour version of this figure, the reader is referred to the online version of this book.)

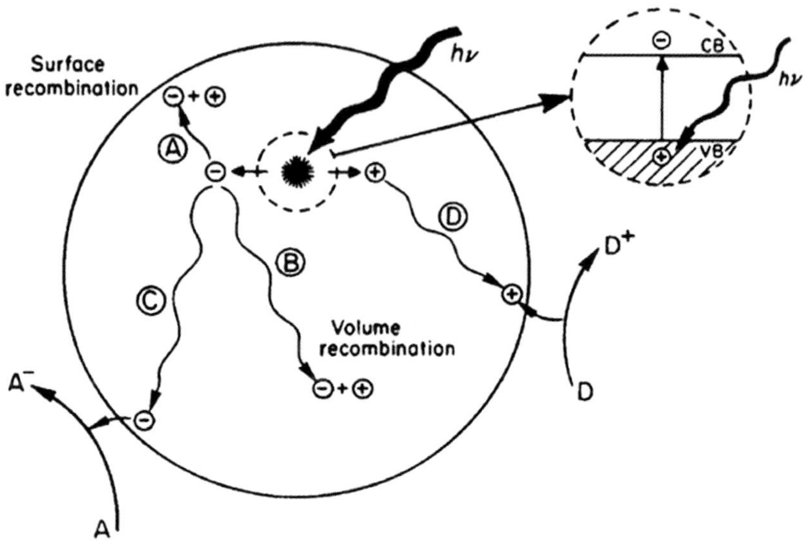

FIGURE 4.2 **Photoexcitation in a semiconductor particle followed by deexcitation events.** *Reprinted with permission from Linsebigler et al. (1995).*

donor (D), reaction (D). The reactions (C) and (D) are the two basic processes in photocatalysis. The electron transfer process is more efficient if the species is preadsorbed on the surface (Linsebigler et al., 1995).

The efficiency of a photocatalytic process is measured as a quantum yield which is defined as the number of events occurring per photon absorbed (Eqn (4.1)) (Mills and LeHunte, 1997).

$$\Phi = \frac{\text{Rate of reaction}}{\text{Rate of absorption of radiation}} \tag{4.1}$$

In heterogeneous semiconductor photocatalysis, the measurement of Φ is difficult and not used very often (Mills and LeHunte, 1997). The reason for this is, that the rate of absorption of photons with energy greater than the band gap is very difficult to assess, since the semiconductor particles will absorb, scatter and transmit light. It has been reported that the fraction of light scattered or reflected by the semiconductor surface may be between 13% and 76% of the incident light intensity (Schiavello et al., 1991).

For this reason the formal quantum efficiency or photonic efficiency (PE) is often used, which is a measure of the rate of reaction divided by the incident light intensity (Eqn (4.2)).

$$\delta = \frac{\text{Rate of reaction (molecules/s)}}{\text{Incident light intensity (photons/s)}} \tag{4.2}$$

For any photochemical process the formal quantum efficiency is lower than the quantum yield.

According to the mechanism illustrated above the overall quantum efficiency is determined by the competition between the recombination and trapping of the electron-hole pair. An increase in the lifetime of the electron-hole pair results in a higher quantum yield (Hoffmann et al., 1995). Retardation of the electron-hole pair recombination is therefore an important aspect of photocatalysis.

The synthesis of the semiconductors may result in surface and bulk irregularities or defects. These irregularities are associated with surface electronic states which differ in their energy from the energy bands present in the bulk semiconductor and these energy states may act as electron-hole recombination traps (Mills and LeHunte, 1997; Bickley et al., 1991) have suggested that the two crystalline phases of TiO_2 anatase and rutile found in Degussa P25 promotes separation of the electron-hole pair and inhibits recombination. The rutile structure of Degussa P25 has been proposed to trap the electrons and thereby suppress the recombination of the electron-hole pair. (Bickley et al., 1991)

4.2.1. The Band Edge Position's Influence on the Photocatalytic Process

The electron transfer processes between the semiconductor and adsorbed species depends on the position of the band gap and redox potential of the adsorbed species. The redox potential of the acceptor molecule is thermodynamically required to be below the conduction band potential of the semiconductor to accept an electron. Similarly, the redox potential of a donor molecule needs to be above the valence band position of the semiconductor in order to donate an electron to the vacant hole. The more positive the valence band, the more oxidising the valence band holes. The free energy of the electron-hole pair is smaller than the energy of the band gap due to the translational entropy of the electrons and holes in the conduction and valence bands, respectively (Hagfeldt and Gratzel, 1995; Gratzel, 2001).

The band edge position of different semiconductors vs the standard hydrogen electrode (SHE) is illustrated in Figure 4.3. The band edge potential follows a Nernstian pH dependence, decreasing 59 mV per pH unit (Carp et al., 2004; Ward et al., 1983), thus the potential of the photogenerated electrons and holes can be controlled by changes in the pH.

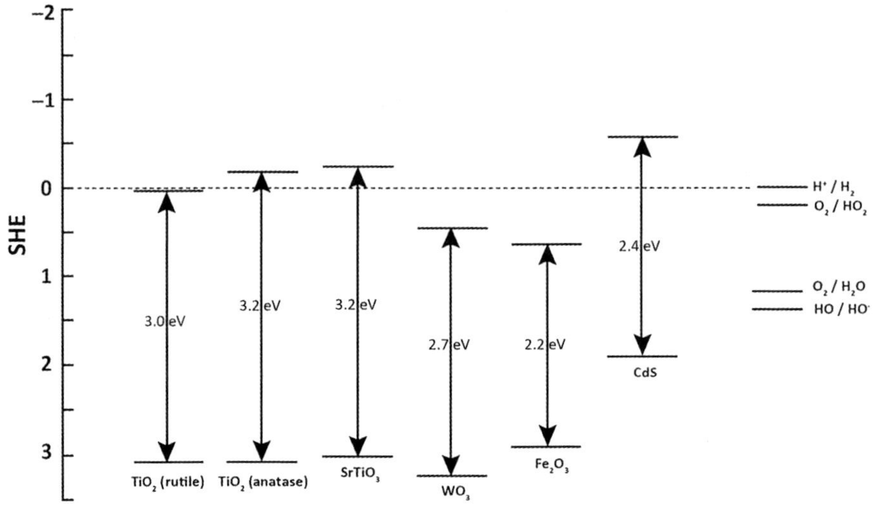

FIGURE 4.3 **The band edge position of various semiconductors vs the standard hydrogen electrode (SHE) at pH 0.** *Reproduced after Mills and LeHunte (1997) with permission from Elsevier.*

The band gap of different semiconductors varies considerably ranging from 2.2 to 3.2 eV. Hence, the wavelength of the light required to photo-activate the semiconductor varies from 564 to 388 nm. In photocatalysis visible light (>400 nm) activation is preferred; however, investigations of CdS ($\leq$517 nm) and other low-band gap semiconductors have shown loss of photocatalytic activity after a period of time due to photocorrosion (oxidation of the semiconductor) (Mills and LeHunte, 1997; Hagfeldt and Gratzel, 1995). In contrast, TiO_2 has been found to be photocatalytically active, cheap, nontoxic, chemically stable over a wide pH range, and is not subject to photocorrosion (Mills and LeHunte, 1997). The photocatalytic mineralisation feature of TiO_2 has been well documented in literature (Carp et al., 2004). However, the band gap of TiO_2 is 3.2 eV (anatase) corresponding to a wavelength of 388 nm which lies in the UV region.

4.3. TIO_2 PHOTOCATALYSIS

4.3.1. Titanium Dioxide

Titanium dioxide is found in three different crystal structures; anatase, brookite and rutile (Gopal et al., 1997). The crystal structure of anatase, rutile and brookite can be described in terms of (TiO_6) octahedrals (Figure 4.4). Each Ti atom is surrounded by an octahedron of six O atoms. The distance between the titanium atoms in anatase is greater than

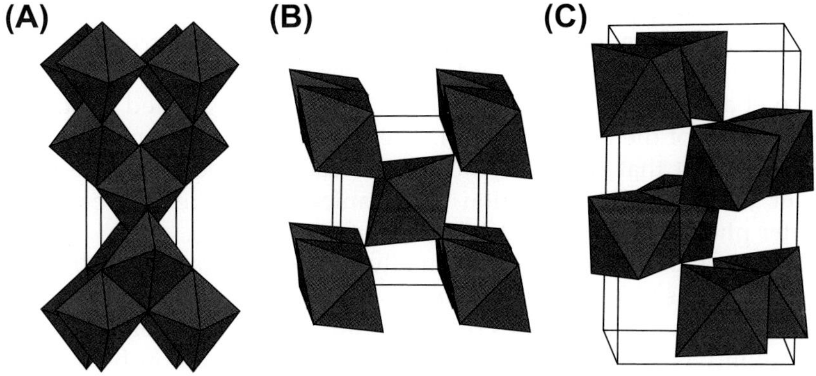

FIGURE 4.4 **Crystal structures of (A) anatase, (B) rutile, and (C) brookite.** *Reprinted with permission from Elsevier (Carp et al., 2004).*

in rutile and brookite, resulting in anatase having a lower density (3.83 g/cm^3) than rutile (4.24 g/cm^3) and brookite (4.17 g/cm^3) (Carp et al., 2004).

Thermodynamic calculations based on calorimetric data have shown that rutile is the most stable form of TiO_2 (Carp et al., 2004; Gopal et al., 1997). However, the small differences in the Gibbs free energy (4−20 kJ/mol) between the three phases suggest that the metastable crystal structures are almost as stable as rutile at normal pressures and temperatures (Carp et al., 2004). In addition investigations have shown that the particle size may influence the stability (Zhang and Banfield, 2000). At very small crystal sizes anatase has been found to be the most stable form due to surface-energy effects (surface free energy and surface stress, which depend on particle size), the critical size being around 11 nm in diameter. Furthermore, brookite has been reported to be the most stable form in the range from 11 to 35 nm, at sizes greater than 35 nm rutile is the most stable form (Zhang and Banfield, 2000).

The enthalpy of the transformation of anatase to rutile is low. Carp et al. 2004 have reported values from −1.3 to −6.0 kJ/mol. The transformation of anatase to rutile at room temperature is very slow and thus anatase is normally regarded to be stable at room temperature. However, at elevated temperatures the rate of transformation increases. No specific transformation temperature is reported but experimental data suggest that the transformation occurs at 400−1000 °C depending on preparation method/particle properties and heating rate (Ahonen, 2001). Crystal defects, impurities, dopants, crystal size and reaction atmosphere have been reported to affect the transformation.

The band gap of anatase is 3.2 eV, which is 0.2 eV greater than the band gap of rutile. The wavelength of the light required to promote an electron over the band gap is ≤388 nm for anatase and ≤414 nm for rutile. Thus, more energy is required to photoactivate the anatase structure. In the literature the activity of anatase has been reported to be greater than rutile (Carp et al., 2004; Hashimoto et al., 2005). This is suggested to be due to the position of the band potential. The band potential of anatase is more negative than that of rutile, thus more reactive species can be formed under the photomineralisation of organics and therefore results in a greater photocatalytic activity (Hashimoto et al., 2005).

The crystallinity of TiO_2 is normally determined by X-ray diffraction (XRD) analysis from the relative intensities of the peaks corresponding to the anatase and rutile phases (Jensen et al., 2004). However, the absolute crystallinity is seldom determined, which means the amorphous fraction of the sample often is neglected (Jensen et al., 2004). This could have a huge impact on the interpretation of the results since it has been shown that the amorphous fraction of TiO_2 has no or little photocatalytic activity (Mills and LeHunte, 1997).

The crystallite size can be determined from the broadening of the corresponding X-ray peak by Scherrer's formula (Eqn (4.3)).

$$\tau = \frac{K \cdot \lambda}{\beta_\tau \cdot \cos\theta} \tag{4.3}$$

where K is the form factor (0.9), β_τ the width of the peak at half the maximum intensity after subtraction of the instrumental noise and θ the diffraction angle.

Besides crystallinity several other factors affect the photocatalytic activity including the particle size, amount and the identity of defects, and surface area and charge (Carp et al., 2004; Jensen et al., 2004; Kang et al., 2001; Yu et al., 2005), many of which are related to the synthesis of the TiO_2 particles.

Degussa P25 and Hombikat UV100 which often are used as reference materials when comparing the photocatalytic activity are prepared at relatively high temperatures by a chloride and sulphate method, respectively. In the chloride method the precursor ($TiCl_4$), is oxidised in the gaseous phase at high temperature (985 °C) in a H_2/O_2 flame (Jensen et al., 2005). In the sulphate process, ilmenite is transformed into iron and titanium sulphates by reaction with sulphuric acid. Titanium hydroxide is precipitated by hydrolysis, filtered and calcinated at 900 °C (Carp et al., 2004).

Other methods of preparation include chemical vapour deposition, physical vapour deposition, and sol-gel methods (Carp et al., 2004; Mills et al., 2003). The sol-gel method has some advantages over other preparation techniques in the many parameters which can be varied in order to obtain TiO_2 particles with specific properties (tailored materials). The sol-gel method leads to the formation of TiO_2 by hydrolysis and condensation of titanium alkoxides. The most commonly used titanium alkoxides are

Ti(O-Et)$_4$, Ti(O-i-Pr)$_4$ and Ti(O-n-Bu)$_4$. The properties of the final TiO_2 particles are strongly affected by the kinetics of the hydrolysis and condensation reactions of the alkoxide precursor (Simonsen and Sogaard, 2010). Acid–base catalysis can be used to control the rate of the hydrolysis and condensation steps (Livage et al., 1988). The obtained TiO_2 particles are usually calcinated at 450–600 °C to remove residual carbon and to obtain crystalline material (anatase/rutile). The calcination process will cause a decrease in the surface area (due to sintering and crystal growth) and loss of surface hydroxyl groups (Carp et al., 2004). Recent work conducted by different groups has shown that it is possible to obtain crystal material at low temperatures (<100 °C) and thus avoid post-preparation heat treatment (Gopal et al., 1997; Li et al., 2004; Jensen et al., 2006; Simonsen, 2009).

A comparison of the particle properties of two commercially available TiO_2 powders (Degussa P25 and Hombikat UV100) and two powders produced by sol-gel methods at relative low temperatures (MW220 and SC134) are shown in Table 4.1. The MW220 is produced by

TABLE 4.1 Properties of the Comparable Commercial and In-house Prepared TiO_2 Catalytic Materials

	Degussa P25	Hombikat UV100	SC134	MW220
SEM				
SEM primary particle size (nm)	50	40	25	20
SEM secondary particle size (nm)	50	800	800	
DLS				
Particle size in suspension (nm)	600 ± 15	818 ± 24	1274 ± 36	270 ± 2
BET				
Surface area powder (m^2/g)	50	360	221	117
XRD				
Anatase (%)	75.6	86.2	65.6	69.1
Rutile (%)	21.6	–	–	–
Amorphous (%)	2.8	13.8	34.4	30.9
Crystallite size (nm)	18	12	7	10

SEM, scanning electron microscope; DLS, dynamic light scattering. *Reproduced with permission from Elsevier (Simonsen et al., 2008).*

a microwave-assisted sol-gel method (Simonsen, 2009) and the SC134 is produced in supercritical carbon dioxide (Jensen et al., 2006).

XRD analysis of the TiO_2 photocatalysts shows that P25 is the only sample consisting of both anatase and rutile (Figure 4.5). P25 consists of approximately 75.6% anatase and 21.6% rutile. In comparison UV100 consists of about 86.2% anatase and the SC134 and MW220 powders about 70% anatase. In addition to the crystallinity the XRD pattern also provides information on the crystallite size. From Table 4.1 it is seen that the crystallite size of the TiO_2 powders vary between 7 and 23 nm.

The specific surface area (SSA) is usually measured by nitrogen adsorption. The Brunauer–Emmett–Teller (BET) adsorption isotherms provide information not only on the SSA but also on porosity and pore size. The BET adsorptions isotherm for the UV100 samples show evidence of a microporous structure leading to the high surface area observed for this powder. The BET adsorptions isotherm for the MW220 and SC134

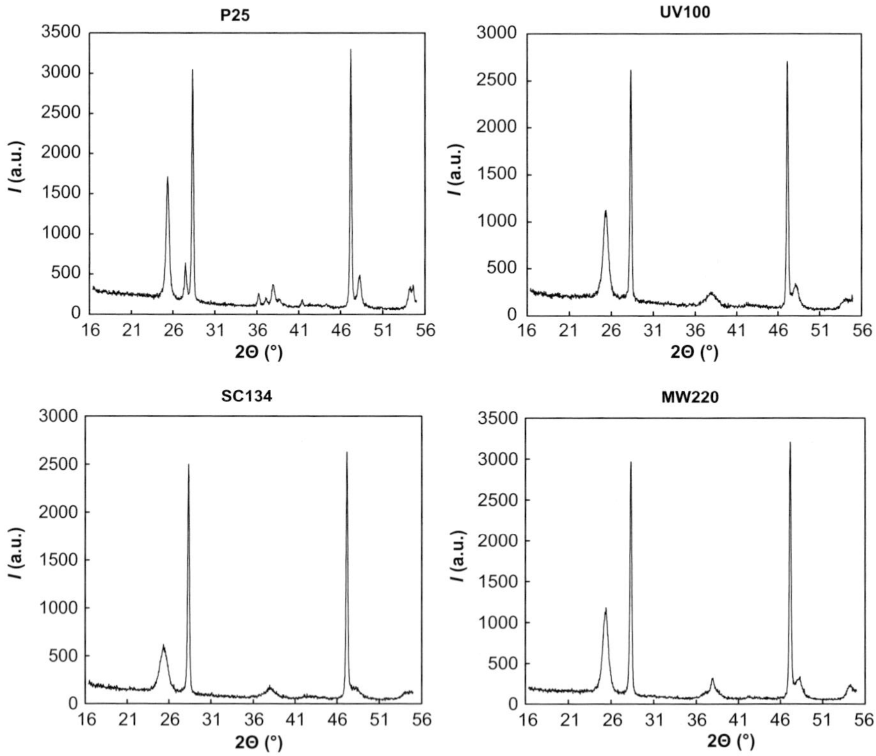

FIGURE 4.5 XRD spectra of the different prepared TiO_2 particles. Anatase (101) = 25.3°, rutile (110) = 27.5°, and CaF_2 (111) = 47.2°. *Reprinted with permission from author (Simonsen, 2009).*

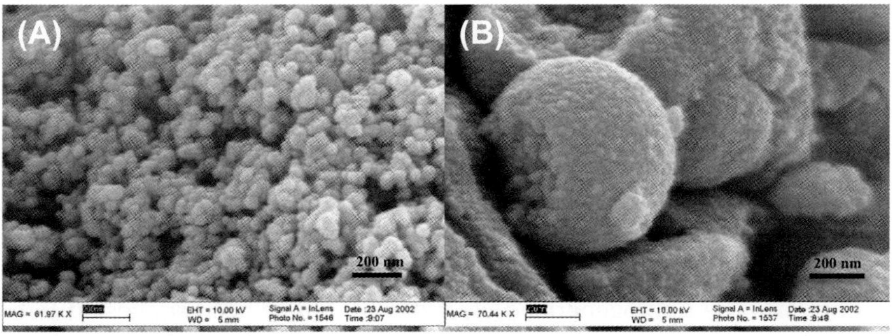

FIGURE 4.6 Scanning electron microscope (SEM) images of the (A) nonporous (P25) and (B) mesoporous (SC134) TiO_2 powders. *Reprinted with permission from Springer (Jensen et al., 2004).*

photocatalysts shows a hysteresis effect suggesting that the particles are mesoporous. In comparison the adsorption isotherms for P25 suggest that the particles are nonporous.

Scanning electron microscope (SEM) images of the nonporous (P25) and mesoporous (SC134) TiO_2 samples (Figure 4.6) show that the primary particle size of both catalysts is around 40 nm. However, in the case of the SC134 photocatalyst these primary particles have aggregated to form larger secondary particles in the size order of 800–1000 nm giving raise to the mesoporous structure.

Great efforts have been made to improve the photocatalytic activity of TiO_2. Two different approaches have been investigated to increase the photocatalytic activity: (1) inhibition of the recombination by increasing the charge separation and thus enhancing the efficiency of the photocatalytic process and (2) increasing the wavelength response range (i.e. excitation of wide band gap semiconductors by visible light) (Linsebigler et al., 1995). Doping of the semiconductor with various metal ions may lead to an enhanced efficiency of the photocatalytic systems due to better charge separation (Carp et al., 2004; Hoffmann et al., 1995). In addition metal ion doping of TiO_2 has been used to shift the absorption into the visible region. The extent of the shift depends on the amount and type of metal ions ($V > Cr > Mn > Fe > Ni$) (Carp et al., 2004). Another approach used to induce visible light activation is substitution of oxygen with N, C, S, and Br which leads to a band gap narrowing in TiO_2 (Carp et al., 2004).

The development of efficient visible light-sensitive photocatalysts is of great interest for both indoor and outdoor self-cleaning application and for water remediation using solar heterogeneous photocatalysis.

4.3.2. Photocatalytic Oxidation

As outlined above the initial step in heterogeneous photocatalysis is the generation of an electron-hole pair (Eqn (4.4)).

$$TiO_2 \xrightarrow{hv < 388\ \text{nm}} TiO_2\left(e^- + h^+\right) \tag{4.4}$$

The redox potential of the photogenerated electron-hole pair can be estimated from the band edge position of the valence and conduction band. The band gap diagram for TiO_2 at pH 7 vs the SHE is shown in Figure 4.7. The photogenerated electrons (e^-) have been suggested to be trapped by oxygen or surface defect sites, which reduce the recombination of the photogenerated electron-hole pair and hence enhance the photocatalytic activity. Liu et al. (2003) have, by means of X-ray photoelectron spectroscopy (XPS) analysis, shown the formation of Ti^{3+} defect sites on the TiO_2 surface through a surface reduction process of Ti^{4+} to Ti^{3+} during UV irradiation. The reduction potential for the conduction band electrons is −0.52 V; however, trapped electrons may loose some of their reducing power. A significant number of the photogenerated electrons will even after trapping possess the reducing power to produce superoxide from dioxygen (Fujishima et al., 2000).

The reduction potential of the photogenerated holes is +2.53 V. The photogenerated holes possess the oxidising power to directly oxidise most organics or to react with surface hydroxyl groups or H_2O to produce hydroxyl radicals (Eqns (4.5) and (4.6)). The highly reactive hydroxyl radicals may react with either preadsorbed organic components/pollutants or in the near proximity of the catalyst surface.

$$h^+ + OH^-_{ad} \rightarrow {}^{\bullet}OH_{ad} \tag{4.5}$$

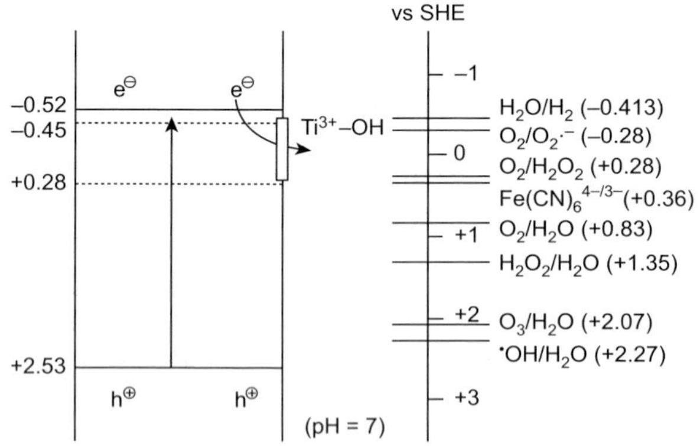

FIGURE 4.7 Potentials for various redox processes occurring on the TiO_2 surface at pH 7. *Reprinted with permission from Elsevier (Fujishima et al., 2000).*

$$h^+ + H_2O \rightarrow {}^{\bullet}OH_{ad} + H^+ \quad (4.6)$$

The species generated from the electron-hole pair may include hydroxyl radicals ($HO^{\bullet}$), superoxide (O_2^-) and hydrogen peroxide (H_2O_2).

Photocatalytical oxidation of soot on TiO_2 films has indirectly shown the formation of reactive species which are liberated from the TiO_2 surface. Mills et al. have shown that soot deposited on a quartz slide placed directly above a TiO_2 film slowly disappears under UV irradiation (Lee et al., 2004). These observations are supported by Lee and Choi who found that soot deposited alongside a TiO_2 film was destroyed after prolonged UV irradiation (Lee and Choi, 2002). Furthermore, studies have shown that the amount of hydroxyl groups on the TiO_2 surface correlates with the observed photocatalytic activity (Figure 4.8), suggesting that the hydroxyl groups and production of hydroxyl radicals play an important role in the photocatalytic mineralisation mechanism (Simonsen et al., 2009).

4.3.3. Photoinduced Superhydrophilicity

In addition to photocatalytic oxidation (PCO) it has been found that UV irradiation induces superhydrophilicity. The PCO mechanism is well recognised; however, two theories have been proposed for the mechanism leading to superhydrophilicity. One model for the photoinduced superhydrophilicity (PSH) is based on surface structural changes occurring by

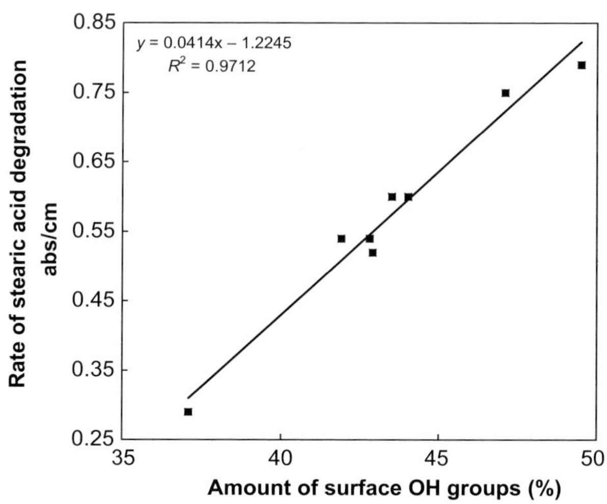

FIGURE 4.8 **Correlation between the photocatalytic activity and the percentage of OH groups on the surface of the TiO_2 films.** *Reprinted with permission from Elsevier (Simonsen et al., 2009).*

UV irradiation. A second model suggests PSH is solely a result of photocatalytic removal of organics on the surface and hence PCO and PSH are directly related (Zubkov et al., 2005; Mills and Crow, 2007). Evidence supporting both models for the PSH has been reported.

Investigation of the wettability of a range of semiconductors suggests that the ability to undergo PSH is connected to surface structural changes, as only a few of the investigated semiconductors showed both photocatalytic and superhydrophilic properties (Miyauchi et al., 2002). Fujishima et al. (2000), Wang et al. (1999), Sun et al. (2001), Sakai et al. (2003, 2001) and Watanabe et al. (1999) have shown that water is reversibly adsorbed on TiO_2 during UV irradiation by help of Fourier transform infrared spectroscopy and XPS analysis. In addition high-resolution XPS studies have shown that UV light activation of TiO_2 leads to increased amounts of hydroxyl groups on the TiO_2 surface (Simonsen et al., 2008, 2009; Yu et al., 2002). The XPS spectra of the O 1s peak of a typical TiO_2 film before and after UV irradiation are shown in Figure 4.9. The spectra have been deconvoluted by addition of two Gaussian-shaped peaks centred at 529.9 and 531.9 eV corresponding to Ti–O in the lattice network of TiO_2 and hydroxyl groups (–OH), respectively (Yu et al., 2002). Figure 4.9 shows the change in the O 1s XPS spectrum of TiO_2 after 1 h of UV irradiation. After UV illumination the peak area assigned to hydroxyl groups increases; indicating that chemisorption of water on the surface of TiO_2 is enhanced by UV irradiation.

The increase in hydroxyl groups is proposed to be due to dissociative adsorption of water at oxygen vacancies. Photogenerated holes have been suggested to be trapped and weaken the bond between titanium and oxygen in the TiO_2 surface. As a result oxygen is liberated to create oxygen vacancies. Water is thought to be dissociatively adsorbed at these defect sites creating a more hydroxylated surface (Sakai et al., 2001; Fujishima and Rao, 1998). The creation of oxygen vacancies in the surface of TiO_2 during UV irradiation under ultrahigh vacuum conditions has been reported by Shultz et al. (1995). Subsequent water exposure was found to regenerate the TiO_2 surface by production of surface hydroxyl groups (Wang et al., 1995). Atomic force microscopy studies have shown roughening of the TiO_2 surface as a result of UV irradiation presumably due to formation of defect sites (Katsumata et al., 2006). Water has been reported to be easily adsorbed on TiO_2 and interact with vacancies even if it is present in almost negligible amounts (Wendt et al., 2005). The interaction between the vacancy and the water molecule is assumed to result in diffusion of a proton to a neighbouring oxygen atom thereby creating two hydroxyl groups.

After the superhydrophilic state has been obtained prolonged irradiation will not lead to any changes in this property. However, if the TiO_2 film is stored in the dark, the TiO_2 surface converts back to its initial more

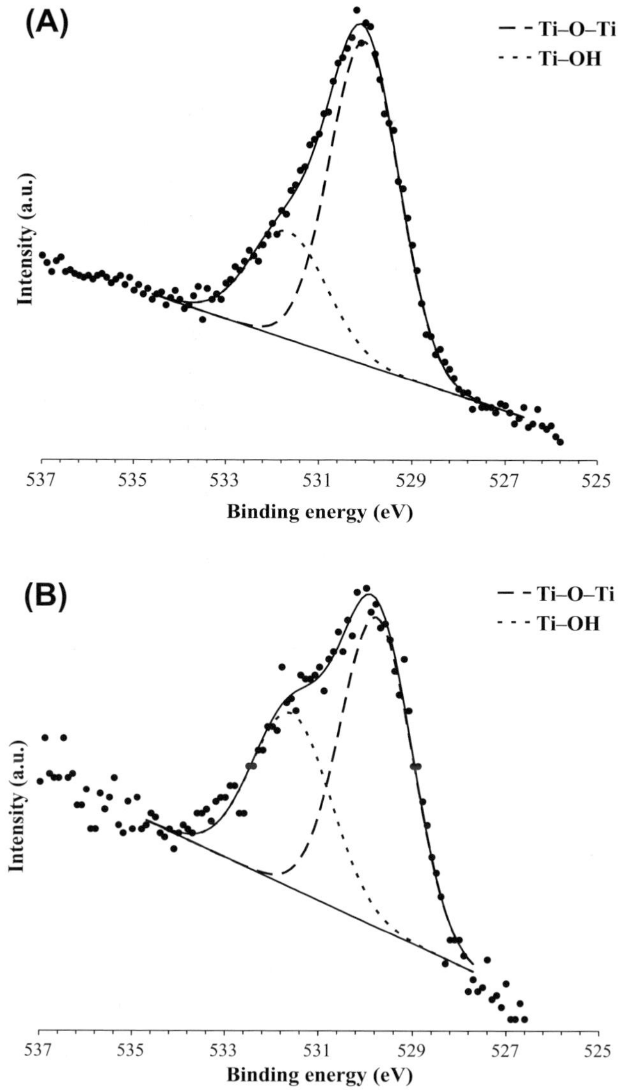

FIGURE 4.9 O 1s spectra obtained from XPS analysis of sol-gel films before (A) and after (B) UV irradiation at 50% Relative Humidity (RH) (25 °C). The TiO_2 films were UV irradiated using UVC light with an intensity of 10 mW/cm^2 measured at the surface of the TiO_2 films. *Reprinted with permission from Elsevier (Simonsen et al., 2009).*

hydrophobic state. Previously UV-irradiated TiO_2 samples have been found to recover their hydrophobic characteristics at increased rates in vacuum (Irie and Hashimoto, 2005). Faster recovery has also been observed at elevated temperature 150 °C suggesting desorption of hydroxyl groups (Hashimoto et al., 2005). The recovery of the TiO_2 surface is suggested to be due to relaxation of the surface where the weakly bonded hydroxyl groups desorb and the initial surface is recreated (Sakai et al., 2003).

In contrast, other groups have reported evidence for the PCO mechanism causing the change in wettability. White et al. (2003) and Zubkov et al., (2005) have studied the hydrophobic—hydrophilic conversion during PCO of a model organic layer of trimethyl acetate and hexane, respectively, and reported the existence of an induction period, during which the organic layer was photooxidised, followed by rapid wetting of the surface. Mills and Crow (2008) has reported that Pilkington Activ™ and plain glass both showed superhydrophilic characteristics after heat treatment at 550 °C suggesting that the superhydrophilicity is attributed to removal of organic contaminants. In the same study similar results were obtained by UVC/O_3 treatment and chemical oxidation using aqua regia. In addition sum frequency generation analysis has shown that the intensity of the C—H stretches due to hydrocarbons decreases upon UV irradiation of an apparent pristine but hydrophobic TiO_2 sample, and regenerates under ambient, dark conditions due to recontamination of the surface by air-borne hydrocarbons (Wang et al., 2003).

Recent work has shown that the amount of OH groups on the TiO_2 surface highly influence the photocatalytic activity and the PSH and that the two mechanisms may be closely related (Simonsen et al., 2009). It is suggested that UV irradiation of the TiO_2 increases the amount of OH groups on the surface of TiO_2. The OH groups can react with photogenerated holes leading to the formation hydroxyl radicals which will mediate the destruction of organics on the TiO_2 surface (Simonsen et al., 2009). This mechanism suggests that the superhydrophilicity is obtained through a combination of the photocatalytic and surface structural change model. Lee et al. (2007) have arrived at similar conclusions in their investigation of the role of water adsorption on the PSH of TiO_2 films.

4.4. CASE STUDY 1 – PHOTOCATALYTIC OXIDATION OF DISINFECTION BY-PRODUCTS IN SWIMMING POOL WATER

Chlorine is the most common disinfectant used in water disinfection practice due to its high efficiency and low cost (Judd and Black, 2000; Chiang et al., 1997; White, 1972). Traditional sodium hypochlorite (NaOCl) has been used for this purpose due to its stability (Kim et al., 2002). The advantage of chlorination in comparison to other disinfection methods such as UV light, active carbon and ozone is that it is present in the water at all times. Other disinfection methods only treat the water which is recycled thus allowing microbial growth of potential pathogen bacteria in the pool water. However, the use of chlorine compounds has been reported to produce various halogenated organic compounds, since organic materials of various forms (perspiration, urine, mucus, spin particles, hair, lotion, etc.) are released into swimming pool water by swimmers (Judd and Black, 2000; Kim et al., 2002; Judd and Bullock, 2003). The two key disinfection by-products in swimming pools are trihalomethanes (THMs) and chloramines, the health impacts of which have been subject to much study over the years (Judd and Bullock, 2003). The most common of these THMs is chloroform (Kim et al., 2002; Aggazzotti et al., 1998). The maximum allowed THM concentration in swimming pool water is 0.1 mg/l.

The presence of these disinfection by-products even in small amounts is a potential health issue, since chloroform has been identified as a possible carcinogen, and hence classified in group 2B by International Agency for Research on Cancer (Aggazzotti et al., 1995).

THMs are volatile substances that can vapourise from water to environmental air depending on many variables, such as vapour pressure, water solubility and the water/air contact area. Exposure of THM in indoor swimming pools occurs by three different routes: inhalation, dermal adsorption and ingestion. Aggazzotti et al. have investigated the exposure to THM in indoor swimming pools by measuring the chloroform concentrations in alveolar air. The investigation showed that it was possible to measure chloroform in alveolar air up to 10 h after 1 h of swimming (Aggazzotti et al., 1995).

Studies have shown that the formation of THM is affected by several parameters, such as dose of chlorine, concentration of bromide and ammonia, pH, temperature and content and type of the natural organic matter. However, these studies have largely been limited to studies on potable water and wastewater (Judd and Black, 2000; Koukouraki and Diamadopoulos, 2003). Some of the changes in chlorination practice commonly suggested are decreasing chlorine dose, changing the point of chlorination, substituting an alternative disinfectant and removing precursors.

4.4.1. Photocatalytic Degradation of Chloroform

The photocatalytic degradation of chloroform and other chlorinated compounds using TiO_2 photocatalysis has been extensively studied (Chen and Jenq, 1998; Choi and Hoffmann, 1997; Martin et al., 1996a,b; Hsiao et al., 1983; Pruden and Ollis, 1983; Yamazaki et al., 2005). The overall reaction scheme for the PCO of chloroform is shown in Eqn (4.7):

$$2CHCl_3 + 2H_2O + O_2 \xrightarrow{h\nu,\ TiO_2} 2CO_2 + 6HCl \qquad (4.7)$$

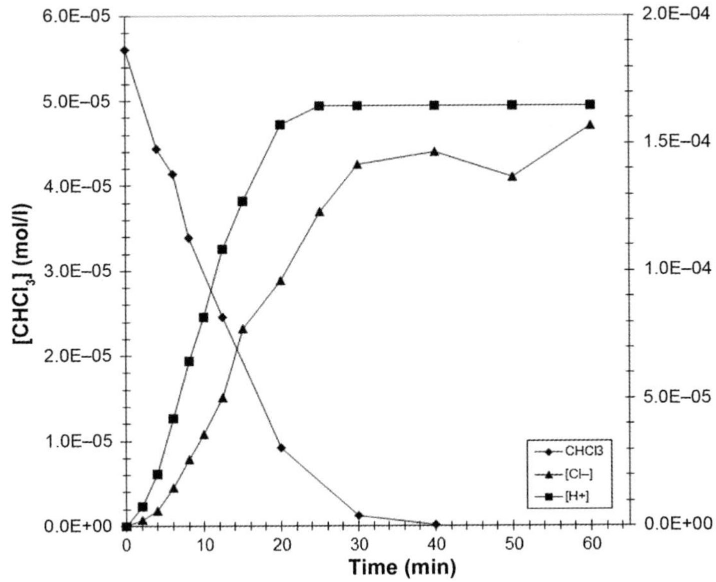

FIGURE 4.10 Degradation profile of chloroform in TiO_2 (Degussa P25) suspension reactor (relationship between chloroform, Cl^- and H^+). *Reprinted with permission from author (Simonsen, 2005).*

The degradation profile for chloroform in a TiO_2 (P25) suspension reactor is shown in Figure 4.10. The correlation between the decrease in the chloroform concentration and the increase in the Cl^- and H^+ concentrations in Reaction (4.7) suggest that no major long-lived intermediates are formed during the degradation.

The photocatalytic degradation of halogenated compounds has been suggested to proceed through a series of radical reactions (Martin et al., 1996a). Investigations conducted by Choi and Hoffmann (1997) have shown that $^{\bullet}CCl_3$ radicals are formed on the surface of TiO_2 through direct oxidation of adsorbed $CHCl_3$. Alternatively, the photogenerated holes can be transferred to adsorbed H_2O or hydroxyl groups on the surface of TiO_2 resulting in formation of hydroxyl radicals which in turn can react with chloroform to produce $^{\bullet}CCl_3$ radicals (Martin et al., 1996a). The $^{\bullet}CCl_3$ radicals have been proposed by several authors to undergo decomposition through reaction with oxygen (Choi and Hoffmann, 1997; Martin et al., 1996a).

The influence of the TiO_2 loading (0–100 mg TiO_2/l) in the suspension reactor is shown in Figure 4.11. The rate of degradation increases with increasing amounts of catalyst up to a certain level. These findings can be explained by two processes: (1) increased amounts of TiO_2 catalyst in the suspension reactor may lead to a shielding effect

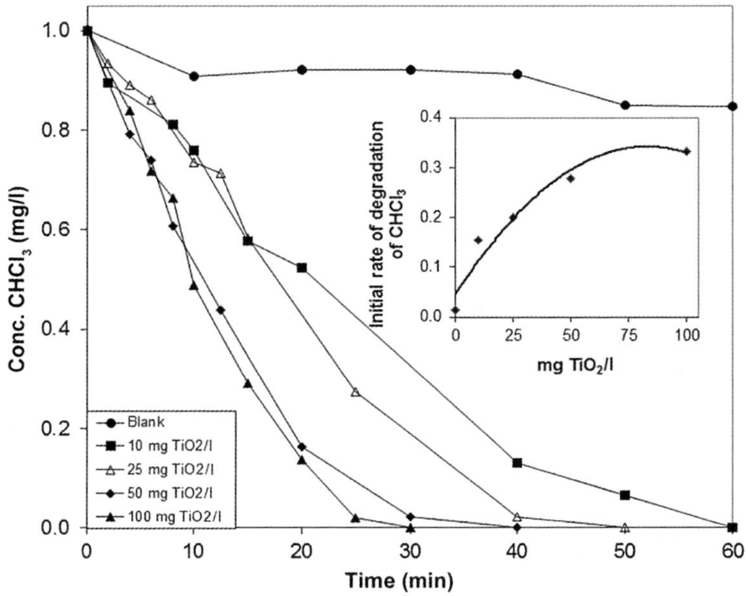

FIGURE 4.11 Photocatalytic degradation of chloroform in TiO_2 (P25) suspension reactor at different TiO_2 loads. Insert: Initial rate of degradation as a function of the catalyst load. *Reprinted with permission from author (Simonsen, 2005).*

resulting in not all the TiO_2 being activated or (2) all the UV light is absorbed by the TiO_2 catalyst and thus the intensity of the UV lamp is the limiting factor.

Not only the amount of catalyst but also the type of TiO_2 catalyst highly influences the rate of photocatalytic degradation of chloroform (see Table 4.2). The initial rate of degradation of chloroform was found to vary from 0.40×10^{-3} to 3.01×10^{-3} dependent on the TiO_2 catalyst used (50 mg TiO_2/l). Degussa P25 was found to be the most efficient catalyst. The difference in the activity of the different TiO_2 catalysts is related to the crystallinity, SSA, particle size in suspension and number of active sites on the surface. XRD investigations of the TiO_2 catalysts show that there were considerable amounts of amorphous TiO_2 present in all the catalysts except Degussa P25 which is found to be nearly completely crystalline (Table 4.1). The preparation of the TiO_2 may also result in surface and bulk irregularities or defects. The irregularities are associated with surface electron states which differ in their energy from the energy bands present in the bulk semiconductor and these energy states can act as electron–hole recombination traps. Bickley et al. (1991) have suggested that the anatase/rutile structure of TiO_2 Degussa P25 promotes charge-pair separation and inhibits recombination. The rutile structure of

TABLE 4.2 Calculated Initial Photonic Efficiencies[1] for the Photocatalytic Degradation of Chloroform in Different Photocatalytic Reactors

			Incident Light Intensity		Initial Rate of Chloroform Degradation	Photonic Efficiencies
		TiO_2 Load	**[mW/cm^2]**	**[Photon/s]**	[n_{CHCl_3}/s]	10^{-3}[n_{CHCl_3}/**photon**]
BI photocatalytic reactor		0.15 mg/cm^2	16	1.00×10^{18}	6.04×10^{15}	1.51
Slurry reactor	Degussa P25	10 mg/l	186	3.90×10^{19}	6.47×10^{16}	1.66
		25 mg/l	186	3.90×10^{19}	8.34×10^{16}	2.15
		50 mg/l	186	3.90×10^{19}	1.47×10^{17}	3.01
		100 mg/l	186	3.90×10^{19}	1.40×10^{17}	3.58
	UV100	50 mg/l	186	3.90×10^{19}	6.95×10^{16}	1.78
	SC134	50 mg/l	186	3.90×10^{19}	1.59×10^{16}	0.41
Multi-tubular[2] photoreactor	Anatase	200 mg/l	Varies depending on experimental conditions	Varies depending on experimental conditions	Varies depending on experimental conditions	0.3–32.8

[1]*Photonic effiency = rate of photochemical process* (n_{CHCl_3}/s)*/intensity of incident light (photons per second).*
[2]*Data extracted from Martin et al. (1996a).*
Reprinted with permission from author (Simonsen, 2009).

Degussa P25 has also been proposed to trap the electrons and thereby suppress the recombination of the electron-hole pair.

4.4.2. Kinetic Investigation of the Degradation of Chloroform

Investigations conducted by Martín et al. have shown that the rate-determining step in the photocatalytic degradation of chloroform is the formation of trichloromethyl radical adsorbed on the surface of TiO_2 (Martin et al., 1996a). The formation of trichloromethyl radical is limited by the amount of chloroform adsorbed on the surface of TiO_2 and the generation of holes. Investigations of several photocatalytic reactions have shown the existence of a linear relation between the rate of reaction and the level of substance adsorbed on the surface of the catalyst. In these cases a Langmuir–Hinshelwood (LH) kinetic model (Eqn (4.8)) has been used to fit the data (Mills and LeHunte, 1997; Kaneko and Okura, 2002; Schiavello, 1997).

$$\frac{\mathrm{d}C}{\mathrm{d}t} = -r = -\frac{k_r K_1 [C]}{1 + K_1 [C]} \tag{4.8}$$

where k_r is the rate constant and K is the adsorption constant.

The rate constant and the adsorption constant can be estimated by plotting the inverse rate of reaction against the inverse concentration which yields a straight line. The linearisation is given by Eqn (4.9).

$$\frac{1}{r} = \frac{1}{k_r} + \frac{1}{k_r K} \frac{1}{[C]} \tag{4.9}$$

The photocatalytic degradation profiles for different initial concentrations of chloroform are shown in Figure 4.12. The insert in Figure 4.12 shows the inverse rate of reaction vs the inverse concentration of chloroform. The straight line indicates that the photocatalytic degradation of chloroform indeed can be described by LH kinetics. The rate and adsorption constant is estimated to 0.4465 mg/ l/min and 0.5431 l/mg, respectively. The constants estimated from the linearisation of LH were used to fit the degradation profile in Figure 4.13. A good correlation between the measured and the predicted rate of degradation of chloroform by the LH kinetic model is seen. The kinetic model also shows that the rate of degradation increases to a certain level after which the rate is constant. These findings can be explained by the fact that the number of active sites on the surface of the TiO_2 particles is constant and thus the rate of degradation (Box 4.1).

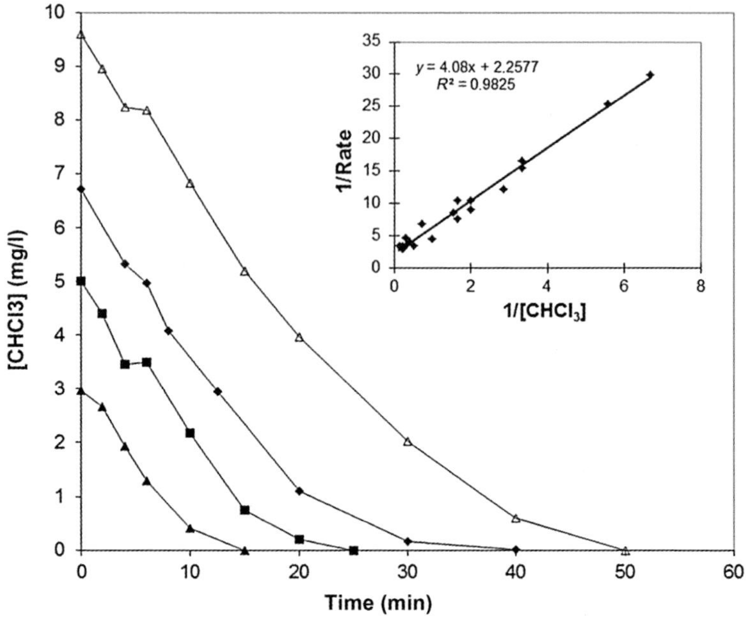

FIGURE 4.12 **Photocatalytic degradation of different initial concentrations of chloroform in TiO_2 (P25) suspension reactor.** *Reprinted with permission from author (Simonsen, 2005).*

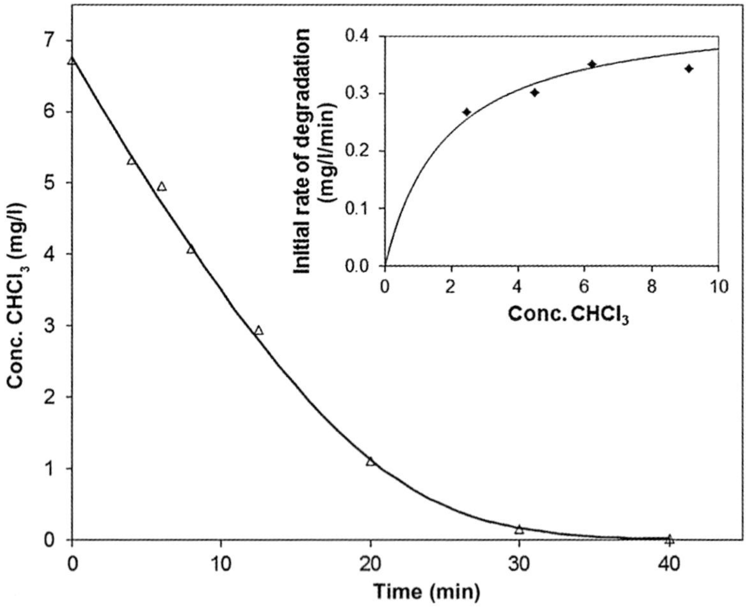

FIGURE 4.13 **Fit to the experimental data using the LH kinetic model.** Insert: Correlation between the experimental determined initial rates of degradation and the proposed LH model. *Reprinted with permission from author (Simonsen, 2005).*

BOX 4.1

SIMPLIFIED REACTION SCHEME FOR THE PHOTOCATALYTIC DEGRADATION OF CHLOROFORM (MARTIN ET AL., 1996A).

Activation: $TiO_2 \xrightarrow{h\nu} e^- + h^+$

Recombination: $e^- + h^+ \xrightarrow{k_2} \text{heat}$

Adsorption: $Ti^{IV} + H_2O \xleftrightarrow{k_{3a},k'_{3a}} Ti^{IV} - H_2O$

$O_L^{2-} + Ti^{IV} + H_2O \xleftrightarrow{k_{3b},k'_{3b}} O_LH^- + Ti^{IV} - OH^-$

$CHCl_3 + S \xleftrightarrow{k_4,k'_4} [CHCl_3]_{ads}$

$O_2 + S_2 \xleftrightarrow{k_5,k'_5} [O_2]_{ads}$

Oxidation: $Ti^{IV} - H_2O + h^+ \xleftrightarrow{k_{6a},k'_{6a}} Ti^{IV} - \bullet OH + H^+$

$[CHCl_3]_{ads} + h^+ \xrightarrow{k_{6b}} [\bullet CCl_3]_{ads} + H^+$

$[CHCl_3]_{ads} + Ti^{IV} - \bullet OH \xrightarrow{k_{7a}} [\bullet CCl_3]_{ads} + Ti^{IV} + H_2O$

$CHCl_3 + Ti^{IV} - \bullet OH \xrightarrow{k_{7b}} \bullet CCl_3 + Ti^{IV} + H_2O$

$[\bullet CCl_3]_{ads} \xleftrightarrow{k_8,k'_8} \bullet CCl_3 + S_1$

$\bullet CCl_3 + O_2 \xrightarrow{k_{9a}} \bullet CCl_3O_2$

$\bullet CCl_3O_2 \xrightarrow{k_{9b}} \bullet CCl_3O + \tfrac{1}{2}O_2$

Reduction: $Ti^{IV} + e^- \xleftrightarrow{k_{10},k'_{10}} Ti^{III}$

$Ti^{III} + [O_2]_{ads} \xleftrightarrow{k_{11},k'_{11}} Ti - \bullet O_2^- + S_2$

$Ti - \bullet O_2^- + H^+ \xleftrightarrow{k_{12},k'_{12}} Ti^{IV} + \bullet HO_2$

Decomposition: $\bullet CCl_3O + \bullet HO_2 \xrightarrow{k_{13}} CCl_3OH + O_2$

$CCl_3OH \xrightarrow{k_{14}} CCl_2O + H^+ + Cl^-$

$CCl_2O + H_2O \xrightarrow{k_{15}} CO_2 + 2Cl^- + 2H^+$

4.4.3. Effect of Oxygen on the Photocatalytic Degradation

The photocatalytic degradation of chloroform in the absence of oxygen is found to be considerably slower compared to the degradation which takes place in the presence of oxygen as shown in Figure 4.14. The

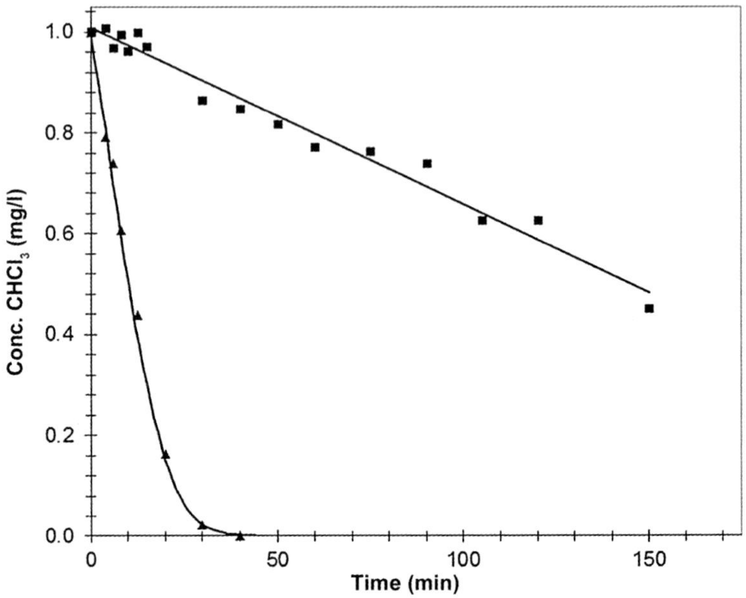

FIGURE 4.14 **Effect of oxygen on the photocatalytic oxidation of chloroform.** Oxidation of chloroform in oxygen-saturated solution and in the absence of oxygen. *Reprinted with permission from author (Simonsen, 2005).*

degradation of chloroform in the absence of oxygen can be described by zero-order kinetics compared the to the LH kinetics proposed for the degradation in presence of oxygen. Zero-order kinetics is usually proposed for reactions on surfaces where all the active sites are occupied. The observed difference in the rate of degradation is most likely due to a higher recombination of the electron-hole pair in the absence of oxygen since oxygen acts as an electron acceptor in most photocatalytic reactions. Moreover, additionally reactive oxygen species (O_2^-) can be formed as a result of the reaction with surface electrons and thereby enhance the rate of degradation.

4.4.4. Influence of Ions and By-Products on Photocatalytic Degradation

Investigations have shown that sulphate, chloride and phosphate ions found in concentrations greater than 10^{-3} mol/l can reduce the rate of degradation by 20–70% due to competitive adsorption at the active sites (Mills and LeHunte, 1997). In another study Chen et al. have investigated the influence of competitive adsorption of various anions during degradation of 1,2-dichloroethane (DCE). The results of the study showed that the order by which the ions inhibited the degradation of DCE was $NO_3^- < Cl^- < HCO_3^- < SO_4^{2-} < HPO_4^{2-}$ (Chen et al., 1997).

The adsorption on TiO_2 depends to a great extent on the surface charge, which in turn varies as a function of pH of the solution. The point of zero charge for TiO_2 has been reported to be in the pH interval 6.0–6.8 (Doll and Frimmel, 2004). The TiO_2 surface is positively charged in acidic media ($pH < 6$), whereas it is negatively charged under neutral and alkaline conditions ($pH > 6$).

In the case of degradation of chloroform in pool water it has been found that the degradation is 4–5 times slower than the degradation observed in distilled water (Figure 4.15). The pH of the solution decreases from around 7 to 5.5, making the TiO_2 surface positively charged. As a result the observed decrease in degradation rate of chloroform is most likely caused by competitive adsorption of chloride and other degradation by-products from chlorination. This is also seen when comparing the LH adsorption constants, the adsorption constant observed for the degradation in pool water was around a factor of five lower ($K = 0.092$ l/mg) than what was observed in distilled water ($K = 0.54$ l/mg^{-1}).

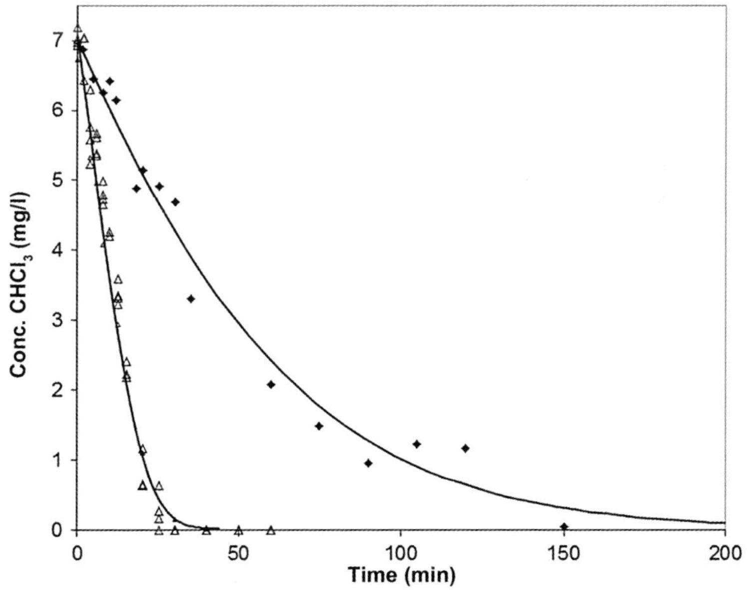

FIGURE 4.15 Comparison between the degradation profiles of chloroform in spiked pool water and distilled water. *Reprinted with permission from author (Simonsen, 2005).*

4.4.5. Photocatalytic Reactor Considerations

Early work on water purification using heterogeneous photocatalysis mainly focused on the photomineralisation of organics dissolved in aqueous solution usually employing the semiconductor in the form of a powder suspension. The use of TiO_2 in suspension is efficient due to the large surface area of the catalyst and the absence of mass transfer limitations (Mills and LeHunte, 1997; Mukherjee and Ray, 1999; Doll and Frimmel, 2004). However, the use of suspensions requires an additional separation step which is both time consuming and expensive. In order to overcome these obstacles immobilisation of the TiO_2 catalyst in the photoreactor seems to be an important step of improvement. There have been made numerous attempts to immobilise TiO_2 on different substrates in order to maximise the surface area of the catalyst as the reaction occurs at the liquid–solid interface (Horikoshi et al., 2002; Lee et al., 2002; Na et al., 2004; Noorjahan et al., 2003). However, many of these photoreactors suffer from nonuniform illumination. A new type of photocatalytic reactor in which the TiO_2 catalyst is immobilised on the surface of the quartz tubes surrounding the UV lamps has been developed. As the TiO_2 catalyst is irradiated from the backside uniform irradiation is achieved (Simonsen and Soegaard, 2008). The backside irradiated (BI) photocatalytic reactor has another advantage as no UV irradiation is lost due to absorption by organic compounds or particles present in the wastewater. Especially in high-particular waters the UV penetration depth is limited. Based on the conducted experiments the new type of photoreactor appears to be an interesting alternative which in time could be implemented in water treatment.

A comparison of the photonic efficiencies for the BI photocatalytic reactor and the traditional photocatalytic powder suspension reactors is seen in Table 4.2. The PE for the BI photocatalytic reactor is 1.5×10^{-3} $CHCl_3$ molecules per photon. In comparison the PE for the degradation of chloroform in a traditional suspension reactor is found to vary from 0.40×10^{-3} to 3.58×10^{-3} depending on the TiO_2 catalyst and the catalyst load. In the literature the photonic efficiencies of different photoreactors used for chloroform degradation ranges from 0.3×10^{-3} to 32.8×10^{-3} (Martin et al., 1996), suggesting that the PE of the BI photocatalytic reactor has a comparable efficiency. The efficiency of the photocatalytic reactor may, however, be improved. The backside irradiation may increase the recombination as the electron-hole pair has to migrate to the surface in order to mediate the photocatalytic degradation. The extent of this recombination has not been investigated. However, an optimal thickness of the TiO_2 coating is believed to exist. UV measurements have shown that 99% of the emitted UV light was absorbed by the TiO_2-coated quartz tubes.

4.5. CASE STUDY 2 – COMPARISON OF THE OXIDATION EFFICIENCY OF UV-ACTIVATED AOP TECHNIQUES

In this chapter a comparison of the efficiency for some UV-activated advanced oxidation process (AOP) techniques often encountered in water treatment is performed. These techniques include heterogeneous photocatalysis (UV/TiO_2), UV-activated hydrogen peroxide (UV/H_2O_2) and persulphate ($UV/S_2O_8^{2-}$), vacuum UV irradiation (VUV) and combinations thereof, all of which generate hydroxyl or sulphate radicals.

UV activation of hydrogen peroxide generates hydroxyl radicals through homolytic splitting of the oxygen–oxygen bond in H_2O_2. Hydrogen peroxide absorbs UV light between 200 and 280 nm with an absorption maximum around 220 nm (Tuhkanen, 2004). Theoretically, the photolysis of H_2O_2 generates two hydroxyl radicals per photon absorbed. However, high concentration of H_2O_2 will scavenge a part of the radicals through Reactions (4.10)–(4.13) resulting in a yield of hydroxyl radical around 0.5 (Tuhkanen, 2004).

$$H_2O_2 + h\nu \rightarrow OH^{\bullet} + OH \tag{4.10}$$

$$OH^{\bullet} + H_2O_2 \rightarrow HO_2^{\bullet} + H_2O \tag{4.11}$$

$$HO_2^{\bullet} + H_2O_2 \rightarrow OH^{\bullet} + H_2O + O_2 \tag{4.12}$$

$$HO_2^{\bullet} + HO_2^{\bullet} \rightarrow H_2O_2 + O_2 \tag{4.13}$$

Persulphate is a strong chemical oxidant with a redox potential of 2.05 and thus can oxidise organic pollutants through direct chemical oxidation. Chemical oxidation using persulphate is generally slow at normal temperatures. The oxidation is enhanced by thermal and UV activation of the persulphate (Reaction (4.14)) (Salari et al., 2008; Bennedsen et al., 2012).

$$S_2O_8^{2-} + h\nu \rightarrow 2SO_4^{\bullet -} \tag{4.14}$$

In aqueous solutions some of the sulphate radicals react with water and form bisulphate and hydroxyl radicals (Reaction (4.14)), initiating Reactions (4.15)–(4.18) (Salari et al., 2008). Overall, these reactions lead to a decrease in the number of sulphate radicals.

$$SO_4^{\bullet -} + H_2O \rightarrow HSO_4^{-} + OH^{\bullet} \tag{4.15}$$

$$HSO_4^{-} \rightarrow H^{+} + SO_4^{2-} \tag{4.16}$$

$$OH^{\bullet} + S_2O_8^{2-} \rightarrow HSO_4^- + SO_4^{\bullet -} + \frac{1}{2}O_2 \tag{4.17}$$

$$SO_4^{\bullet -} + OH^{\bullet} \rightarrow HSO_4^- + \frac{1}{2}O_2 \tag{4.18}$$

VUV irradiation (<200 nm) produces hydroxyl radicals from photolysis of water (Tasaki et al., 2009). Under VUV irradiation water itself is split into a hydrogen atom and hydroxyl radical, other oxidative species such as hydrogen peroxide may also be formed simultaneously (Tasaki et al., 2009).

4.5.1. Measure of Efficiency of UV-Activated AOPs

The dye *p*-nitrosodimethylaniline (RNO) has proved to be a very suitable model compound for evaluation of the oxidation power and PE of AOPs (Bennedsen et al., 2012; Simonsen et al., 2010; Muff et al., 2011; Holt et al., 2003; Bors et al., 1979; Comninellis, 1994; Kraljić and Trumbore, 1965; Wabner and Grambow, 1985; Simonsen et al., 2013). In aqueous solution RNO forms a strong yellow colour due to the conjugated double bonds and nitroso group. The bleaching of RNO can be monitored by UV/Vis absorption measurements ($\lambda_{max} = 440$ nm) (Figure 4.16). The evaluation of the efficiency of the UV-activated AOPs was carried out in a

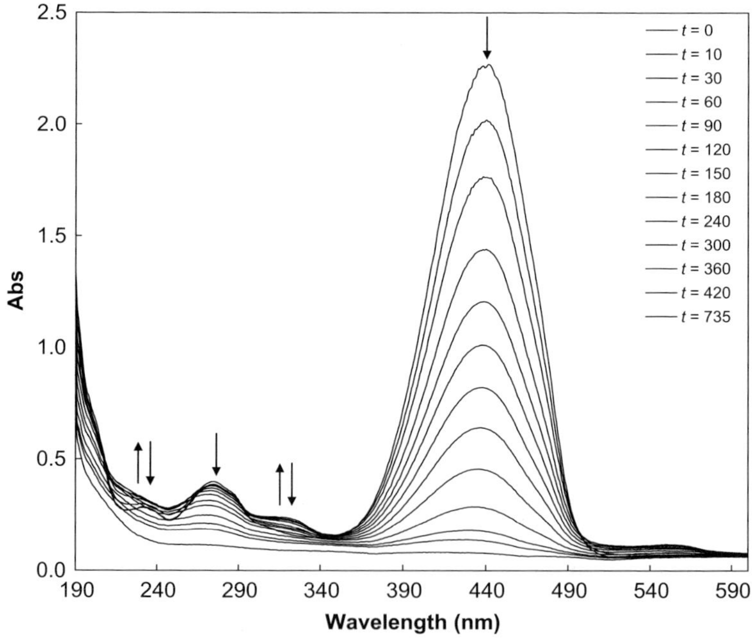

FIGURE 4.16 Bleaching of RNO in oxygen-saturated solution at pH 6 using a photocatalytic reactor with TiO_2 immobilised on liner (Simonsen et al., 2010).

TABLE 4.3 Irradiation Intensity in the Photoreactor, Initial Rate of Bleaching of RNO, Calculated PE for the Bleaching of RNO for the Different UV-Activated Systems and Calculated PE for the Mineralisation of RNO for the Different UV-Activated Systems

	Irradiation Intensity		Rate of RNO Bleaching	Photonic Efficiency RNO Bleaching	Photonic Efficiency RNO Mineralisation	
Photoreactor[2] System	**W**	**Photons/s**	**Molecules/s**	**$*10^{-2}$ N_{RNO}/photon**	**$*10^{-2}$ N_{RNO}/photon**	**%[1]**
UV	2.4	3.06×10^{18}	1.28×10^{15}	0.042	0.0033	0.17
UV/TiO_2 TiO_2-coated liner[3]	2.4	3.06×10^{18}	1.11×10^{16}	0.363	0.052	2.70
UV/TiO_2 (suspension) 50 mg/l (Degussa P25)	2.4	3.06×10^{18}	9.03×10^{16}	2.95	–	–
UV/$S_2O_8^{2-}$ $[S_2O_8^{2-}] = 1 \times 10^{-3}$ M	2.4	3.06×10^{18}	8.99×10^{16}	2.93	0.295	15.3
UV/$S_2O_8^{2-}$/TiO_2 $[S_2O_8^{2-}] = 1 \times 10^{-3}$ M, TiO_2 coated liner	2.4	3.06×10^{18}	4.39×10^{16}	1.43	0.196	10.1
UV/H_2O_2 $[H_2O_2] = 1 \times 10^{-3}$ M	2.4	3.06×10^{18}	9.12×10^{16}	2.98	0.349	18.1
UV/H_2O_2/TiO_2 $[H_2O_2] = 1 \times 10^{-3}$ M, TiO_2-coated liner	2.4	3.06×10^{18}	7.40×10^{16}	2.41	0.229	11.9
VUV	3.9	5.49×10^{18}	6.57×10^{16}	1.10	0.134	6.96
VUV/TiO_2 TiO_2-coated liner	3.9	5.49×10^{18}	6.51×10^{16}	1.09	0.0878	4.56
VUV/H_2O_2 $[H_2O_2] = 1 \times 10^{-3}$ M	3.9	5.49×10^{18}	1.83×10^{17}	3.06	–	–

[1]*Percentage of the maximum quantum yield (2×10^{-2}) for the mineralisation.*

[2]*Tubular photoreactor with centred UV lamp (l = 31 cm, d = 6 cm), total volume 3 l, flow rate 300 l/h, initial RNO concentration 10 mg/l (6.66×10^{-5} M), pH 6.5, and temperature $20 \pm 1\,°C$.*

[3]*Twenty-six TiO_2-coated stainless steel discs (d = 5.5 cm) installed perpendicular to the UV lamp.*

Reproduced with permission from Simonsen et al. (2013).

batch system, where the RNO solution was recycled from a water-cooled reservoir through the photoreactor. The bleaching of RNO is suggested to occur through an oxidation mechanism in which hydroxyl radicals react with the conjugated double bonds at the nitroso group (Simonsen et al., 2010; Muff et al., 2011). Further oxidation of RNO will lead to complete mineralisation through Reaction (4.19).

$$2C_8H_{10}N_2 + 26O_2 \rightarrow 16CO_2 + 4HNO_3 + 8H_2O \tag{4.19}$$

The photonic efficiencies determined by bleaching of RNO for the different UV-activated AOP techniques are shown in Table 4.3. The highest PE determined for the bleaching of RNO was observed for the VUV/H_2O_2 system (3.06%). Comparable photonic efficiencies were observed for the UV/H_2O_2 (2.98%), UV/TiO_2 (suspension) (2.95%) and UV/$S_2O_8^{2-}$ (2.93%) processes. These numbers suggest that only around 3% of the emitted photons are used in the bleaching of RNO (Simonsen et al., 2013). However, RNO is not only bleached but also further oxidised in the process. Prolonged treatment by the UV-activated AOP methods will lead to complete mineralisation of RNO. The degree of mineralisation can be estimated for total organic carbon (TOC) measurements. TOC data

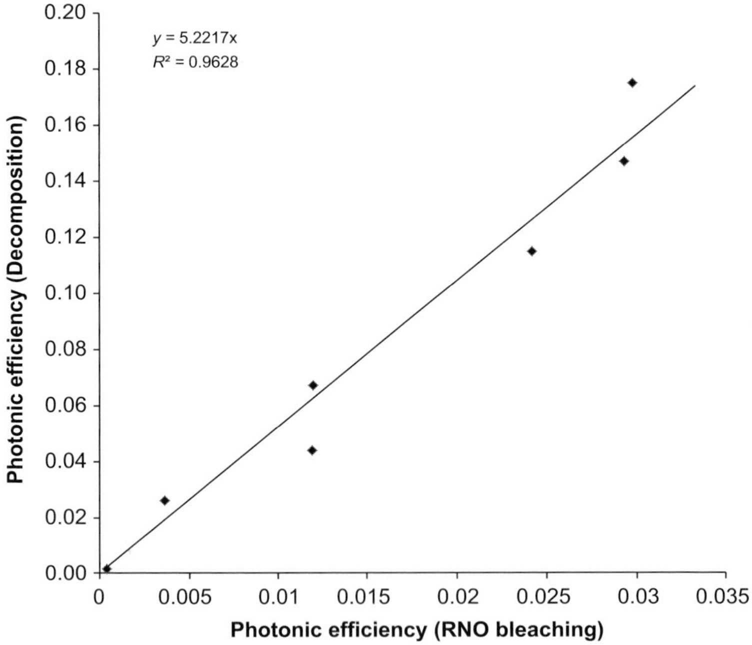

FIGURE 4.17 **Correlation between the photonic efficiencies determined by RNO bleaching and RNO decomposition measured by TOC.** *Reprinted with permission from Simonsen et al. (2013).*

in Table 4.3 suggest that the PE based on mineralisation of RNO is around 20% for the UV/H_2O_2 process.

Comparison of PE for the photocatalytic system applying immobilised TiO_2 in the form of a TiO_2-coated liner and the TiO_2 suspension reactor shows that the TiO_2 suspension is around eight times more efficient. The significant higher PE found in the case of the TiO_2 suspension reactor is due to the larger surface area of the catalyst available for reaction and the absence of mass transfer limitations.

The correlation between the PEs determined from bleaching and mineralisation is shown in Figure 4.17. The correlation between the PEs suggests that the bleaching of RNO can be used to evaluate the oxidation power of the UV-activated AOP techniques.

4.6. SUMMARY

The initial step in photocatalysis is the generation of an electron-hole pair within the photocatalyst particle by light activation. The photogenerated electron-hole pair can take part in different redox reactions which lead to the formation of reactive species, i.e. hydroxyl radicals which can mediate the degradation of organic components. In this process complete mineralisation of organic pollutants to carbon dioxide and mineral acids can be achieved. Literature reviews have shown that a wide range of organic components such as aliphatics, aromatics, dyes and surfactants can be completely mineralised by semiconductor photocatalysis.

The activity of the photocatalyst is determined by the crystallinity, crystallite size, particle size, SSA, amount and identity of defects and surface hydroxyl groups, many of which are related to the synthesis of the catalyst.

In spite of the potential of this promising technology, large-scale water treatment system is not that common even though commercial units are available on the market. In the last few years, a large number of publications have appeared based on laboratory-scale studies with generally positive results for decomposition of different toxic compounds in water. In order to make this approach cost alternative to current technologies, more fundamental research is still needed in order to develop more active photocatalysts and new reactors in which the photocatalysts can be immobilised or efficiently recovered.

References

Aggazzotti, G., Fantuzzi, G., Righi, E., Predieri, G., 1995. Environmental and biological monitoring of chloroform in indoor swimming pools. J. Chromatogr. A 710, 181–190.

Aggazzotti, G., Fantuzzi, G., Righi, E., Predieri, G., 1998. Blood and breath analyses as biological indicators of exposure to trihalomethanes in indoor swimming pools. Sci. Total Environ. 217, 155–163.

Ahonen, P, 2001. Aerosol Production and Crystallization of Titanium Dioxide from Metal Alkoxide Droplets.

Bennedsen, L.R., Muff, J., Søgaard, E.G., 2012. Influence of chloride and carbonates on the reactivity of activated persulfate. Chemosphere 86, 1092–1097.

Bickley, R.I., Gonzalezcarreno, T., Lees, J.S., Palmisano, L., Tilley, R.J.D., 1991. A structural investigation of titanium-dioxide photocatalysts. J. Solid State Chem. 92, 178–190.

Bors, W., Michel, C., Saran, M., 1979. On the nature of biochemically generated hydroxyl radicals. Studies using the bleaching of *p*-nitrosodimethylaniline as a direct assay method. Eur. J. Biochem. 95, 621–627.

Carp, O., Huisman, C.L., Reller, A., 2004. Photoinduced reactivity of titanium dioxide. Prog. Solid State Chem. 32, 33–177.

Chen, P.H., Jenq, C.H., 1998. Kinetics of photocatalytic oxidation of trace organic compounds over titanium dioxide. Environ. Int. 24, 871–879.

Chen, H.Y., Zahraa, O., Bouchy, M., 1997. Inhibition of the adsorption and photocatalytic degradation of an organic contaminant in an aqueous suspension of TiO_2 by inorganic ions. J. Photochem. Photobiol. A Chem. 108, 37–44.

Chiang, P.C., Chang, E.E., Ko, Y.W., Lou, J.C., 1997. Balancing disinfection efficiency and THM formation during chlorination: theoretical considerations. Can. J. Chem. Eng. 75, 892–898.

Choi, W.Y., Hoffmann, M.R., 1997. Novel photocatalytic mechanisms for $CHCl_3$, CHBr3, and CCl_3CO_2- degradation and the fate of photogenerated trihalomethyl radicals on TiO2. Environ. Sci. Technol. 31, 89–95.

Comninellis, C., 1994. Electrocatalysis in the electrochemical conversion/combustion of organic pollutants for waste water treatment. Electrochim. Acta 39, 1857–1862.

Doll, T.E., Frimmel, F.H., 2004. Development of easy and reproducible immobilization techniques using TiO2 for photocatalytic degradation of aquatic pollutants. Acta Hydrochim. Hydrobiol. 32, 201–213.

Frank, S.N., Bard, A.J., 1977. Heterogenous photocatalytic oxidation of cyanide ion inaqueous solutions at titanium dioxide powder. J. Am chem. Soc. 99, 303–304.

Fujishima, A., Honda, K., 1972. Electrochemical photolysis of water at a semiconductor electrode. Nature 238, 37–38.

Fujishima, A., Rao, T.N., 1998. Interfacial photochemistry: fundamentals and applications. Pure Appl. Chem. 70, 2177–2187.

Fujishima, A., Zhang, X.T., 2006. Titanium dioxide photocatalysis: present situation and future approaches. C. R. Chim. 9, 750–760.

Fujishima, A., Rao, T.N., Tryk, D.A., 2000. Titanium dioxide photocatalysis. J. Photochem. Photobiol. C Photochem. Rev. 1, 1–21.

Goodeve, C.F., Kitchener, J.A., 1938. The Mechanism of photosensitization by solids. Trans. Faraday. Soc. 34, 902–908.

Gopal, M., Chan, W.J.M., DeJonghe, L.C., 1997. Room temperature synthesis of crystalline metal oxides. J. Mater. Sci. 32, 6001–6008.

Gratzel, M., 2001. Photoelectrochemical cells. Nature 414, 338–344.

Hagfeldt, A., Gratzel, M., 1995. Light-Induced redox reactions in nanocrystalline systems. Chem. Rev. 95, 49–68.

Hashimoto, K., Irie, H., Fujishima, A., 2005. TiO_2 photocatalysis: a historical overview and future prospects. Jpn. J. Appl. Phys. 44, 8269–8285.

Hoffmann, M.R., Martin, S.T., Choi, W.Y., Bahnemann, D.W., 1995. Environmental applications of semiconductor photocatalysis. Chem. Rev. 95, 69–96.

Holt, K.B., Forryan, C., Compton, R.G., Foord, J.S., Marken, F., 2003. Anodic activity of boron-doped diamond electrodes in bleaching processes: effects of ultrasound and surface states. New J. Chem. 27, 698–703.

Horikoshi, S., Watanabe, N., Onishi, H., Hidaka, H., Serpone, N., 2002. Photodecomposition of a nonylphenol polyethoxylate surfactant in a cylindrical photoreactor with TiO_2 immobilized fiberglass cloth. Appl. Catal. B Environ. 37, 117–129.

Hsiao, C.Y., Lee, C.L., Ollis, D.F., 1983. Heterogeneous photocatalysis – degradation of dilute solutions of dichloromethane (CH2Cl2), chloroform ($CHCl_3$), and carbon-tetrachloride (CCl4) with illuminated TiO_2 photocatalyst. J. Catal. 82, 418–423.

Irie, H., Hashimoto, K., 2005. In: Environmental Photochemistry Part II: Photocatalytic Active Surfaces and Photo-induced High Hydrophilicity/High Hydrophobicity. Springer-Verlag, Duesseldorf, Germany.

Jensen, H., Joensen, K.D., Jorgensen, J.E., Pedersen, J.S., Sogaard, E.G., 2004. Characterization of nanosized partly crystalline photocatalysts. J. Nanopart. Res. 6, 519–526.

Jensen, H., Soloviev, A., Li, Z.S., Sogaard, E.G., 2005. XPS and FTIR investigation of the surface properties of different prepared titania nano-powders. Appl. Surf. Sci. 246, 239–249.

Jensen, H., Joensen, K.D., Iversen, S.B., Sogaard, E.G., 2006. Low temperature synthesis of metal oxides by a supercritical seed enhanced crystallization (SSEC) process. Ind. Eng. Chem. Res. 45, 3348–3353.

Judd, S.J., Black, S.H., 2000. Disinfection by-product formation in swimming pool waters: a simple mass balance. Water Res. 34, 1611–1619.

Judd, S.J., Bullock, G., 2003. The fate of chlorine and organic materials in swimming pools. Chemosphere 51, 869–879.

Kaneko, M., Okura, I., 2002. Photocatalysis – Science and Technology. Springer-Verlag, Berlin, Heidelberg, New York.

Kang, M., Lee, S.Y., Chung, C.H., Cho, S.M., Han, G.Y., Kim, B.W., et al., 2001. Characterization of a TiO_2 photocatalyst synthesized by the solvothermal method and its catalytic performance for CHCl3 decomposition. J. Photochem. Photobiol. A Chem. 144, 185–191.

Kato, S., Mashio, F., 1956. Autooxidation by TiO_2 as a photocatalyst. Abtr. Book Annu. Meet Chemical Society of Japan. 223.

Katsumata, K., Nakajima, A., Shiota, T., Yoshida, N., Watanabe, T., Kameshima, Y., et al., 2006. Photoinduced surface roughness variation in polycrystalline TiO_2 thin films. J. Photochem. Photobiol. A Chem. 180, 75–79.

Kim, H., Shim, J., Lee, S., 2002. Formation of disinfection by-products in chlorinated swimming pool water. Chemosphere 46, 123–130.

Koukouraki, E., Diamadopoulos, E., 2003. Modelling the formation of THM (tri-halomethanes) during chlorination of treated municipal wastewater. Water Recycling Mediterr. Region 3, 277–284.

Kraljić, I., Trumbore, C.N., 1965. *p*-Nitrosodimethylaniline as an OH radical scavenger in radiation chemistry. J. Am. Chem. Soc. 87, 2547–2550.

Lee, N.C., Choi, W.Y., 2002. Solid phase photocatalytic reaction on the soot/TiO2 interface: the role of migrating OH radicals. J. Phys. Chem. B 106, 11818–11822.

Lee, J.C., Kim, M.S., Kim, B.W., 2002. Removal of paraquat dissolved in a photoreactor with TiO_2 immobilized on the glass-tubes of UV lamps. Water Res. 36, 1776–1782.

Lee, S.K., McIntyre, S., Mills, A., 2004. Visible illustration of the direct, lateral and remote photocatalytic destruction of soot by titania. J. Photochem. Photobiol. A Chem. 162, 203–206.

Lee, F.K., Andreatta, G., Benattar, J., 2007. Role of water adsorption in photoinduced superhydrophilicity on TiO_2 thin films. Appl. Phys. Lett. 90, 181928.

Li, Y., White, T.J., Lim, S.H., 2004. Low-temperature synthesis and microstructural control of titania nano-particles. J. Solid State Chem. 177, 1372–1381.

Linsebigler, A.L., Lu, G.Q., Yates, J.T., 1995. Photocatalysis on TiO_2 surfaces – principles, mechanisms, and selected results. Chem. Rev. 95, 735–758.

Liu, H.M., Yang, W.S., Ma, Y., Cao, Y., Yao, J.N., Zhang, J., et al., 2003. Synthesis and characterization of titania prepared by using a photoassisted sol-gel method. Langmuir 19, 3001–3005.

Livage, J., Henry, M., Sanchez, C., 1988. Sol-gel chemistry of transition-metal oxides. Prog. Solid State Chem. 18, 259–341.

Martin, C.A., Baltanas, M.A., Cassano, A.E., 1996a. Photocatalytic reactors. 3. Kinetics of the decomposition of chloroform including absorbed radiation effects. Environ. Sci. Technol. 30, 2355–2364.

Martin, C.A., Baltanas, M.A., Cassano, A.E., 1996b. Photocatalytic decomposition of chloroform in a fully irradiated heterogeneous photoreactor using titanium oxide particulate suspensions. Catal. Today 27, 221–227.

Matsunaga, T., Tomato, R., Nakajima, T., Wake, H., 1985. Photoelectrochemical sterilization of microbial cells by semiconductor powders. FEMS Microbiol. Lett. 29, 211–214.

Mills, A., Crow, M., 2007. In situ, continuous monitoring of the photoinduced superhydrophilic effect: influence of UV-type and ambient atmospheric and droplet composition. J. Phys. Chem. C 111, 6009–6016.

Mills, A., Crow, M., 2008. A study of factors that change the wettability of titania films. Int. J. Photoenergy, 470670.

Mills, A., Lee, S.K., 2002. A web-based overview of semiconductor photochemistry-based current commercial applications. J. Photochem. Photobiol. A Chem. 152, 233–247.

Mills, A., LeHunte, S., 1997. An overview of semiconductor photocatalysis. J. Photochem. Photobiol. A Chem. 108, 1–35.

Mills, A., Hill, G., Bhopal, S., Parkin, I.P., O'Neill, S.A., 2003. Thick titanium dioxide films for semiconductor photocatalysis. J. Photochem. Photobiol. A Chem. 160, 185–194.

Miyauchi, M., Nakajima, A., Watanabe, T., Hashimoto, K., 2002. Photocatalysis and photoinduced hydrophilicity of various metal oxide thin films. Chem. Mater. 14, 2812–2816.

Muff, J., Bennedsen, L., Søgaard, E., 2011. Study of electrochemical bleaching of *p*-nitrosodimethylaniline and its role as hydroxyl radical probe compound. J. Appl. Electrochem. 41, 599–607.

Mukherjee, P.S., Ray, A.K., 1999. Major challenges in the design of a large-scale photocatalytic reactor for water treatment. Chem. Eng. Technol. 22, 253–260.

Na, Y.S., Kim, D.H., Lee, C.H., Lee, S.W., Park, Y.S., Oh, Y.K., et al., 2004. Photocatalytic decolorization of rhodamine B by fluidized bed reactor with hollow ceramic ball photocatalyst. Korean J. Chem. Eng. 21, 430–435.

Noorjahan, M., Reddy, M.P., Kumari, V.D., Lavedrine, B., Boule, P., Subrahmanyam, M., 2003. Photocatalytic degradation of H-acid over a novel TiO_2 thin film fixed bed reactor and in aqueous suspensions. J. Photochem. Photobiol. A Chem. 156, 179–187.

Pruden, A.L., Ollis, D.F., 1983. Degradation of chloroform by photoassisted heterogeneous catalysis in dilute aqueous suspensions of titanium-dioxide. Environ. Sci. Technol. 17, 628–631.

Sakai, N., Fujishima, A., Watanabe, T., Hashimoto, K., 2001. Enhancement of the photoinduced hydrophilic conversion rate of TiO2 film electrode surfaces by anodic polarization. J. Phys. Chem. B 105, 3023–3026.

Sakai, N., Fujishima, A., Watanabe, T., Hashimoto, K., 2003. Quantitative evaluation of the photoinduced hydrophilic conversion properties of TiO_2 thin film surfaces by the reciprocal of contact angle. J. Phys. Chem. B 107, 1028–1035.

Salari, D., Daneshvar, N., Niaei, A., Aber, S., Rasoulifard, M.H., 2008. The photo-oxidative destruction of C.I. Basic yellow 2 using UV/S2O8 2–process in an annular photoreactor. J. Environ. Sci. Health Part A 43, 657–663.

Schiavello, M., 1997. Heterogeneous Photocatalysis. John Wiley & sons.

Schiavello, M., Augugliaro, V., Palmisano, L., 1991. An experimental method for the determination of the photon flow reflected and absorbed by aqueous dispersions containing polycrystalline solids in heterogeneous photocatalysis. J. Catal. 127, 332–341.

Shultz, A.N., Jang, W., Hetherington, W.M., Baer, D.R., Wang, L., Engelhard, M.H., 1995. Comparative second harmonic generation and X-ray photoelectron spectroscopy studies of the UV creation and O_2 healing of Ti3+ defects on (110) rutile TiO_2 surfaces. Surf. Sci. 339, 114–124.

Simonsen, M.E., 2005. Characterization of Nano-crystalline Titania and Photocatalytic Degradation of Chloroform.

Simonsen, M.E., 2009. Sol-gel Processes for Development of Nanomaterials for Ceramic and Photocatalytic Applications.

Simonsen, M.E., Soegaard, E.G., 2008. Photocatalytic degradation of chloroform using a new type of photo-reactor. In: Sol-Gel Methods for Materials Processing – Focusing on Materials for Pollution Control, Water Purification, and Soil Remediation, pp. 421–426.

Simonsen, M.E., Sogaard, E.G., 2010. Sol-gel reactions of titanium alkoxides and water: influence of pH and alkoxy group on cluster formation and properties of the resulting products. J. Sol Gel Sci. Technol. 53, 485–497.

Simonsen, M.E., Jensen, H., Li, Z., Sogaard, E.G., 2008. Surface properties and photocatalytic activity of nanocrystalline titania films. J. Photochem. Photobiol. A Chem. 200, 192–200.

Simonsen, M.E., Li, Z., Sogaard, E.G., 2009. Influence of the OH groups on the photocatalytic activity and photoinduced hydrophilicity of microwave assisted sol-gel TiO_2 film. Appl. Surf. Sci. 255, 8054–8062.

Simonsen, M.E., Muff, J., Bennedsen, L.R., Kowalski, K.P., Søgaard, E.G., 2010. Photocatalytic bleaching of p-nitrosodimethylaniline and a comparison to the performance of other AOP technologies. J. Photochem. Photobiol. A 216, 244–249.

Simonsen, M.E., Jensen, C.V., Søgaard, E.G., 2013. Comparison of different UV-activated AOP methods. J. Adv. Oxid. Technol. 16, 179–187.

Sun, R.D., Nakajima, A., Fujishima, A., Watanabe, T., Hashimoto, K., 2001. Photoinduced surface wettability conversion of ZnO and TiO2 thin films. J. Phys. Chem. B 105, 1984–1990.

Tasaki, T., Wada, T., Fujimoto, K., Kai, S., Ohe, K., Oshima, T., et al., 2009. Degradation of methyl orange using short-wavelength UV irradiation with oxygen microbubbles. J. Hazard. Mater. 162, 1103–1110.

Tuhkanen, T.A., 2004. In: Parsons, S. (Ed.), first ed. IWA Publishing, London, p. 356.

Wabner, D., Grambow, C., 1985. Reactive intermediates during oxidation of water lead dioxide and platinum electrodes. J. Electroanal. Chem. Interf. Electrochem. 195, 95–108.

Wang, L.Q., Baer, D.R., Engelhard, M.H., Shultz, A.N., 1995. The adsorption of liquid and vapor water on TiO_2(110) surfaces: the role of defects. Surf. Sci. 344, 237–250.

Wang, R., Hashimoto, K., Fujishima, A., Chikuni, M., Kojima, E., Kitamura, A., Shimohigoshi, M., Watanabe, T., 1998. Light-induced amphiphilic surfaces. Nature 388, 431–432.

Wang, R., Sakai, N., Fujishima, A., Watanabe, T., Hashimoto, K., 1999. Studies of surface wettability conversion on TiO_2 single-crystal surfaces. J. Phys. Chem. B 103, 2188–2194.

Wang, C.Y., Groenzin, H., Shultz, M.J., 2003. Molecular species on nanoparticulate anatase TiO2 film detected by sum frequency generation: trace hydrocarbons and hydroxyl groups. Langmuir 19, 7330–7334.

Ward, M.D., White, J.M., Bard, A.J., 1983. Electrochemical investigation of the energies of particulate titanium dioxide photocatalysts. The methyl viologen-acetate system. J. Am. Chem. Soc. 105, 27–31.

Watanabe, T., Nakajima, A., Wang, R., Minabe, M., Koizumi, S., Fujishima, A., et al., 1999. Photocatalytic activity and photoinduced hydrophilicity of titanium dioxide coated glass. Thin Solid Films 351, 260–263.

Wendt, S., Schaub, R., Matthiesen, J., Vestergaard, E.K., Wahlstrom, E., Rasmussen, M.D., et al., 2005. Oxygen vacancies on TiO_2(110) and their interaction with H_2O and O-2: a combined high-resolution STM and DFT study. Surf. Sci. 598, 226–245.

White, G.C., 1972. Handbook of Chlorination – For Potable Water, Wastewater, Cooling Water, Industrial, Processes, and Swimming Pools. Van Nostrand Reinhold Company.

White, J.M., Szanyi, J., Henderson, M.A., 2003. The photon-driven hydrophilicity of titania: a model study using TiO_2(110) and adsorbed trimethyl acetate. J. Phys. Chem. B 107, 9029–9033.

Yamazaki, S., Yoshida, A., Abe, H., 2005. Photocatalytic degradation of chloroform in the gas phase on the porous TiO2 pellets: effect of Cl accumulated on the catalyst surface. J. Photochem. Photobiol. A Chem. 169, 191–196.

Yu, J.G., Yu, J.C., Ho, W.K., Jiang, Z.T., 2002. Effects of calcination temperature on the photocatalytic activity and photo-induced super-hydrophilicity of mesoporous TiO_2 thin films. New J. Chem. 26, 607–613.

Yu, J.G., Xiong, J.F., Cheng, B., Liu, S.W., 2005. Fabrication and characterization of Ag-TiO_2 multiphase nanocomposite thin films with enhanced photocatalytic activity. Appl. Catal. B Environ. 60, 211–221.

Zhang, H.Z., Banfield, J.F., 2000. Understanding polymorphic phase transformation behavior during growth of nanocrystalline aggregates: insights from TiO_2. J. Phys. Chem. B 104, 3481–3487.

Zubkov, T., Stahl, D., Thompson, T.L., Panayotov, D., Diwald, O., Yates, J.T., 2005. Ultraviolet light-induced hydrophilicity effect on TiO_2(110)(1x1). Dominant role of the photooxidation of adsorbed hydrocarbons causing wetting by water droplets. J. Phys. Chem. B 109, 15454–15462.

CHAPTER

5

Near- and Supercritical Water

Assistant Professor Rudi P. Nielsen

Section of Chemical Engineering, Department of Biotechnology, Chemistry and Environmental Engineering, Aalborg University Esbjerg, Niels Bohrs Vej 8, DK 6700 Esbjerg, Denmark

When subjecting water to high pressures and temperatures it behaves nothing like water at ambient conditions. Solubilities change dramatically at these conditions, to the point where water and oil become fully miscible and salts are no longer soluble. Steel, and steel alloys, will corrode at a higher rate and the risk of hydrogen embrittlement becomes an actual risk when processing with water at these conditions. The focus of this text is on these changes and the reasons for them.

Fluids at high pressures and temperatures exhibit significant changes in properties, such as solubilities and reactivity. These changes in properties can be utilised for a large number of applications ranging from extraction of chemicals from a complex matrix to conversions of biomass into bio-oil or syngas or destruction of hazardous chemicals by oxidation. In this work, however, focus shall be on the fundamentals of the near- and supercritical states. That water is the medium of choice does not mean that water is the only substance that is utilised for supercritical technology; however, water is an interesting substance for supercritical technology due to changes in ion dissociation constants and dielectric constant, which shall be described in detail in the following sections.

5.1. PROPERTIES OF THE SUPERCRITICAL PHASE

A supercritical fluid is a substance or a mixture of substances that is kept with temperature and pressure above the critical values of these. Beyond this temperature and pressure the substance only exists in a single phase, i.e. the concept of a gas and liquid phase with an interface between them does not exist once the supercritical region is entered. Here only one phase exists which exhibits characteristics in between that of the gas and

Chemistry of Advanced Environmental Purification Processes of Water
http://dx.doi.org/10.1016/B978-0-444-53178-0.00005-5

liquid phase. As described by Heidemann and Khalil the critical point is a point for temperature, pressure and volume, where the composition and density of the equilibrium phases are identical (Heidemann and Khalil, 1980). To understand the location of the supercritical region a general understanding of the phase diagram is needed. The phase diagram shows the various phases (solid, liquid and gas) as a function of temperature and pressure, for water this is shown in Figure 5.1.

The phase diagram shows several characteristics of water, which may be described as the curves separating the gas, liquid and solid phases, and two distinct points. The curves themselves are where transitions between phases occur, a list of these transitions and their names are given in Table 5.1.

In the phase diagram the regions given by these phase transition curves are single phase regions, meaning that within a region only a single phase will exist. Two phases, such as liquid and gas will only exist on points on the vaporization/condensation curve, since the curves mark the transition between phases. It appears that the melting/freezing curve is vertical meaning that high pressures are required to achieve an ice to liquid water transition by increasing pressure alone. However, the curve is in fact sloping, as illustrated in Figure 5.2.

Here it is evident that with increasing pressure ice will change into liquid water, although it will take a pressure increase of approximately 1.5 MPa to decrease the melting point by 1 K.

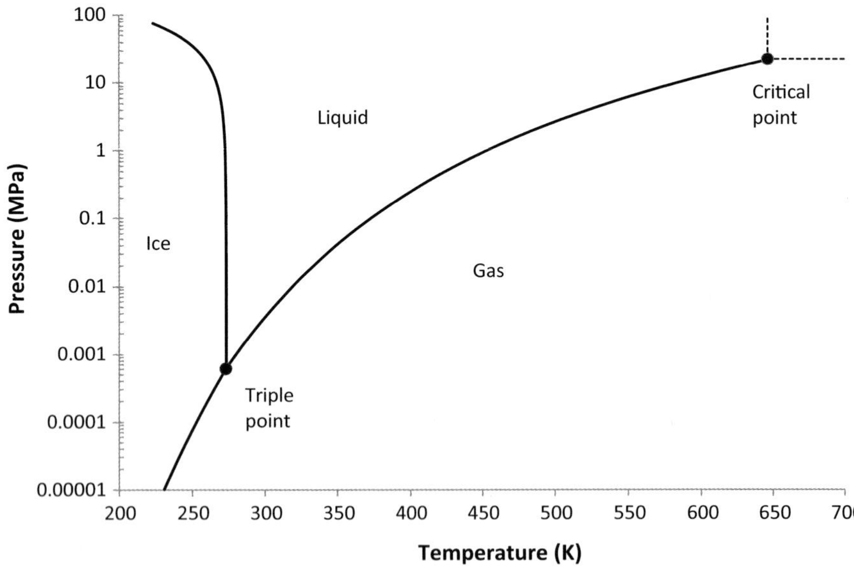

FIGURE 5.1 **Phase diagram of water.** Curves indicate transitions between phases with triple and critical points indicated. *Data from Hardy (1998), Choukroun and Grasset (2007), Wagner and Pruss (2002), IAWPS (2011).*

TABLE 5.1 Transitions between Phases in the Phase Diagram and Their Denominations

Transitions	Phase Change Occurring
Sublimation	Solid to gas
Deposition	Gas to solid
Freezing	Liquid to solid
Melting	Solid to liquid
Vaporization	Liquid to gas
Condensation	Gas to liquid

Besides the curves that define the borders of the phases two points of interest are present in the phase diagram (Figure 5.1). These are the critical point and the solid–liquid–gas triple point which are located at the pressure and temperature conditions given in Table 5.2.

In reality, water has several triple points, although not shown in Figure 5.1, where three phases coexist in equilibrium. The triple point is a point in pressure and temperature where gas, liquid and solid phases exist simultaneously. At this point changes from one phase to another may occur very rapidly with only slight variations in temperature and

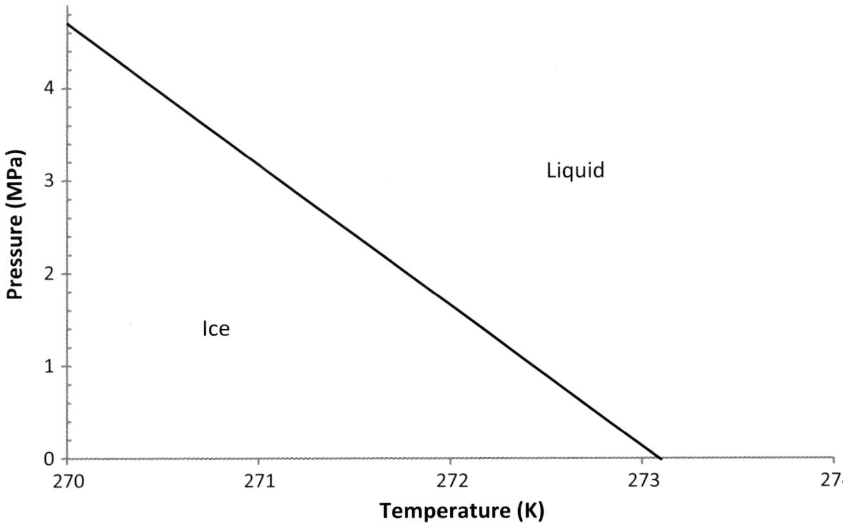

FIGURE 5.2 **Phase diagram of water showing ice and liquid water phases in a narrow temperature range.** *Data from Hardy (1998), Choukroun and Grasset (2007), Wagner and Pruss (2002), IAWPS (2011).*

TABLE 5.2 Temperature and Pressure Values for Triple and Critical Points of Water

Point	Temperature (K)	Pressure (Pa)
Triple point	273.16	611.73
Critical point	647.10	$22.06 \cdot 10^6$

pressure. Most often when referring to the triple point of water the reference is to the solid–liquid–gas triple point; however, due to the large number of ice structures that water may form, triple points exists as solid–solid–gas, solid–solid–liquid or solid–solid–solid triple points.

One interesting feature about the supercritical region is that by entering this region it is possible to move from a liquid to a gas without crossing the boiling/condensation line of the phase diagram. This shows that it is possible to change from one phase to another without subjecting the fluid to the mechanisms of boiling or condensation.

The critical point is the point beyond which the supercritical phase exists. The critical point may be defined as done by Reid and Beegle (1977) 'A stable point on the limit of stability'. Although this may seem a rather contradictory statement, it refers to the placement of the critical point on the phase envelope. Figure 5.3 illustrates a phase envelope with the critical point added.

The border of the phase envelope is what is referred to by Reid and Beegle (1977) as the limit of stability. At the border a split from a single phase into two phases occurs, which means that the single phase is not stable anymore, i.e. the limit of stability. However, at the critical point this

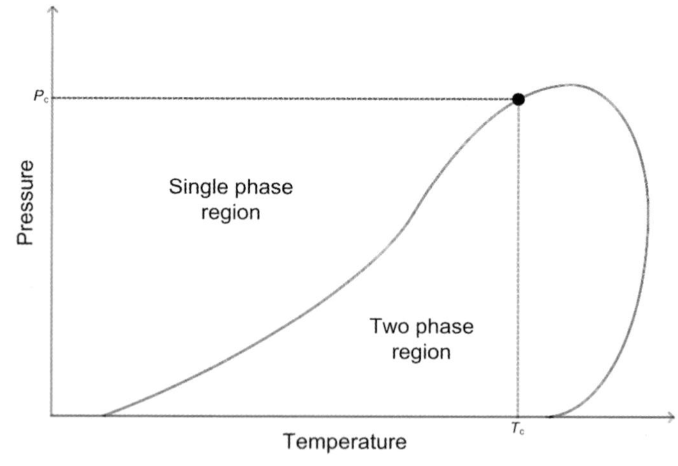

FIGURE 5.3 **Phase envelope with critical point shown.**

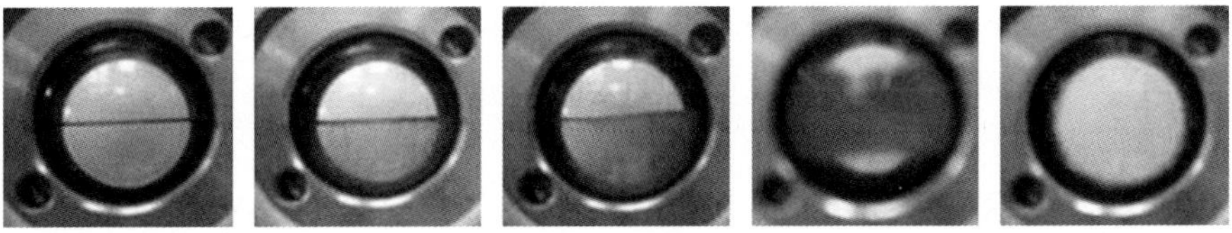

FIGURE 5.4 **Transition into the supercritical region by increasing temperature.** (For colour version of this figure, the reader is referred to the online version of this book.)

split into two phases does not occur, since here the supercritical phase forms, meaning that a single stable phase is present.

Knowing that the critical point is present is one thing; however, most times one needs to determine the exact values for critical temperature and pressure. This is necessary in those cases where processing is carried out at elevated temperatures and pressures and knowledge is needed about the phase distribution in the process. A method for determining the critical point of a given mixture is that of Heidemann and Khalil (1980), using the modification of the method proposed by Michelsen (1980) for part of the calculation. This method iteratively solves a set of equations with temperature and volume as independent variables to determine the critical parameters of a multicomponent mixture.

One question that always arises is 'what does the supercritical phase look like?' Figure 5.4 shows the transition from a liquid/gas system into a supercritical system by increasing temperature in the reaction vessel.

As can be seen from the figure the interface between gas and liquid gradually expands, until one single phase is achieved. This phase, which can be seen, is the supercritical phase, which is stable as long as temperature and pressure remain above their critical values.

Although this chapter deals exclusively with water, every substance has a critical point. Table 5.3 lists the critical temperatures and pressures for some selected substances.

Of these substances especially CO_2 and methanol are often used in processing where supercritical fluids are utilised. Compared to water most substances have relatively low critical properties, making it easier to

TABLE 5.3 Critical Temperatures and Pressures of Selected Substances

Substance	Critical Temperature (K)	Critical Pressure (MPa)
CO_2	304.2	7.38
Methane	190.4	4.60
Methanol	512.6	8.10
Acetone	308.1	4.70

reach the supercritical phase, and thus utilise the properties exhibited by this phase.

5.2. PROPERTIES OF SUPERCRITICAL WATER

Water is a benign medium for processing; meaning the harmful effects that are connected with using organic solvents is eliminated. Furthermore, water is a low-cost solvent and use of water as a processing medium also yields opportunities for clean processing and pollution prevention (Patrick et al., Dec. 2001; Nolen et al., 2003; Savage et al., 1995; Savage, 1999).

One thing of special interest when using supercritical water for processing is the changes in several key parameters. Table 5.4 gives the density, viscosity and diffusivity of gas, liquid and supercritical fluids.

It is evident that the properties of the supercritical fluid are somewhere between those of gasses and liquids. Based on these properties and coupled with the fact that supercritical fluids do keep the solvating power of fluids it becomes evident that the changed properties allow for a wide range of applications such as extraction processes.

INFORMATION BOX

Example Application of Supercritical Fluid: Extraction

Since the supercritical fluid retains the solvating power of a fluid while having lower viscosity but higher diffusivity than a liquid it may be used to extract components from a sample matrix.

During processing the supercritical fluid will enter the sample matrix with relative ease due to lower viscosity than a conventional liquid, such as CCl_4, which is often used for extractions. Once inside the sample matrix the components are then soluted in the supercritical fluid, and may subsequently be removed from the matrix with the supercritical fluid.

Although processing with supercritical fluids may be relatively expensive due to the temperature and pressure requirements, the benefit of using a solvent that is far less hazardous to the health of the operator than conventional extraction solvents can in many cases outweigh the cost.

Experimental work using CO_2 to remove components from disposed car tyres have shown results of similar degrees of extraction compared to conventional Soxhlet extractions using CCl_4.

TABLE 5.4 Comparison of Properties

	Density kg/m^3	Viscosity $\mu Pa \cdot s$	Diffusivity mm^2/s
Gas	1	0.01	1–10
Supercritical	100–800	0.05–0.1	0.01–0.1
Liquid	1000	0.5–1.0	0.001

The properties of the supercritical phase can be quite beneficial; Savage (2000) compiles a list of benefits from using supercritical water for processes with heterogeneous catalysis:

- Providing concentrations of reactant gases (e.g. H_2, O_2) higher than those that can be obtained in conventional gas–liquid systems.
- Eliminating interphase transport limitations that would exist in multiphase reacting systems (e.g. hydrogenation, partial oxidation).
- Providing easier product separation because solubility can be a strong function of pressure and temperature in the critical region.
- Allowing in situ extraction of coke precursors and hence a longer time on stream for hydrocarbon processing catalysts.
- Optimizing the strongly pressure-dependent properties of the supercritical reaction medium for a given reaction.
- Providing higher diffusivities than liquids and better heat transfer than gases.

Although this list is focused on processes incorporating heterogeneous catalysts, the points specified by Savage (2000) are still valid in most cases where supercritical water is used. Some of these points may seem less obvious, and subsequently they will be treated in turn and explained further.

At near-critical conditions the tendency of water to autodissociate increases significantly with K_W several orders of magnitude larger than at ambient conditions (this shall be treated in greater detail later). This means that the concentrations of H^+ and OH^- will be significantly higher, which is beneficial in processes where these reactants are needed.

Since the supercritical phase is a single stable phase there is no need to move components from one phase to another. This means that any hindrances that would be associated with the phase transitions, such as mobilization from water to gas, are not present when processing in supercritical water.

As shall be seen later the solubility of polar and nonpolar components in supercritical fluids can change significantly. Since temperature and pressure are the main parameters for controlling the properties in the supercritical phase and since the properties, as shall be seen later, can change drastically with only slight variations in pressure and temperature it is possible to finely tune the properties of the supercritical fluid based on the parameters.

A more specific application of this is in the case of processing using heterogeneous catalysts. Especially in hydrocarbon processing formation of coke occurs due to deposition of the coke precursors on the catalyst surface, which will lead to coverage of the catalyst surface or poisoning. Since as stated before, solubilities can be controlled by variation of pressure and temperature it is possible to keep the coke precursors in solution, and thus prevent deposition on the catalyst surface and subsequently poisoning of the catalyst (Savage et al., 1995).

In case of reactions which are strongly pressure dependent the use of a supercritical fluid for processing may yield some benefits. In reactions where products are favoured by high pressure the supercritical fluid is an excellent reaction media.

Higher diffusivities than liquids and better heat transfer than gasses, is stated as a beneficial property and should be quite self-explanatory. However, the higher diffusivities means that the supercritical fluid can better enter a sample matrix thus resembling gas properties. The better heat transfer means that heat exchanging on a supercritical medium requires less surface area than a gas heat exchanger, this is again where the properties 'between gas and liquid' and their benefits are evident.

In his list Savage (2000) does not include the changes in ion dissociation constant. This constant is an indication of the amount of dissociation of water molecules into hydrogen and hydroxide ions.

$$H_2O \rightleftarrows H^+ + OH^- \tag{5-I}$$

The ion dissociation constant is subsequently given as

$$K_W = \frac{[H^+][OH^-]}{[H_2O]} \tag{5.1}$$

The ion dissociation constant changes significantly around the critical point and in the near-critical region. This means that the amount of H^+ and OH^- ions are quite high, actually approximately three orders of magnitude higher than at ambient conditions, as the critical point is approached. This is evident from Figure 5.5 which shows the temperature dependence of the ion dissociation constant of water at various pressures.

Besides changes in the ion dissociation constant, also the dielectric constant changes significantly from ambient to supercritical temperature as seen in Figure 5.6.

At ambient temperature, although at high pressure, water is a polar substance; however, with the decrease in dielectric constant water will behave like a nonpolar substance at supercritical conditions. Since at low densities in the supercritical region the water molecules are highly

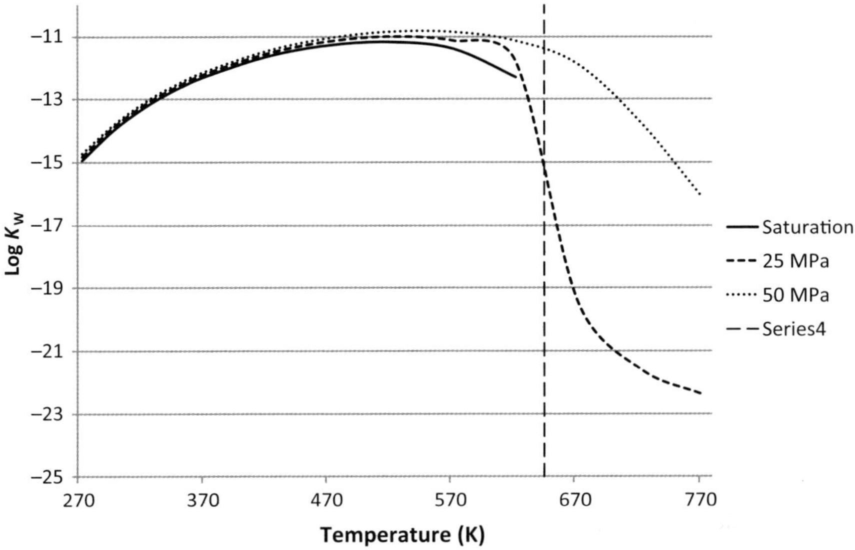

FIGURE 5.5 Ion dissociation constant of water at saturation pressure, 25 and 50 MPa as a function of temperature. *Data from Marshall and Franck (1981).*

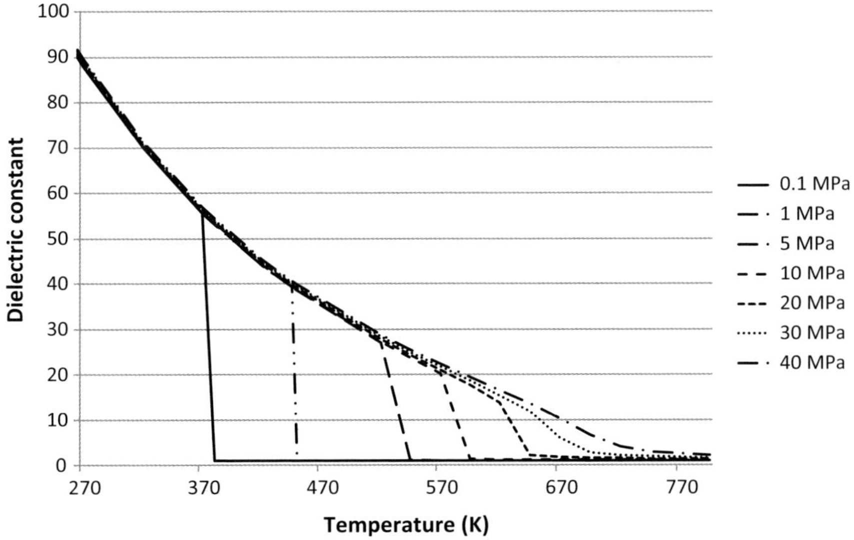

FIGURE 5.6 Dielectric constant of water at various pressures and temperatures. *Data from Archer and Wang (1990).*

spatially distributed and due to temperature they have a high kinetic energy, the hydrogen bonding of the molecules decreases, which will be shown later. This means that water will be completely miscible with nonpolar organic components at these conditions (Savage et al., 1995; Savage, 1999). This is an important property of supercritical water, and supercritical fluids in general, that polar substances will behave as nonpolar substances once they are near or in the supercritical region.

However, the dielectric constant does also influence the ability of the substance to act as a solvent for salts. With decreasing dielectric constant salts will precipitate from the supercritical substance (Archer and Wang, 1990; Peterson et al., 2008a) which may be quite problematic. This reduction of solubility is due to decreased shielding of ions, since the dielectric constant (ε) is a measure of this effect as it is related to the dipole moment and the amount of molecules per volume unit. The high spatial distribution at low densities causes the number of molecules per volume to decrease leading to a significant decrease in dielectric constant (Seminario et al., 1994).

Several studies have been carried out regarding solubility of sodium carbonate and sulphate in near- and supercritical water (Kahn and Rogak, 2004; Rogak and Teshima, 1999; Shvedov and Tremaine, 2000; Armellini and Tester, 1993; Armellini et al., 1994). Figure 5.7 shows the solubility of

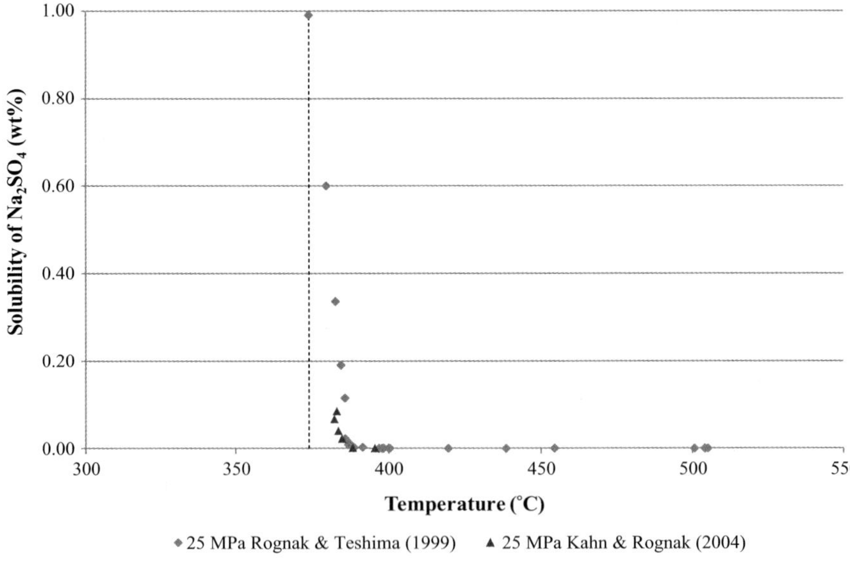

FIGURE 5.7 Solubility of sodium sulphate at 25 MPa as a function of temperature. The dashed line indicates the critical temperature. *Data by Kahn and Rogak (2004), Rogak and Teshima (1999).* (For colour version of this figure, the reader is referred to the online version of this book.)

sodium sulphate at 25 MPa and various temperatures in the supercritical region.

It is evident from Figure 5.7 that solubility decreases significantly once the supercritical region is entered in accordance with the decrease in the dielectric constant (Savage, 1999; Peterson et al., 2008a).

INFORMATION BOX

Processing with Supercritical Water Issue: Precipitation of Salts

As precipitation of salts may occur when using supercritical water for processing, due to the properties of the supercritical phase, quite serious issues may arise. One issue is abrasion of valves and piping, since the precipitated salt particles will be carried along in the process stream and when impacting on pipe walls, valves and other process equipment may cause abrasion on these. This could possibly lead to a decrease in lifespan of the unit which is subject to the abrasion, thus increasing the costs of keeping the process running due to being forced to replace units with increased frequency.

To remediate this effect precipitated salts should preferably be removed from the processing stream continuously. This can be accomplished in several ways, such as a sedimentation unit, where flow rates are reduced significantly and the precipitate may settle into the bottom of the sedimentation unit. To remove as much of the precipitate, or rather, to ensure precipitation does not occur further down the process line, this unit should preferably be placed where the dielectric constant of the water is at its minimum.

In some cases density is used as a measure when dealing with supercritical fluids. The reason for this is that density is a combination of temperature and pressure, and holds besides the information of these parameters also information regarding solubilities of nonpolar compounds in the case of a polar solvent. Figure 5.8 shows the pressure/temperature/density relationship for water.

From Figure 5.8 it is observed that density is strongly dependent on temperature and pressure. It may be utilised to exert a great deal of control on the density of a fluid in its supercritical state and thus also varying the properties of this fluid.

Why do these changes in water occur then? According to both Nakahara et al. (2001) and Bellissent-Funel (2001) the changes in water,

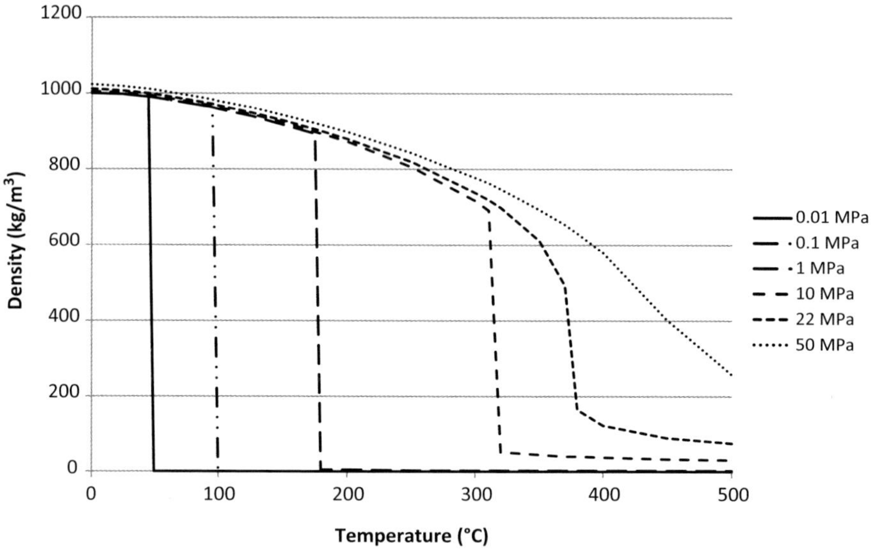

FIGURE 5.8 Temperature, pressure and density relationship for water. *Data from Harvey (1998).*

near and above critical conditions is due to the change in hydrogen bonding. How the bonding occurs in water and how the molecules rearrange is subject to some debate (Ball, 2008); however, there is an agreement that the amounts of hydrogen bonds is influenced by temperature and pressure (Gupta et al., 1992; Gorbaty and Kalinichev, 1995; Hoffmann and Conradi, 1997; Martí, 1999; Stokely et al., 2010; Wernet et al., 2004; Salzmann et al., 2006).

This has been examined by Nakahara et al. (2001), which reveals that the amount of hydrogen bonding decreases above the critical point. This corresponds to what was determined by Martí (1999), these results are shown in Figure 5.9.

Investigations of the number of hydrogen bonds using nuclear magnetic resonance spectroscopy to investigate the amount of hydrogen bonding suggest that the reason for the diminishing amount of hydrogen bonding with diminishing density is due to the freedom of the water molecule (Nakahara et al., 2001). At low densities the water molecule that has broken a hydrogen bond will have more rotational and translational energy, but less vibration and the distribution of water molecules will be less spatially homogeneous. This means that new hydrogen bonds to replace broken bonds are not as easily formed as at higher densities where the molecules are closer packed and thus hydrogen bonding with another molecule occurs more readily.

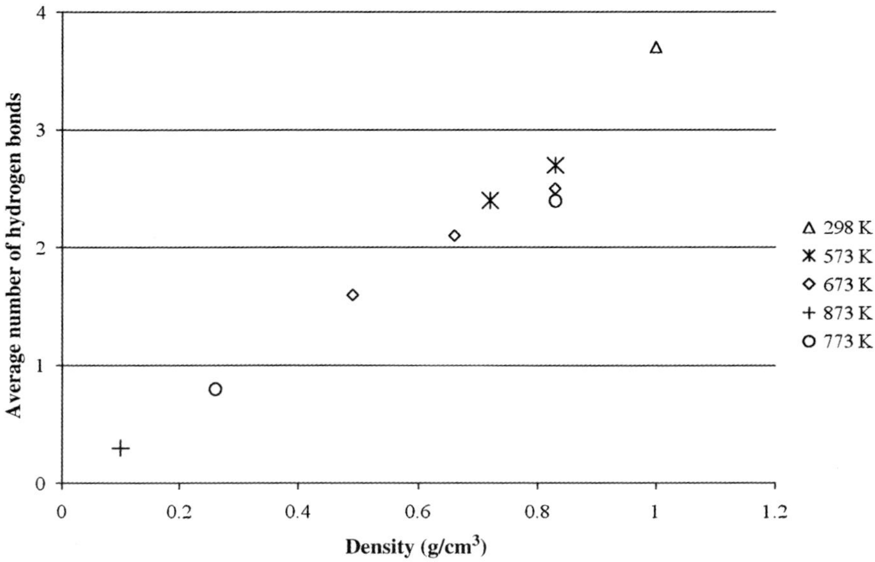

FIGURE 5.9 **Average number of hydrogen bonds in water as a function of density and temperature.** *Data by Martí (1999).*

Also, Gorbaty and Gupta (1998) have studied the hydrogen bonding in water at elevated temperatures and generated results for the temperature dependence of the degree of hydrogen bonding χ. The degree of hydrogen bonding is the amount of hydrogen bonds in water at given conditions compared to the amount of hydrogen bonds in ice, which is the state with most hydrogen bonds per water molecule, i.e. four. Based on their measurements and the data of several others (Gupta et al., 1992; Gorbaty and Kalinichev, 1995; Hoffmann and Conradi, 1997; Franck and Roth, 1967; Gorbaty and Demianets, 2000) they created an expression for the degree of hydrogen bonding in water.

$$\chi = -8.68 \cdot 10^{-4} T + 0.61 \tag{5.2}$$

Since this expression is only dependent on temperature it is only valid for a small range of pressures or densities. Gorbaty and Gupta (1998) claim that this model is accurate to within ± 0.1 in the temperature range of 280−800 K and density range of 0.5−1.1 g/cm^3. Plotting this expression yields the graph shown in Figure 5.10.

This expression is shown by Gorbaty and Gupta (1998) to be quite well related to the experimental measurements, which were made in pressure ranges from 0.1 to 100 MPa. This shows that the temperature dependence of the degree of hydrogen bonding is significantly stronger than the pressure dependence. One could interpret this as that the kinetic energy of

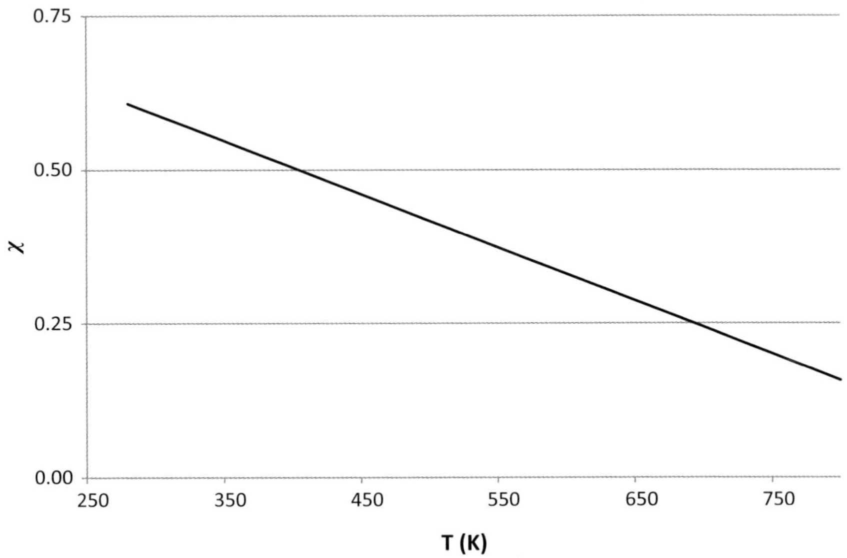

FIGURE 5.10 **Degree of hydrogen bonding in water.** A value of 1.0 corresponds to ice.

the molecules, which is increased with temperature, is what plays the most significant role in how many hydrogen bonds a water molecule will form.

5.3. APPLICATIONS OF SUPERCRITICAL WATER

Supercritical fluids have a large number of applications. In this section some case studies that relate to the use of supercritical water for processing are discussed. These cases will both involve purely processing issues, such as salt precipitation and high pressure measurements, while others will focus more on environmental applications or the environmental impacts of supercritical fluids as processing media.

5.3.1. Wet Air Oxidation

Although wet air oxidation is not a reaction in supercritical media it may operate at conditions close to the critical point. Furthermore, wet air oxidation is a well-known technology that is being utilised in full-scale operations, which makes it a good example of the application of hot compressed water for water treatment.

Oxidation in supercritical media is also used, however, to a lesser degree than oxidation at subcritical conditions since the latter is a well-known and proven technology. Most of the development of the wet air oxidation process took place in the 1950s. F.J. Zimmermann and his group worked with treatment of pulp mill liquor and patented the wet air oxidation process. This process is known as Zimpro, a shortening of Zimmermann process.

There are several types of wet oxidation, using various media for processing as listed in Table 5.5.

There are many applications of wet air oxidation; however, as it is an oxidation process they all centre on treatment of a process stream to oxidize components in this stream. Applications are mainly treatment of wastewater stream or sludge to decompose organic contaminants in these streams. Often wet air oxidation is used for waste streams that are too toxic for biodegradation or with too low (<100 g/l) chemical oxygen demand (COD) to be used for incineration (Kim and Ihm, 2011). COD is a measure of the amount of oxygen needed to fully oxidize an organic substance or a mixture of substances expressed in milligrams per litre. The general premise of COD is that any organic molecule can be fully oxidized to CO_2 and water using a strong oxidizing agent, such as oxygen. In very dilute streams the amount of COD in the process stream may be too low for the process to be economically viable, due to the amount of energy used, making the cost per unit contaminant too high. Klavarioti et al. (2009) state that wet air oxidation is applicable in the COD range of 10–100 g/l. In these cases wet air oxidation provides another possibility for the treatment of these waste streams.

Wet air oxidation may be used both as precursor to other treatments (Kim and Ihm, 2011; Oller et al., 2011; Banerjee et al., 2009; Heimbuch and Wilhelmi, 1985) or as a single-step treatment (Heimbuch and Wilhelmi, 1985; Zalouk et al., 2009; Pintar, 2003; Debellefontaine et al., 1996), depending on process and requirements for the effluent. Typical parameters for wet air oxidation are given in Table 5.6.

TABLE 5.5 Types of Wet Oxidation

Process	Description
Wet oxidation (WO)	Oxidation in aqueous solution of organics. The oxidizing media can be air, oxygen, peroxide, nitric acid, etc.
Wet air oxidation (WAO)	Wet oxidation using air for generation of radicals.
Wet peroxide oxidation (WPO)	Wet oxidation using hydrogen peroxide (H_2O_2) for the oxidation.
Supercritical water oxidation (SCWO)	Wet oxidation carried out in supercritical water.
Catalytic wet air oxidation (CWAO)	Wet air oxidation where either supported or unsupported catalysts are utilised.

TABLE 5.6 Parameters of Wet Air Oxidation

Parameter	Range
Temperature	120–325 °C
Pressure	0.3–20 MPa
Residence time	3–240 min.
COD removal	75–90%

Data summarized from Luck (1999), Kolaczkowski et al. (1999), Gasso et al. (1995), Debellefontaine and Foussard (2000).

A wide range of commercial wet air oxidation processes are available, ranging from low pressure and temperature (e.g. LOPROX process) to those operating in the near-critical region (e.g. ATHOS process). A summary of commercial processes is given in Table 5.7.

Several of the commercial wet air oxidation processes utilises catalysts as either supported or unsupported. If the catalyst is supported or unsupported denotes if the active part of the catalyst is in suspension (unsupported) or fixed on a solid media (supported), such as pellets, in the reactor bed.

The principle of wet air oxidation is that the feedstock is pressurized and mixed with air or oxygen before entering a reactor where the oxidation occurs. Prior to entering the process stream is heated through one or more heat exchangers to attain process temperature. Following

TABLE 5.7 Commercial Wet Air Oxidation Processes and Their Parameters

Process	Catalyst	P (MPa)	T (°C)	Residence Time (min)
ATHOS	Unsupported	7–15	250–300	<60
Ciba–Geigy	Unsupported	16	295	>180
Kenox	None	4.1–4.7	200–240	~40
LOPROX	Unsupported	5–20	<200	60–180
NS-LC	Supported	0.9–8	160–270	~60
ORCAN	Unsupported	0.3	120	–
Osaka gas	Supported	7	250	~30
OXYJET	None	7	140–180	~3
Wetox	None	4	210–250	30–60
Zimpro	None	2–12	140–325	20–240

Data summarized from Luck (1999), Kolaczkowski et al. (1999), Gasso et al. (1995), Debellefontaine and Foussard (2000).

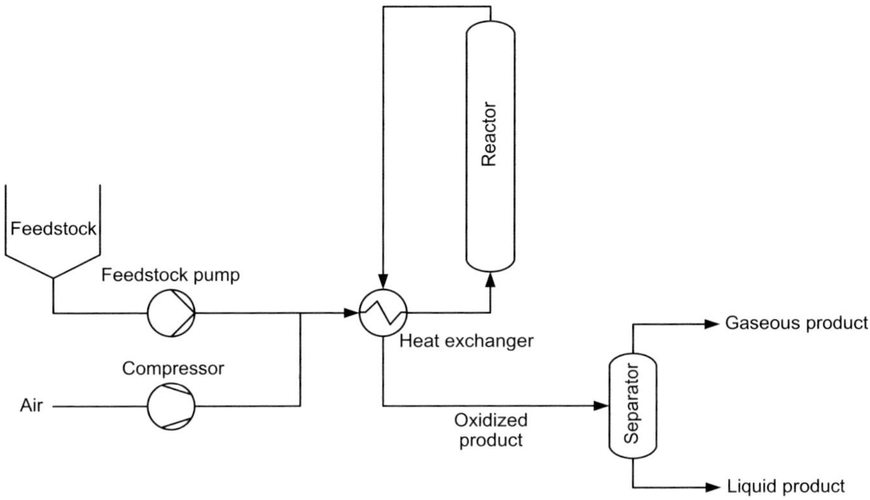

FIGURE 5.11 **Principle of wet air oxidation.**

oxidation the product is separated into a liquid and a gas phase before further processing. The general principle is shown in Figure 5.11.

In the shown principle the hot oxidized product is used for heating the process stream entering the reactor. This is often utilised to optimize the energy balance of the process, i.e. reusing as much of the energy as possible. The reactor itself is shown as a generic column reactor, but in reality, several reactor designs are possible for wet air oxidation (Kolaczkowski et al., 1999; Debellefontaine and Foussard, 2000; Beyrich et al., 1979; Miachon et al., 2003), the simplest, however, is the vertical bubble column.

The wet air oxidation process is described as a free radical oxidation as given in (5-II) through (5-III) with M denoting a homogeneous species (Li et al., 1991).

$$\mathrm{RH} + \mathrm{O_2} \rightarrow \mathrm{R^{\bullet}} + \mathrm{HO_2}^{\bullet} \tag{5-II}$$

$$\mathrm{RH} + \mathrm{HO_2}^{\bullet} \rightarrow \mathrm{R^{\bullet}} + \mathrm{H_2O_2} \tag{5-III}$$

$$\mathrm{H_2O_2} + M \rightarrow 2\mathrm{HO^{\bullet}} \tag{5-IV}$$

The general principle is that oxygen will react with the weakest C–H bond and form the radicals which are responsible for the oxidation of the organics. The reaction pathways in wet air oxidation have been investigated for a long time. A general reaction pathway is described by Li et al. (1991) based on the work by Foussard et al. (1989) who suggested that the main intermediate in the oxidation process is acetic acid. The general

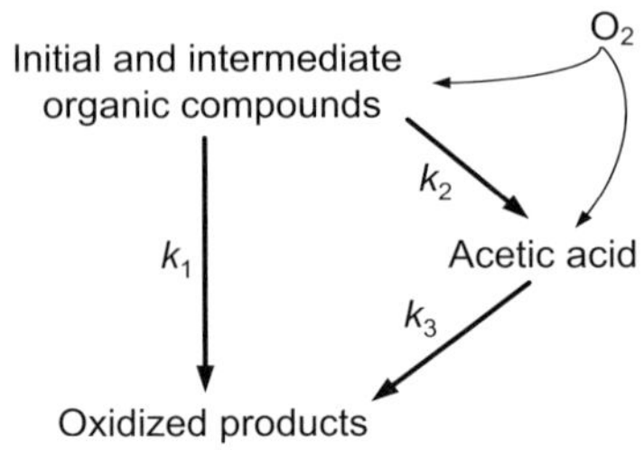

FIGURE 5.12 **General reaction pathway of wet oxidation as proposed by the works of Li et al. (1991) and Foussard et al. (1989).**

reaction pathways may thus be illustrated as in Figure 5.12, where all organic compounds are oxidized either directly or by the acetic acid intermediate.

This oxidation may be described using three reaction constants which are

k_1, oxidation of compounds without acetic acid as an intermediate.
k_2, oxidation of compounds into acetic acid.
k_3, oxidation of acetic acid.

Based on this general reaction pathway (which is applicable in both catalysed and noncatalyzed wet air oxidation (Levec and Pintar, 2007)), where the oxidation of compounds with and without acetic acid as an intermediate are assumed to be parallel reactions, the following general expression for the oxidation has been proposed (Li et al., 1991), which has shown good agreement with wet oxidation in both sub- and supercritical water.

$$\frac{[\text{Organics}]}{[\text{Organics}]_0} = \frac{k_2}{k_1 + k_2 - k_3} e^{-k_3 t} + \frac{(k_1 - k_3)}{k_1 + k_2 - k_3} e^{-(k_1 - k_2)t} \tag{5.3}$$

This expression gives a measure of the degree of oxidation or the ratio between the initial and current concentrations of compounds that can be oxidized as a function of time.

Wet oxidation is a technology that has shown to be applicable in a wide array of situations for wastewater or sludge treatment or as a precursor to other unit operations. There is a significant variation in process parameters for wet oxidation depending on which commercial process is utilised. However, in any case it is the oxidation of organic compounds by radical reactions that is the key. In general two pathways may describe wet air oxidation, that is, directly to the oxidized product or with acetic acid as an intermediate. Of course other intermediates exist but this reaction pathway is capable of describing the process and allows for calculation of kinetics given the rate constants.

5.3.2. Degradation and Conversion Reactions in Near- and Supercritical Water

Oxidation and degradation of substances in near- and supercritical water as a method of cleaning wastewater is not the only option when processing at these conditions. Another option is the use of the near- and supercritical water to produce either drop-in fuels or synthesis gas which may be utilised for production of biodiesel through the Fischer—Tropsch synthesis. This section has a focus on the degradation and conversion aspects of using near- and supercritical water.

When using near- and supercritical water for processing applications a large number of reactions may occur depending on the components in the system. In this case focus shall be on the reactions occurring when lignocellulosic material is treated in supercritical water. This will often lead to degradation of the lignocellulosic material, or biomass, into gasses by what is called supercritical water gasification. Extensive work has been carried out in this field, both with regards to reaction pathways and the effect of alkali salts as catalysts for these reactions (Nielsen, 2010; Peterson et al., 2008b; Schmieder et al., 2000; Kruse and Dinjus, 2005; Kruse et al., 2005; Kruse et al., 2000; Sinag et al. 2003; Dinjus & Kruse, 2004; Kruse and Dinjus, 2003; Boukis et al., 2005).

In some ways supercritical water gasification may resemble steam reforming. Albeit operating at higher temperatures steam reforming is production of hydrogen and CO from feedstock of not only mainly methane, but also other feedstock. Generally the reaction may be given as a set of three reactions.

$$C_nH_m + nH_2O \rightarrow nCO + (n + m/2)H_2 \quad \text{(5-V)}$$

$$CO + H_2O \rightarrow CO_2 + H_2 \quad \text{(5-VI)}$$

$$CO + 3H_2 \rightarrow CH_4 + H_2O \quad \text{(5-VII)}$$

However, when using feedstock other than methane the risk of coke formation increases (Levec and Pintar, 2007). Formation of these contaminants may be by Boudouard reactions or cracking of alkanes due to temperatures in processing. The formation of coke by the Boudouard reaction may be inhibited by alkali metals (Nielsen, 2010; Peterson et al., 2008b). Also, solid carbon, i.e. coke, may react with water and form hydrogen, which also counters the coke formation.

$$2C + 3H_2O \rightarrow CO + CO_2 + 3H_2 \quad \text{(5-VIII)}$$

This may occur at a wide temperature range and yielding CO or CO_2 depending on the temperature (Smith et al., 1954). Low temperatures favour CO production while at high temperatures CO_2 is formed.

Besides formation of coke during processing formation of tar from lignin degradation products also must be considered. Tar is high-molecular weight aromatic compounds, which create a very viscous substance. Commonly components such as naphthalene, pyrene and xylene are considered tar compounds. Padban et al. (2000) report that the formation of tar may be reduced at high temperatures in gasification. This means that fast heating to process conditions is desirable, to avoid as much tar formation as possible, since the high process temperature will reduce tar formation.

At near-critical conditions the dissociation product of water is three orders of magnitude higher than at ambient conditions. This means that the dissociation of water (*I*) becomes an aspect to take into account as well. Effectively, this means that acid- and base-catalysed reactions can occur without the addition of acids or base to the process media. This is of interest as components, possibly stemming from transesterfication reactions, which are acid/base catalysed, have been determined in bio-oil produced at near-critical conditions (Nielsen, 2010).

Besides effects of catalysts and alkali, Antal and Xu (1999) have reported that the reactor wall itself may have an effect on conversion. They suggest that as the reactor wall is subject to process conditions that change the reactor wall and yield an effect on the process. Due to plugs in the flow reactor they cleaned the system with H_2O_2 and afterwards observed decreased presence of hydrocarbons in the produced gas. Hence it was suggested that this effect is caused by the reactor wall, i.e. the process is affected by the history of the reactor, meaning that the reactions will change over time due to changes in the reactor wall such as composition in the surface. The plugging issue with tubular reactors for conversion has been pointed out by both Antal and Xu (1999) as well as Matsumura (2004). Thus it has been shown that a fluidized bed reactor may be used even in supercritical water, with almost the same efficiencies as for previously utilised reactors.

The reaction pathways of this conversion process are not fully defined or experimentally verified. Based on literature studies concerning other thermochemical conversion processes and behaviour close to the critical point (Watanabe et al., 2002, 2003; Bühler et al., 2002; Matsumura et al., 2005; Yoshida et al., 2004; Watanabe et al., 2001) a series of reaction pathways for conversion of cellulose, lignin and fatty acids may be proposed. These are illustrated in Figure 5.13.

As can be seen the conversion of lignocellulosic material may occur through an extensive network of reactions.

Initially cellulose is decomposed into sugars, i.e. fructose and glucose, by hydrolysis of the glycoside bonds. These sugars are expected to be

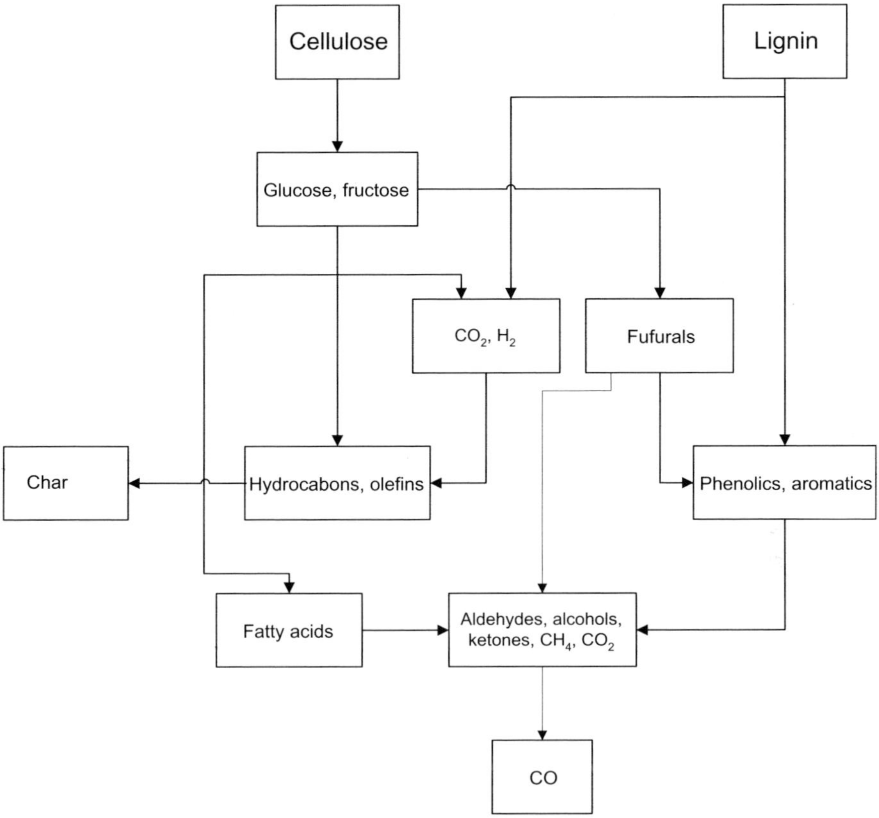

FIGURE 5.13 Possible reaction pathways.

converted into either a gas phase or an oil phase. The oil phase is expected to be mainly carboxylic acids formed from sugars with furfurals being an intermediate. This reaction is competitive to the steam reforming leading to the gas phase products of CO_2 and H_2. The gas phase components will participate in the Water-Gas-Shift(WGS) and the Fischer–Tropsch synthesis of hydrocarbons, which is catalysed by the zirconia catalyst. With regards to the oil phase the components may react further to form char in the system. This would mainly be a product of the Boudouard reaction, although it is inhibited by the alkali added to the feed.

Fatty acids are expected to undergo various changes in the process. This can be hydrogenation of unsaturated bonds, decomposition into methane, ketones, alcohols, etc. Also decarboxylation of the fatty acids forming CO_2 may occur as well as hydrolysis of any fatty acids in the form of fatty acid methyl esters. Generally the changes occurring to fatty acids are reduction of molecular size by removal of one or more groups.

Lignin is converted by formation of mainly aromatics and phenolics. Also, aldehydes, ketones and decomposition into CO, CO_2 and H_2 are expected from the conversion of lignin which may react through the WGS, although to a lesser extent than for the cellulose products due to less formation of the gasses.

These reaction pathways lead to products such as hydrocarbons, H_2, CO, alcohols and carboxylic acids.

In the case of cellulose, the result after several steps of hydrolysis will be formation of sugars as the cellulose network is degraded. Hydrolysis is acid/base catalysed, meaning there are favourable conditions in near-critical water for this type of reaction. The hydrolysis of cellulose must be considered one of the main initial reactions in the conversion process.

Another reaction worth mentioning is the decarboxylation, which is the formation of CO_2 from carboxylic acids. Decarboxylation is a thermal effect; however, Watanabe et al. (2001) have reported that acetic acid is stabilized in supercritical water to some extent. This indicates that some decarboxylation will occur at process conditions, but it is possible that the carboxylic acids are stabilized to some extent and in the presence of the zirconium oxide catalyst.

It is evident that these reaction pathways incorporate both heterogeneous and homogeneous catalyst as well as acidity, pressure and temperature. With the amount of possible reactions, in a system like this it requires extensive investigations to determine the exact reaction pathway for various compounds when catalysis, inhibition and synergies must be taken into account. Figure 5.14 gives a graphic representation of a likely reaction pathway for the conversion of lignocellulosic material in near-critical water.

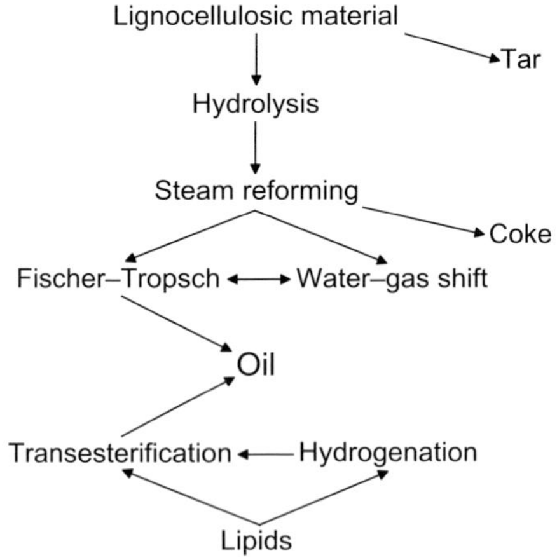

FIGURE 5.14 **Possible reaction pathways.** *Nielsen (2010).*

This shows that lignocellulosic material will be converted to oil compounds through hydrolysis of the glucoside bonds in the cellulose, steam reforming and subsequent water–gas shift and Fischer–Tropsch reactions to form oil compounds. During this tar from degradation of lignin and coke from the Bouduoard reactions may form; however, these are to some extent inhibited by a high heating rate and the alkali salts used as catalyst. Furthermore, lipids are likely to undergo hydrogenation and transesterfication in their conversion into oil compounds. This reaction pathway is a general outline of what is likely to occur during processing; however, with the amounts of compounds present it is only the main steps, as it is likely a multitude of reactions are occurring simultaneously.

5.4. SUMMARY

At high pressures and temperatures the supercritical phase is entered. This phase exhibits properties between those of gasses and liquids, allowing for special uses of these supercritical fluids. The main focus has been supercritical water, which, in addition to the gas/liquid-like properties, also behaves like a nonpolar solvent, meaning that organics and water are fully miscible at these conditions. The reasons for these property changes may be accounted for by the changes in amounts of hydrogen bonds formed by the individual water molecule. In ice a water molecule will form four hydrogen bonds to neighbouring water molecules; however, as temperature increases the amount of kinetic energy in the molecules will cause fewer formations of hydrogen bonds, leading to changes in properties of water.

Supercritical fluids have certain properties that make them useful in wide range of applications. In this chapter water has been the main focus, and in its supercritical state several key properties have been discussed.

- Viscosity, diffusivity and heat conductivity is between that of liquid water and water vapour.
- Water will be fully miscible with nonpolar substances at supercritical conditions
- Salts will not be soluble in supercritical water which may lead to precipitation and depositions
- In near-critical conditions the amount of H^+ and OH^- ions in water will be several orders of magnitude larger than at ambient conditions.

These properties allow for near- and supercritical water to be used in many different applications such as extraction, conversion or oxidization processes.

The wet air oxidation process was described, as an example of how to use water at high pressures and temperatures for treatment of contaminated effluents. Wet air oxidation is an application that is commercially viable and has been in operation since the 1950s, illustrating that using hot compressed water for processing is not new technology, even though a lot of research into the field is still being carried out.

Also, the reactions that may occur when treating lignocellulosic material, such as plant residues, in supercritical water were described. In this case it is evident that the degradation of the lignocellulosic material as well as the formation of fatty acids from the degradation products is part of a very complex network of reactions.

The applications of near- and supercritical water are not limited, a vast number of technologies incorporating the use of this processing media exist. The benefit of using a benign medium for processing and the special properties of the medium means that it is an area that is and will increase in scope.

References

Antal Jr., M.J., Xu, X., 1999. Hydrogen production from high moisture content biomass in supercritical water. U.S. DOE Hydrogen Program Rev.

Archer, D.G., Wang, P., 1990. The dielectric constant of water and Debye-Hückel limiting law slopes. J. Phys. Chem. Ref. Data 19 (2), 371–411.

Armellini, F.J., Tester, J.W., April 1993. Solubility of sodium chloride and sulfate in sub- and supercritical water vapor from 450–550 °C and 100–250 bar. Fluid Phase Equilib. 84, 123–142.

Armellini, F.J., Tester, J.W., Hong, G.T., September 1994. Precipitation of sodium chloride and sodium sulfate in water from sub- to supercritical conditions: 150 to 550 °C, 100 to 300 bar. J. Supercrit. Fluids 7, 147–158.

Ball, P., 2008. Water—an enduring mystery. Nature 452, 291–292.

Banerjee, S., Sen, R., Pandey, R. a., Chakrabarti, T., Satpute, D., Giri, B.S., Mudliar, S., December 2009. Evaluation of wet air oxidation as a pretreatment strategy for bioethanol production from rice husk and process optimization. Biomass Bioenergy 33 (12), 1680–1686.

Bellissent-Funel, M.-C., 2001. Structure of supercritical water. J. Mol. Liq. 90 (1–3), 313–322.

Beyrich, J., Gautschi, W., Regenass, W., Wiedmann, W., 1979. Design of reactors for the wet air oxidation of industrial waste water by means of computer simulation. Comput. Chem. Eng. 3, 161–165.

Boukis, N., Galla, U., D'Jesus, P., Müller, H., Dinjus, E., October 2005. Gasification of wet biomass in supercritical water: results of pilot plant experiments. In: Proceedings of the 14 European Biomass Conference, pp. 964–967.

Bühler, W., Dinjus, E., Ederer, H.J., Kruse, A., Mas, C., 2002. Ionic reactions and pyrolysis of glycerol as competing reaction pathways in near- and supercritical water. J. Supercrit. Fluids 22, 37–53.

Choukroun, M., Grasset, O., September 2007. Thermodynamic model for water and high-pressure ices up to 2.2 GPa and down to the metastable domain. J. Chem. Phys. 127 (12), 124506.

Debellefontaine, H., Foussard, J.N., 2000. Wet air oxidation for the treatment of industrial wastes. Chemical aspects, reactor design and industrial applications in Europe. Waste Manage. 20, 15–25.

Debellefontaine, H., Chakchouk, M., Foussard, J.N., Tissot, D., Striolo, P., 1996. Treatment of organic aqueous wastes: wet air oxidation and wet peroxide oxidation. Environ. Pollut. 92 (2), 155–164.

Dinjus, E., Kruse, A., April; 2004. Hot compressed water – a suitable and sustainable solvent and reaction medium? J. Phys. Condens. Matter 16 (14), 1161–1169.

Foussard, J.N., Debellefontaine, H., Besomes-Vailhé, J., 1989. Efficient elimination of organic liquid wastes: wet air oxidation. J. Environ. Eng. 115, 367–385.

Franck, E.U., Roth, K., 1967. Infra-red absorption of HDO in water at high pressures and temperatures. Discuss. Faraday Soc. 43, 108–114.

Gasso, S., Gonzalez, M., Baldasano, J.M., Lemonnier, J.P., Abatzoglou, N., Chornet, E., 1995. Wet oxidation of refractory organic compounds in industrial aqueous wastes via the Oxyjet technology. Waste Manage. Res. 13 (1), 37–46.

Gorbaty, Y.E., Demianets, Y.N., 2000. The pair-correlation functions of water at a pressure of 1000 bar in the temperature range 25–500 °C. Chem. Phys. Lett. 100 (5), 450–454.

Gorbaty, Y.E., Gupta, R.B., August 1998. The structural features of liquid and supercritical water. Ind. Eng. Chem. Res. 37 (8), 3026–3035.

Gorbaty, Y.E., Kalinichev, A.G., April 1995. Hydrogen bonding in supercritical water. 1. Experimental results. J. Phys. Chem. 99 (15), 5336–5340.

Gupta, R.B., Panayiotou, C.G., Sanchez, I.C., Johnston, K.P., August 1992. Theory of hydrogen bonding in supercritical fluids. AIChE J. 38 (8), 1243–1253.

Hardy, B., April 1998. ITS-90 formulations for vapor pressure, frostpoint temperature, dewpoint temperature, and enhancement factors in the range −100 to +100 °C. In: The Proceedings of the Third International Symposium on Humidity & Moisture, pp. 1–8.

Harvey, A.H., 1998. NISTIR 5078: Thermodynamic Properties of Water: Tabulation from the IAPWS Formulation 1995 for the Thermodynamic Properties of Ordinary Water Substance for General and Scientific Use.

Heidemann, R.A., Khalil, A.M., 1980. The calculation of critical points. AIChE J. 26 (5), 769–779.

Heimbuch, J.A., Wilhelmi, A.R., November 1985. Wet air oxidation—A treatment means for aqueous hazardous waste streams. J. Hazard. Mater. 12 (2), 187–200.

Hoffmann, M.M., Conradi, M.S., April 1997. Are there hydrogen bonds in supercritical water? J. Am. Chem. Soc. 119 (16), 3811–3817.

IAWPS, September 2011. Revised Release on the Pressure along the Melting and Sublimation Curves of Ordinary Water Substance, pp. 1–7.

Kahn, M.S., Rogak, S.N., August 2004. Solubility of Na2SO4, Na2CO3 and their mixture in supercritical water. J. Supercrit. Fluids 30 (3), 359–373.

Kim, K.H., Ihm, S.K., 2011. Heterogeneous catalytic wet air oxidation of refractory organic pollutants in industrial wastewaters: a review. J. Hazard. Mater. 186 (1), 16–34.

Klavarioti, M., Mantzavinos, D., Kassinos, D., 2009. Removal of residual pharmaceuticals from aqueous systems by advanced oxidation processes. Environ. Int. 35 (2), 402–417.

Kolaczkowski, S.T., Plucinski, P., Beltran, F.J., Rivas, F.J., McLurgh, D.B., 1999. Wet air oxidation: a review of process technologies and aspects in reactor design. Chem. Eng. J. 73 (2), 143–160.

Kruse, A., Dinjus, E., 2003. Hydrogen from methane and supercritical water. Angewandte Chemie (International Ed. in English) 42 (8), 909–911.

Kruse, A., Dinjus, E., 2005. Influence of salts during hydrothermal biomass gasification: the role of the catalysed water-gas shift reaction. Zeitschrift für Physikalische Chemie 219 (3), 341–366.

Kruse, A., Krupka, A., Schwarzkopf, V., Gamard, C., Henningsen, T., April 2005. Influence of proteins on the hydrothermal gasification and liquefaction of biomass. 1. Comparison of different feedstocks. Ind. Eng. Chem. Res. 44 (9), 3013–3020.

Kruse, A., Meier, D., Rimbrecht, P., Schacht, M., Dec. 2000. Gasification of pyrocatechol in supercritical water in the presence of potassium hydroxide. Ind. Eng. Chem. Res. 39 (12), 4842–4848.

Levec, J., Pintar, A., 2007. Catalytic wet-air oxidation processes: a review. Catal. Today 124 (3–4), 172–184.

Li, L., Chen, P., Gloyna, E.F., November 1991. Generalized kinetic model for wet oxidation of organic compounds. AIChE J. 37 (11), 1687–1697.

Luck, F., 1999. Wet air oxidation: past, present and future. Catal. Today 53 (1), 81–91.

Marshall, W.L., Franck, E.U., 1981. Ion product of water substance, 0-1000 °C, 1-10000 bar, new international formulation and its background. J. Phys. Chem. Ref. Data 10 (2), 295–305.

Martí, J., 1999. Analysis of the hydrogen bonding and vibrational spectra of supercritical model water by molecular dynamics simulations. J. Chem. Phys. 110 (14), 6876–6886.

Matsumura, Y., 2004. Fundamental design of a continuous biomass gasification process using a supercritical water fluidized bed. Int. J. Hydrogen Energy 29 (7), 701–707.

Matsumura, Y., Minowa, T., Potic, B., Kersten, S.R.A., Prins, W., van Swaaij, W., van de Beld, B., Elliott, D.C., Neuenschwander, G.G., Kruse, A., Antal Jr., M.J., 2005. Biomass gasification in near- and super-critical water: status and prospects. Biomass Bioenergy 29 (4), 269–292.

Miachon, S., Perez, V., Crehan, G., Torp, E., Ræder, H., Bredesen, R., Dalmon, J.-A., 2003. Comparison of a contactor catalytic membrane reactor with a conventional reactor: example of wet air oxidation. Catal. Today 82 (1–4), 75–81.

Michelsen, M.L., 1980. Calculation of phase envelopes and critical properties for multicomponent mixtures. Fluid Phase Equilib. 4, 1–10.

Nakahara, M., Matubayasi, N., Wakai, C., Tsujino, Y., 2001. Structure and dynamics of water: from ambient to supercritical. J. Mol. Liq. 90 (1–3), 75–83.

Nielsen, R.P., 2010. The Physical Chemistry of the CatLiq Process. Aalborg University.

Nolen, S.A., Liotta, C.L., Eckert, C.A., Gläser, R., 2003. The catalytic opportunities of near-critical water: a benign medium for conventionally acid and base catalyzed condensations for organic synthesis. Green Chem. 5 (5), 663.

Oller, I., Malato, S., Sánchez-Pérez, J.A., September 2011. Combination of advanced oxidation processes and biological treatments for wastewater decontamination–a review. Sci. Total Environ. 409 (20), 4141–4166.

Padban, N., Wang, W., Ye, Z., Bjerle, I., Odenbrand, I., 2000. Tar formation in pressurized fluidized bed air gasification of woody biomass. Energy Fuels 14 (3), 603–611.

Patrick, H.R., Griffith, K., Liotta, C.L., Eckert, C.A., Gläser, R., December 2001. Near-critical water: a benign medium for catalytic reactions. Ind. Eng. Chem. Res. 40 (26), 6063–6067.

Peterson, A.A., Vontobel, P., Vogel, F., Tester, J.W., 2008a. In situ visualization of the performance of a supercritical-water salt separator using neutron radiography. Journal of Supercritical Fluids 43 (3), 490–499.

Peterson, A.A., Vogel, F., Lachance, R.P., Fröling, M., Antal Jr., M.J., Tester, J.W., 2008b. Thermochemical biofuel production in hydrothermal media: a review of sub- and supercritical water technologies. Energy Environ. Sci. 1 (1), 32.

Pintar, A., 2003. Catalytic processes for the purification of drinking water and industrial effluents. Catal. Today 77 (4), 451–465.

Reid, R.C., Beegle, B.L., September 1977. Critical point criteria in legendre transform notation. AIChE J. 23 (5), 726–732.

Rogak, S.N., Teshima, P., 1999. Deposition of sodium sulfate in a heated flow of supercritical water. AIChE J. 45 (2), 240–247.

Salzmann, C.G., Radaelli, P.G., Hallbrucker, A., Mayer, E., Finney, J.L., 2006. The preparation and structures of hydrogen ordered phases of ice. Science 311 (5768), 1758–1761.

Savage, P.E., 1999. Organic chemical reactions in supercritical water. Chem. Rev. 99, 63.

Savage, P.E., Gopalan, S., Mizan, T.I., Martino, C.J., Brock, E.E., 1995. Reactions at supercritical conditions: applications and fundamentals. AIChE J. 41 (7), 1723–1778.

Savage, P.E., November 2000. Heterogeneous catalysis in supercritical water. Catal. Today 62 (2–3), 167–173.

Schmieder, H., Abeln, J., Boukis, N., Dinjus, E., Kruse, A., Kluth, M., Petrich, G., Sadri, E., Schacht, M., April 2000. Hydrothermal gasification of biomass and organic wastes. J. Supercrit. Fluids 17 (2), 145–153.

Seminario, J.M., Concha, M.C., Murray, J.S., Politzer, P., 1994. Theoretical analyses of O2/H2O systems under normal and supercritical conditions. Chem. Phys. Lett. 222 (1–2), 25–32.

Shvedov, D., Tremaine, P.R., 2000. The solubility of sodium sulfate and the reduction of aqueous sulfate by magnetite under near-critical conditions. J. Solution Chem. 29 (10), 889–904.

Sinag, A., Kruse, A., Schwarzkopf, V., December 2003. Formation and degradation pathways of intermediate products formed during the hydropyrolysis of glucose as a model substance for wet biomass in a tubular reactor. Eng. Life Sci. 3 (12), 469–473.

Smith, R.N., Pierce, C., Joel, C.D., 1954. The low temperature reaction of water with carbon. J. Phys. Chem. 58 (4), 298–302.

Stokely, K., Mazza, M.G., Stanley, H.E., Franzese, G., 2010. Effect of hydrogen bond cooperativity on the behavior of water. Proc. Natl. Acad. Sci. U.S.A. 107 (4), 1301–1306.

Wagner, W., Pruss, A., 2002. The IAPWS formulation 1995 for the thermodynamic properties of ordinary water substance for general and scientific use. J. Phys. Chem. Ref. Data 31 (no. 2), 387–535.

Watanabe, M., Inomata, H., Smith, R.L., Arai, K., 2001. Catalytic decarboxylation of acetic acid with zirconia catalyst in supercritical water. Appl. Catal. A 219 (1–2), 149–156.

Watanabe, M., Inomata, H., Arai, K., 2002. Catalytic hydrogen generation from biomass (glucose and cellulose) with ZrO2 in supercritical water. Biomass Bioenergy 22 (5), 405–410.

Watanabe, M., Inomata, H., Osada, M., Sato, T., Adschiri, T., Arai, K., 2003. Catalytic effects of NaOH and ZrO2 for partial oxidative gasification of n-hexadecane and lignin in supercritical water. Fuel 82, 545–552.

Wernet, P., Nordlund, D., Bergmann, U., Cavalleri, M., Odelius, M., Ogasawara, H., Näslund, L.Å., Hirsch, T.K., Ojamäe, L., Glatzel, P., Pettersson, L.G.M., Nilsson, A., 2004. The structure of the first coordination shell in liquid water. Science 304, 995–999.

Yoshida, T., Oshima, Y., Matsumura, Y., 2004. Gasification of biomass model compounds and real biomass in supercritical water. Biomass Bioenergy 26 (1), 71–78.

Zalouk, S., Barbati, S., Sergent, M., Ambrosio, M., 2009. Disposal of animal by-products by wet air oxidation: performance optimization and kinetics. Chemosphere 74 (2), 193–199.

CHAPTER

6

Membrane Filtration in Water Treatment – Removal of Micropollutants

Ph.D. fellow Henrik Tækker Madsen

Section of Chemical Engineering, Department of Biotechnology, Chemistry and Environmental Engineering, Aalborg University Esbjerg, Niels Bohrs Vej 8, DK 6700 Esbjerg, Denmark

The global water scarcity is increasing, and as such it has become necessary to take new water sources that we earlier may have thought of as inaccessible into consideration. Examples of such sources could be brackish water, sea water, water sources abandoned because of pollution and waste water, which in this respect is better thought of as used water. The problem with these water sources is that they contain components that in one way or another make the water undrinkable. Examples include high mineral concentrations that cause dewatering of cells, toxic compounds and pathogens. Problems with water are of course not limited to drinking water, but are general for the entire water industry. The main idea to solve these problems is based on the principle of separation. Processes such as sand filtration, extraction, coagulation, flocculation, settling and distillation are all examples of processes based on separation. Often however these processes are very specific. Sand filtration removes iron and manganese, but does not affect calcium and sodium, and a series of different processes may be required to obtain water of sufficient quality. In water treatment, membrane filtration is a relatively new separation process that potentially makes it possible to perform all separations in one or few steps, and as we will see at the end of the chapter there is also great potential in integrating membrane schemes with change to Advanced oxidation process (AOP).

This chapter will not aim at being the complete handbook on membrane filtration; the topic is membrane filtration for removal of micropollutants.

Chemistry of Advanced Environmental Purification Processes of Water
http://dx.doi.org/10.1016/B978-0-444-53178-0.00006-7

Nevertheless, it will be necessary with some background theory on membranes to understand how they work, and should a more detailed description be necessary this can be found in the references. We will begin with answering the question 'what is a membrane?' by building a conceptual model. Then the different types and ways of operating membranes are presented and compared, and with this in mind we will look at models for membrane filtration and use the models to give predictions about the fundamental properties of membranes. As the final part of the theory on membranes, we will look at some of the challenges faced when using membranes.

6.1. INTRODUCTION

Membranes have played a crucial role in human history, although for most of their history they have gone on relatively unnoticed. The human body, and much of nature, relies on the principle of membrane filtration to perform a number of tasks such as to control the cellular environment and to excrete waste products. Without membrane filtration, trees would not stand erect, and human life would be drastically altered simply because we would be unable to create a cellular environment different from the surrounding environment. It is however only recently that engineers have begun using membranes in water treatment. In 1993 there were only eight membrane filtration systems installed in the United States, but since then outbreaks of pathogens[1] and a significant lowering in the production cost of membranes have greatly increased the application of membranes (Crittenden et al., 2012).

A membrane is a physical barrier that either completely rejects or reduces the flux of a given compound so that it may be separated from the rest of the stream. As we will see, a good conceptual model for a membrane is a porous media that will only allow for flow of compounds smaller than the pore diameter. Other factors may influence the retention, and these will make it possible for molecules larger than the pore diameter to move through the membrane, but also for membranes with pore diameters larger than the molecules to reject these. However, for the general understanding of the membrane filtration process, the model of the porous media operating by size exclusion will be sufficient, and we will see that models based on the size exclusion principle will be able to describe the filtration process.

[1]Especially the 1993 cryptosporidiosis outbreak in Milwaukee with over 400,000 caused illnesses and 50 deaths are thought to a changed the view on the use of membranes in drinking water treatment.

6.2. TYPES OF MEMBRANES AND MEMBRANE PROCESSES

Separation with membranes occurs because of the existence of a gradient across the membrane, and membrane processes may be divided into groups based on the specific type of gradient. The most common way of categorising membranes is to divide them into two groups: pressure driven and non-pressure driven. Some of the most notable examples of each group are listed in Table 6.1.

In pressure-driven membrane processes, the filtration occurs because a pressure gradient is created across the membrane. The pressure gradient is called the transmembrane pressure (TMP), and as the driving force of the pressure-driven membrane processes, it plays a crucial role in membrane filtration. In the non-pressure driven processes one of the most notable driving forces is concentration gradients. As seen concentration gradients are the basis of processes such as dialysis and forward osmosis, which are both in effect osmosis processes. The use of electric potentials as the driving force has also gained a lot of attention because the electrical potential enables a more selective separation based on the charge of the solutes. However, in the water treatment industry, it is the pressure-driven processes that has become most widespread, whereas the non-pressure driven processes are used more in specialised operations. As a result, we focus on the pressure-driven membrane processes in this chapter.

As mentioned previously, the primary effect governing membrane filtration is size exclusion, and membranes are therefore divided into four main groups depending on the size of particle they retain: microfiltration (MF), ultrafiltration (UF), nanofiltration (NF) and reverse osmosis (RO) (Kucera, 2010; Peinemann and Nunes, 2010), see Table 6.2 and Figure 6.1.

MF membranes have the largest pores and will retain bacteria, polymers and large colloidal particles, while UF membranes will allow for removal of smaller colloidal particles, viruses and macromolecules such as proteins and some humic acids. Because the solutes that are separated

TABLE 6.1 Membrane Processes Divided into Pressure and Non-pressure Driven

Pressure Driven		Non-pressure Driven	
Membrane Process	**Driving Force**	**Membrane Process**	**Driving Force**
Microfiltration	Pressure	Dialysis	Concentration
Ultrafiltration	Pressure	Electrodialysis	Electrical potential
Nanofiltration	Pressure	Pervaporation	Pressure
Reverse osmosis	Pressure	Forward osmosis	Osmosis
		Membrane distillation	Vapour pressure

TABLE 6.2 Comparison of Pressure-Driven Membranes

	Particle Size (μm)	Molecular Weight (Da)	Particle Characteristics	Pressure (bar)
MF	0.1–1	10^5–10^6	Macromolecular to cellular	0.5–1
UF	0.01–0.1	1000–10^5	Molecular to macromolecular	1–10
NF	0.001–0.01	100–1000	Ionic to molecular	5–70
RO	0.0001–0.001	100	Ionic	10–100

are often in the form of undissolved particles, MF and UF is often called particular filtration. In general MF/UF membranes and NF/RO membranes can be separated based on their operational pressure. MF/UF membranes are operated at relatively low pressure, which distinguish them from NF and RO that operate at higher pressures. The higher operational pressure is because of the small pore size of these membranes, which results in a higher resistance to flow. Because of the small pore size, NF and RO allow for separation of even smaller molecules as well as

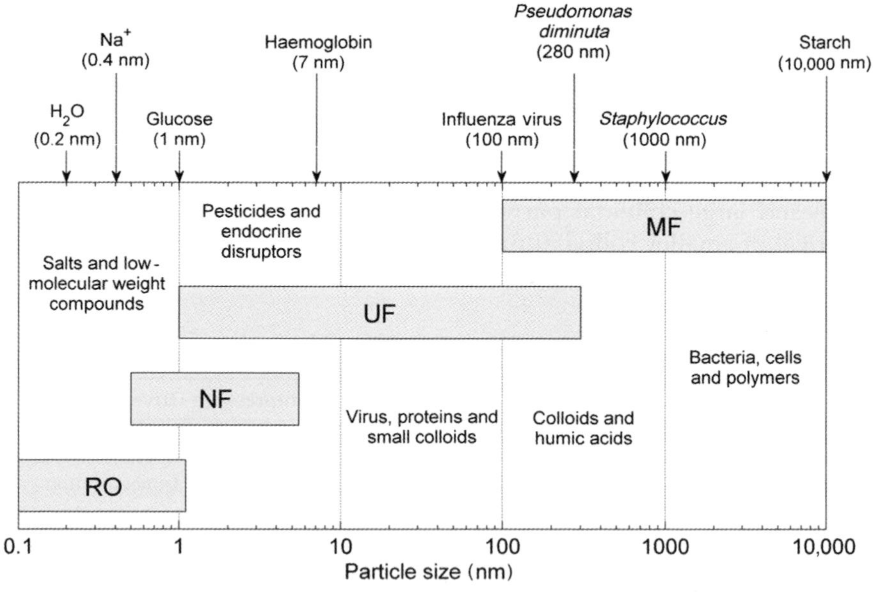

FIGURE 6.1 **Particle size retention for membranes.** (For colour version of this figure, the reader is referred to the online version of this book.)

TABLE 6.3 Comparison of MF/UF and NF/RO (Crittenden et al., 2012)

Process Characteristics	MF/UF	NF/RO
Objective	Particle removal, microorganism removal	Desalination, micropollutant removal, softening, natural organic matter (NOM) removal
Target contaminant	Particles	Dissolved solutes
Most common configuration	Hollow fibre	Spiral wound
Most common flow pattern	Dead end	Cross-flow
Backwash recycle	Yes	No
Materials	Polymer, ceramic	Polymer
Competing process	Granular filtration	Active carbon filtration, ion exchange, precipitative softening, distillation

minerals. RO membranes are in effect nonporous[2] and will retain ~99.7% of the salinity (Kucera, 2010; Peinemann and Nunes, 2010). Because of the effective mineral retention, RO membranes are used in desalination. RO membranes will also remove micropollutants. However, RO membranes are operated at high pressures (10–100 bars), and the high removal of minerals may necessitate a re-mineralisation of the water after treatment, which will add further costs to the process. Instead, NF may be used when extensive desalination is not required. NF membranes are sometimes described as 'loose' RO membranes because of slightly larger pores, with pore diameters around 1 nm. This pore size is the on the molecular scale, and NF membranes can therefore be used to retain micropollutants. As seen from Figure 6.1, there is a substantial overlap between UF and NF membranes, and that based on the displayed particle size domains, NF seems to be unnecessary. NF membranes do however disserve their place in the membrane hierarchy. They represent an intermediate between RO and UF membranes, and they are characterised by having a negative surface charge that affects their rejection of solutes. Species will be removed more efficiently with increasing charge, and NF membranes will as such have a higher rejection of divalent ions such as Ca^{2+}, Mg^{2+} and SO_4^{2-} compared with monovalent ions such as Na^+, K^+ and Cl^-. When there is an imbalance in divalent cations and anions, the difference in rejection of species dependent on their charge also gives rise to another interesting rejection phenomenon known as Donnan exclusion.

[2]This is actually a disputed fact, which we will return to during the discussion of models for membranes, but for the general understanding of RO membranes the idea of a nonporous membrane is suitable.

Often MF/UF ad NF/RO will be treated separately because they operate in fundamentally different ways. MF/UF works almost exclusively by size exclusion, and they may as such achieve very high removal efficiencies with an almost complete removal of solutes. This is obtained regardless of influent pressure and concentration. For NF/RO diffusion of solutes through the membrane makes the removal efficiency dependent on parameters such as influent pressure, solute concentration and water flux rate. In this chapter we will attempt to treat both types of membranes at the same time, and pointing out whenever a distinction is required. In Table 6.3 an overview of the main differences are given.

6.3. FILTRATION TECHNIQUES AND MEMBRANE MODULE CONFIGURATION

The conceptual model of a membrane presented in Figure 6.2 describes a situation in which an influent stream is forced through a semi-permeable membrane to create a filtrate with a reduced concentration of solutes. The influent stream is called the feed, the filtrate is called permeate and the process is called dead-end filtration. Dead-end filtration is well suited for a batch production, but if the process is continuous, the build-up of rejected solutes at the surface will quickly lead to formation of a cake layer that significantly reduces the flux through the membrane. The cake layer can be removed by reversing the flow and performing a backwash, but this means that the production must be stopped.

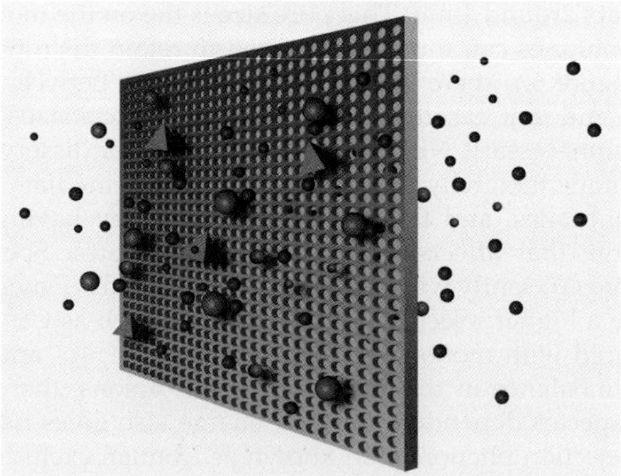

FIGURE 6.2 Conceptual model of a membrane where only particles smaller than the membrane pores penetrates. (For colour version of this figure, the reader is referred to the online version of this book.)

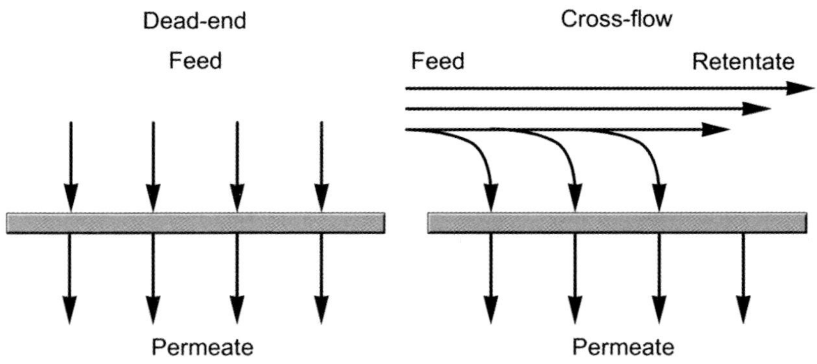

FIGURE 6.3 **Comparison of dead-end and cross-flow filtration.** (For colour version of this figure, the reader is referred to the online version of this book.)

Furthermore, many membranes, especially RO and NF, are not designed for backwash, and the high cost of these membranes makes dead-end filtration less attractive. Sand filtration and coffee brewing represents other examples of dead-end filtration.

Instead of dead-end filtration, cross-flow filtration may be used. Here the feed flows tangential across the membrane surface, and is forced through the membrane through a pressure gradient between the two sides of the membrane. The result is one influent stream, the feed, and two effluent streams, a permeate and a concentrate stream. The concentrate is also called the retentate, reject, waste or brine. Cross-flow filtration allows for a theoretical[3] continuous operation, and lowers the potential for build-up of solutes at the membrane (Figure 6.3). Typical cross-flow rates is between 0.5 and 1 m/s and is around four orders of magnitude greater than the velocity of water towards the membrane surface (Crittenden et al., 2012). This tangential flow across the membrane surface helps to keep the membrane clean by scouring the surface for deposits.

In dead-end filtration the TMP is controlled by varying the mechanical energy deposited by a pump relative to the membrane surface area, but in cross-flow filtration mode an increase in pump work load would only create an increase in the flow across the membrane surface, not through it. To control TMP in cross-flow mode, the filtration is operated with a flow control valve on the retentate stream as can be seen in Figure 6.4. The flow control valve creates the back pressure necessary for forcing the fluid through the membrane, and by opening or closing this valve, the pressure and cross-flow velocity can be adjusted. A completely closed flow control valve will turn the cross-flow membrane into a dead-end filter (Kucera, 2010).

[3]With time the membrane will foul, and it becomes necessary to clean the membrane.

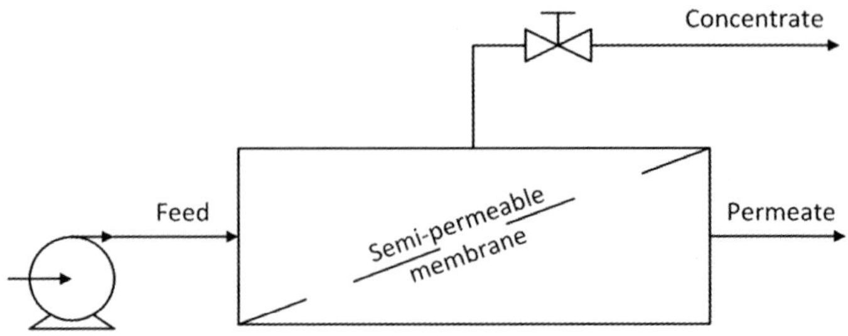

FIGURE 6.4 Schematic of membrane separation process.

6.3.1. Membrane Modules – Application in Real Life

There exist four types of membrane modules that may be divided into two groups (Kucera, 2010):

1. Flat sheet membranes
   a. Plate and frame
   b. Spiral wound
2. Tubular shaped membranes
   a. Tubular
   b. Hollow fibre

Plate and frame and tubular membrane modules are usually used in high solid situation, where ease of cleaning is an important factor.

In plate and frame modules, flat sheet membranes are stacked with spacer elements in between. The spacer elements keep the membranes from getting into contact with each other and function as channels through which the feed/retentate flows. The membranes are placed on support plates, with membranes on either side. The support plates separate the membranes and functions as the permeate channel. Because it is possible to open the module and get into contact with the membranes, membranes stacked in a plate and frame module are relatively easy to clean (Kucera, 2010).

Tubular modules may be compared with a shell and tube heat exchanger, where the feed is introduced on the tube side and the permeate exits on the shell side. The module consists of a perforated steel tube with the membrane placed on the inside, and the feed is forced through the membrane into the shell side. The tubes have a diameter ranging from 1/2 to 1 in, and may therefore be cleaned easily by shooting a sponge ball down the feed tube. Tubular configuration is also often seen when membranes are made of ceramic materials, but these membranes are not widespread yet.

The disadvantage of the two module systems is the low packing density, which is the amount of membrane surface area per unit. As a result a relatively large amount of equipment is necessary per unit membrane area, and this raises costs.

Hollow fibre and spiral wound membranes offer a higher packing density making them more economical. A hollow fibre module consists of a large array of hollow tubes with an inside diameter ranging from about 1 mm down to about 42 μm (Kucera, 2010). With respect to direction of flow, the filtration may be both inside—out, as with the tubular membrane system, or outside—in. Outside—in has the advantage that it creates a larger surface area and may as such treat larger amounts of water. It is also less prone to clogging because of larger particles, but it cannot be operated in cross-flow mode and will as such be a dead-end filtration method. Because the tubes are very flexible, a typical outside—in configuration is to fold the fibres in half and mold the two open ends in an epoxy block. The folded end may then also be molded in an epoxy block, and the entire fibre/epoxy system can finally be inserted in a closed module. Feed is being introduced to the module in one end and the filtration is driven by a pressure gradient over the membrane fibres. The retentate can be collected at the other end of the module, while the permeate is collected from all the fibres. The inside—out scheme may be operated in cross-flow mode, and is therefore less sensitive to high turbid waters. However, the fibres may be clogged by particles and it gives a larger pressure drop depending on the fibre diameter. Inside—out modules may be designed with the fibres elongated in the module, with feed entering one end of the fibres and the retentate coming out of the other end. Permeate is then collected from the shell side (Crittenden et al., 2012). Hollow fibre configuration is especially often used for MF/UF membranes, because they have a high packing factor compared with other configurations and are relatively easily backwashed.

Spiral wound membranes are constructed as a collection of membrane leafs wrapped around a permeate collection tube to form a cylindrical module. Each leaf is constructed by attaching membranes on both sides of a porous material, called the permeate carrier spacer, and gluing the membranes together along three of the four sides. The side left open is then attached to the permeate collection tube. Between leafs are placed feed spacer elements, which functions by separating membranes on adjacent leafs and allowing feed to stream in between the membranes. By rolling the leafs and feed spacers around the permeate collection tube, a spiral structure is formed. The feed may then move through the membrane to become permeate, which is collected from the permeate collection tube, and retentate is collected at the end of the membrane element opposite to where the feed was introduced (Crittenden et al., 2012; Kucera, 2010; Peinemann and Nunes, 2010). One of the advantages of spiral membranes is that they are standardised in size, and typically vary from 1 m (40 in) to 1.5 m (60 in) in length and 0.1 m (4 in) to 0.46 m (18 in) in diameter. The most common is the 8 in diameter (Crittenden et al., 2012). In practice four to seven elements are placed in series in a pressure vessel to produce one membrane unit. The pressure vessel is called the membrane housing and is typically either constructed from fibre glass or stainless steel. Spiral wound configurations are especially used for NF and RO because the feed spacers may be designed to create turbulent flow conditions that will reduce problems with ions building up concentration at the membrane surface, which may lead to scaling.

6.4. FILTRATION THEORY – CONCEPTS AND MODELS

So far we have discussed membranes in words and described membranes as processes based purely on straining. However, to be able to make calculations and predictions on membrane performance in a given situation, a more fundamental approach is necessary. In this section we will start by introducing some key concepts for the description of a given membrane filtration process and then move on to models describing MF/UF and NF/RO membranes.

6.4.1. Key Concepts – Rejection and Recovery

Rejection is the term used to describe how large a percentage of a given solute is retained by the membrane. As an example a 95% rejection of atrazine means that the membrane will retain 95% of the atrazine in the influent stream, while allowing for passage of 5%. Rejection is calculated with Eqn (6.1).

$$\text{Rejection} = 1 - \frac{C_P}{C_F} \tag{6.1}$$

where C_P is the concentration of the solute in the permeate and C_F is the concentration in the feed.

Recovery describes the percentage of the feed flow that is turned into permeate, and hereby recovered. It is calculated with Eqn (6.2).

$$\text{Recovery} = \frac{F_P}{F_F} \tag{6.2}$$

where F_P is the permeate flow and F_F is the feed flow. To obtain a high recovery is one of the primary objectives of a membrane filtration process, because this is equal to a high amount of permeate relative to concentrate, which will most often be an additional cost to the process since it must be handled in one way or another. Also, a low recovery means that a lot of

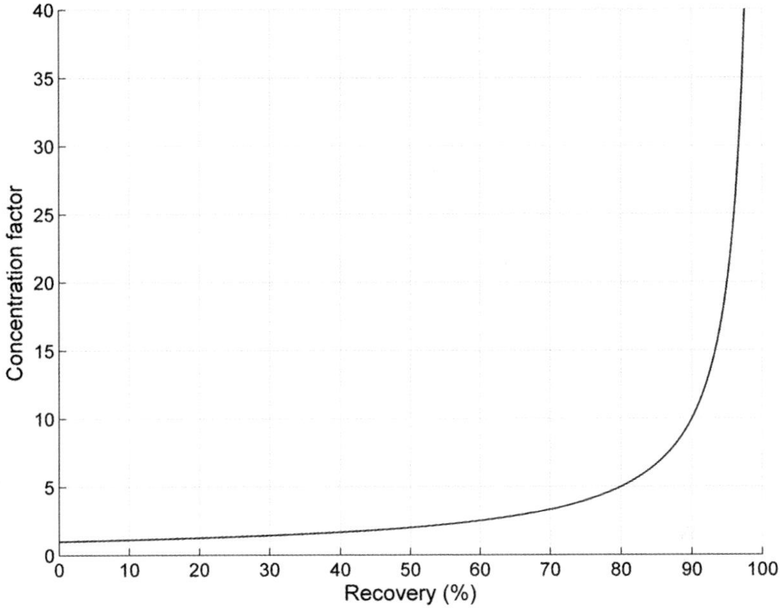

FIGURE 6.5 Concentration factor as a function of recovery. (For colour version of this figure, the reader is referred to the online version of this book.)

water is transported through the system without being converted into product.[4] However, because recovery is inversely connected to the concentration factor as seen in Figure 6.5, a high recovery may lead to significant increases in concentration in the concentrate. In Section 6.5 the effect of increasing concentration on the filtration process is discussed.

6.4.2. Steric Model for Membrane Filtration

As mentioned previously MF/UF membranes work by steric hindrance, meaning that particles larger than the pore size do not permeate the membrane, while NF/RO membranes are more complex since they are used to filtrate dissolved solutes that may diffuse inside the membrane matrix. Therefore, especially RO membranes are often thought of as being nonporous. However, pores have been directly observed directly by use of scanning electron microscopy and atomic force microscopy, and

[4]Here we assume that the permeate is the desired product, but in some cases the concentrate may be the product. An example is the use of membranes to concentrate a given product such as a specific antibiotic from a fermentation reaction. Still, this does not change the conclusion that a low recovery is not desirable.

even NF/RO membranes may as such be modelled by use of a pore flow model (Kiso et al., 2010, 2011). Although when applied on NF/RO membranes, it is done under the assumption that no or only very little interaction occurs between the solute and the membrane.[5] During the description of the pore flow model, we will use the term 'solute' to denote the species being rejected by the membrane. For MF/UF membranes a more accurate term would be particle since the species rejected by these membranes are often not dissolved in the water. In the following description the reader is free to exchange the word 'solute' with 'particle'.

In a pore flow model, the first factor that affects the rejection of a given solute is how easily it enters a pore, and this is determined by the ratio of solute radius to pore radius,[6] λ.

$$\lambda = \frac{r_\mathrm{s}}{r_\mathrm{p}} \tag{6.3}$$

For solutes with a radius larger than the pore radius, the solute will be completely rejected by the membrane. If a solute has a radius smaller than the pore radius, the ease of entering the pore is given by the steric partitioning factor, Φ. Φ accounts for the solubility equilibrium at the interface between pore and bulk solution, and it depends on the geometry of the solute. For spherical solutes Φ is given by (Deen, 1987; Kiso et al., 2010, 2011; Otero et al., 2008)

$$\Phi = (1 - \lambda)^2 \tag{6.4}$$

For solutes that enter the pores, the flow will consist of both a diffusive and a convective flow, and these will be influenced by the solute size relative to the pore radius. If the solute is much smaller than the pore radius, it is not affected by the walls of the pore and the solute flux follows Stokes' law (Deen, 1987), but this is a borderline case. As the solute radius relative to the pore radius increases, the solute will experience a hydrodynamic force affecting both the convective transport and the diffusive transport of the solute through the pore. The effect is described with the convective and diffusive hindrance factors (K_c and K_d), which are defined as (Kiso et al., 2011; Otero et al., 2008)

$$K_\mathrm{c} = \frac{v}{v_\mathrm{w}}, \quad K_\mathrm{d} = \frac{D_\mathrm{p}}{D_\infty} \tag{6.5}$$

[5] To lower the interaction between solute and membrane the solute and membrane should have different polarity and we will also assume the solute to be uncharged.

[6] The pore is usually modelled as a cylinder. Although this is not required, it is also the approach used in this description of the model.

where v is the solute velocity in a pore and v_w is the velocity of the solvent. D_p is the solute diffusion coefficient in a pore, and D_∞ is the bulk diffusivity. To determine these, two empirical coefficients are used: the enhanced drag coefficient, K^{-1}, and the lag coefficient, G (Deen, 1987). These can be correlated to λ with the empirical expressions

$$K^{-1}(\lambda) = a + b\lambda + c\lambda^2 + d\lambda^3 \tag{6.6}$$

$$G(\lambda) = e + f\lambda + g\lambda^2 + h\lambda^3 \tag{6.7}$$

where the values of the constants have been reported by (Bowen and Sharif, 1994; Bowen et al., 1997; Deen, 1987) to be

$$a = 1.0,\ b = -2.30,\ c = 1.154 \text{ and } d = 0.224$$

$$e = 1.0,\ f = 0.054,\ g = -0.988 \text{ and } h = 0.441$$

For $0 < \lambda \leq 0.8$, and

$$a = -0.105,\ b = 0.318,\ c = -0.213 \text{ and } d = 0$$

$$e = -6.830,\ f = 19.348,\ g = -12.518 \text{ and } h = 0$$

For $0.8 < \lambda \leq 1$. It is possible to use the first set of values for the entire range of $0 < \lambda \leq 1.0$ as shown by (Kiso et al., 2011) and still obtain satisfactory results.

The enhanced drag coefficient describes the effect of the pore wall on the convective transport of the solute, and the lag coefficient describes the effect on diffusive transport. The hindrance factors can be found by setting them equal to the coefficients, but when the pores are relatively long and narrow, the lag coefficient should also be multiplied by $(2-\Phi)$ (Bowen and Sharif, 1994; Otero et al., 2008).

$$\begin{aligned} K_c &= (2-\Phi)G(\lambda) \\ K_d &= K^{-1}(\lambda) \end{aligned} \tag{6.8}$$

Under the assumption of no interaction between the solute and membrane, the solute flux only consists of convective and diffusive flows (Bowen et al., 1997; De Pinho et al., 2002; Deen, 1987; Nakao and Kimura, 1981; Otero et al., 2008).

$$J_S = -D_p\frac{dC}{dx} + K_c C\frac{J_W}{A_k} = -K_d D_\infty\frac{dC}{dx} + K_c C\frac{J_W}{A_k} \tag{6.9}$$

where J_W is the solvent flux and A_k is the membrane porosity given as a percentage of open area per unit membrane surface. For dilute solutions, J_W may be approximated to the solution flux J_V. The solute flux can also be evaluated by permeate concentration (C_p)

$$J_S = C_p\frac{J_V}{A_k} \tag{6.10}$$

By using Eqn (6.10) and the approximation of J_W to J_V, Eqn (6.9) may be rewritten to

$$\frac{dC}{dx} = \frac{J_V}{A_k D_\infty}\left(\frac{K_c}{K_d}C - \frac{1}{K_d}C_p\right) \tag{6.11}$$

This equation can be solved by integrating across the pore with the boundary conditions $x = 0$ and $C = C_i$ at the entrance and $x = \Delta x$ and $C = C_o$ at the exit.

$$\int_{C_i}^{C_o} \frac{1}{\left(\frac{K_c}{K_d}C - \frac{1}{K_d}C_p\right)} dC = \int_0^{\Delta x} \frac{J_V}{A_k D_\infty} dx \Updownarrow \tag{6.12}$$

$$\left| \frac{\ln\left(\frac{K_c}{K_d}C - \frac{1}{K_d}C_p\right)}{\frac{K_c}{K_d}} \right|_{C_i}^{C_o} = \left| \frac{J_V x}{A_k D_\infty} \right|_0^{\Delta x} \Updownarrow \tag{6.13}$$

By using that $C_i = C_f \Phi$ and $C_o = C_p \Phi$ we get

$$\ln\left(\frac{\frac{K_c}{K_d}\Phi C_p - \frac{1}{K_d}C_p}{\frac{K_c}{K_d}\Phi C_f - \frac{1}{K_d}C_p}\right) = \frac{J_V \Delta x}{A_k D_\infty}\frac{K_c}{K_d} \Updownarrow \tag{6.14}$$

It can be seen that the right side of the equation is unit less, and this is called the Peclet number (Pe). By inserting the Peclet number and multiplying both numerator and denominator with K_d/C_p Eqn (6.14) becomes

$$\frac{K_c \Phi - 1}{K_c \Phi \frac{C_f}{C_p} - 1} = \exp(\text{Pe}) \Updownarrow, \quad \text{Pe} = \frac{J_V \Delta x}{A_k D_\infty}\frac{K_c}{K_d} \tag{6.15}$$

The water flux (J_V) in a pore may be calculated with the Hagen–Poiseuille equation (Kiso et al., 2011).

$$J_V = \frac{r_p(\Delta P - \Delta \pi)}{8 \eta \Delta x} \tag{6.16}$$

where ΔP and $\Delta \pi$ are the applied and osmotic pressures, respectively, and η is the viscosity of the water. The viscosity will however not affect the rejection. It is also included in the equation for bulk diffusivity (Kiso et al., 2011)

$$D_\infty = \frac{kT}{6\pi\eta}\frac{1}{r_s} \tag{6.17}$$

where T is the temperature and k is the Boltzmann constant, and by inspection it can be seen that the two viscosity terms cancel.

Through rearrangement and by using the definition for rejection from Eqn (6.1), an equation for the rejection of a given solute as predicted by the steric pore flow model is reached.

$$R = 1 - \frac{K_c\Phi}{1 - (1 - K_c\Phi)\exp(-\mathrm{Pe})} \tag{6.18}$$

This model shows the rejection (R) to be a function of solute radius (r_s), pore radius (r_p), the porosity of the membrane (A_k), temperature (T), applied pressure (ΔP) and osmotic pressure ($\Delta\pi$). If all other factors are held constant and the rejection of compounds with varying solute radius is measured, the pore size of the membrane may be determined. In Example 1 a calculation is made to determine the pore size of a membrane.

EXAMPLE 1: CALCULATION OF PORE SIZE FOR AN NF/RO MEMBRANE

A series of filtration experiments have been made with five compounds of varying solute radius where the rejection for each compound has been measured (R_m). The experiments were carried out at 25 °C and 10 bar, with solute feed concentration of 0.001 M. The data from the experiments are

r_s (nm): 0.2511, 0.2580, 0.2615, 0.2776, 0.3130, 0.3614

R_m: 0.7561, 0.8431, 0.8985, 0.9283 0.9746, 0.9971

To calculate the pore size, a script file in Matlab was constructed. The basic principle in the solution was for each (r_s, R_m) data value to calculate the rejection (R_c) for a series of r_p values, and then calculate the squared difference between each R_c and the R_m value. Then the sum of squares was calculated, and the pore radius was found as the r_p value leading to the smallest sum of squares.

```
clear
clc
x=[0.2:0.001:1];   % Investigated span for pore radius
(nm)
r_p=x*10.^-9;          % Pore radius (m)
Rm=[0.7561 0.8431 0.8985 0.9283 0.9746 0.9971];
r_s=[0.2146 0.2244 0.2293 0.2522 0.3025 0.3712]*10^-9;

for i=1:length(Rm)
n=1.002 * 10^-3;            % Viscosity (Pa s)
t=298;                     % Temperature (K)
P=1*10^6;                   % Applied pressure (Pa)
R=8.314;                        % Gas constant (J/mol*K)
```

(Continued)

EXAMPLE 1: CALCULATION OF PORE SIZE FOR AN NF/RO MEMBRANE *(cont'd)*

```
c = 1.00;                          % Concentration (mol/m^3)
P_pi = c*R*t;                      % Osmotic pressure (Pa)
k = 1.3806*10^-23;                 % Boltzmann constant
D = k*t/(6*pi*n*(r_s(i)));         % Diffusion coefficient
L = r_s(i)./r_p;                   % Lambda value

% Calculation of steric partition factor, T
F      = (1-L).^2;
QQ     = find(L>1);                % Selection rule
F(QQ) = 0;
T      = F;

G      = 1+0.054*L-0.988*L.^2+0.441*L.^3;   % Lag
coefficient
Kc     =   (2-T).*G;                             % Hindrance
factor for convection
Kd     = 1-2.30*L+1.154*L.^2+0.224*L.^3;    % Hindrance
factor for diffusion
D_p    = Kd*D;                                % Diffusivity
in a pore
u      = T.*Kc;
v      = Kc./Kd;
Pe     = ((r_p.^2.*(P-P_pi)*6*pi.*r_s(i))/(8*k*t)).
*v; % Peclet number
Rc     = 1 - u./(1 - (1 - u).*exp(-Pe)); % Calculated
rejection
Delta = (Rm(i)-   Rc).^2;      %   Squared   difference
between measured and calculated rejection
DELTA(i,:) = Delta;          %  Save  calculated  delta
value in a matrix
end
SUM = sum(DELTA);    % Sum of all delta values for each r_p
SSy = (SUM./(length(Rm)-1)).^(1/2);        %    Sum    of
squares as a function of r_p
[minVal minInd] = min(SSy);      % Identification of index
for minimum SSy
RP = x(minInd)                   % Pore radius (nm) with
minimum SSy
```

RP = 0.302 nm

To check the calculated pore radius and the validity of the model, the radius may be used to calculate a new series of rejection values that can subsequently be plotted against the measured rejection values.

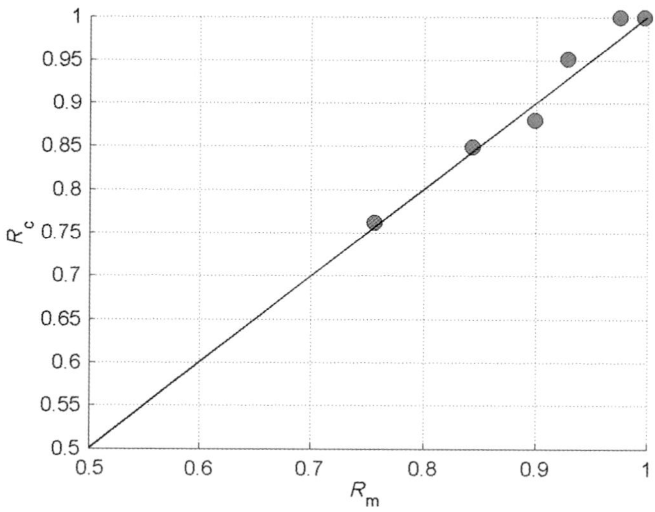

FIGURE 6.6 Rejection values calculated with the pore flow model plotted against measured rejection values. (For colour version of this figure, the reader is referred to the online version of this book.)

The model compounds used in Example box 1 are on the nanoscale, meaning that we are in the area of NF/RO membranes, and we see that there is a fine correlation between the measured and the calculated rejection values (Figure 6.6). This shows that the rejection of solutes by NF/RO membranes may be described by purely steric models such as the pore flow model, and this has also been found in experiments (Kiso et al., 2011). The model compounds used in the example were however alcohols, which due to their polarity was not expected to interact significantly with the hydrophobic membrane. For compounds that do interact with the membrane another type of model may be necessary to describe the rejection.

If the solute and membrane have similar polarity/hydrophobicity, the solute may obtain a lower free energy by adsorbing and diffusing into the membrane. Also, as the pore radius decreases, solvent/solute molecules moving through pores in the membrane will be less likely to be surrounded by other solvent/solute molecules, and will instead 'see' membrane molecules, which is the same situation as for diffusion through the membrane. In this way transport through very narrow pores may also be considered as diffusion through the membrane. Rejection of a solute then becomes the result of a competing transport process between the diffusion of the solvent and the solute through the membrane. The model describing this kind of rejection mechanism is called the solution diffusion model.

6.4.3. Solution Diffusion Model

For a diffusion-based model, the transport of a species 'i' through a membrane may be described by a difference in chemical potential on the two sides of the membrane (Wijmans and Baker, 1995).

$$J_i = -H_i \frac{d\mu_i}{dx} \tag{6.19}$$

where H_i is a constant of proportionality. The difference in chemical potential describes a driving force, and as seen in Table 6.1, the two most important driving forces in membrane filtration are pressure and concentration gradient. To account for these two gradients, the chemical potential may be expressed as

$$d\mu_i = RT d\ln(\gamma_i c_i) + V_{m,i} dP \tag{6.20}$$

where γ_i is the activity coefficient of species i, $V_{m,i}$ is the molar volume and concentration is given as the molar fraction (mol/mol). Integration with respect to concentration and pressure, under assumption of a constant molar volume gives

$$\mu_i = \mu_i^0 + RT\ln(\gamma_i c_i) + V_{m,i}(P - P_i^0) \tag{6.21}$$

Here μ_1^0 is the chemical potential of pure i at a reference pressure P_i^0.

In the solution diffusion model, the pressure in the membrane is considered constant,[7] and the chemical potential in Eqn (6.19) may therefore be expressed as only a function of the concentration gradient. Here we assume that the activity coefficient is constant.[8]

$$J_i = -\frac{H_i RT}{c_i}\frac{dc_i}{dx} \tag{6.22}$$

By integrating across the membrane from $(0, c_{i,0})$ to $(l, c_{i,l})$,[9] an expression of flux as a function of activity be obtained, and it is seen that this is equal to Fickian diffusion, where l is the thickness of the membrane.

[7]Because the membrane is considered practically nonporous, pressure is not transmitted through the liquid in the pores, but through the solid. Since we do not need the existence of a gradient in pressure through a solid, as we would have needed if the pressure was transmitted by the liquid, the pressure can be assumed constant through the membrane.

[8]$\frac{du}{dx} = RT\frac{d\ln(\gamma c)}{dx} = \frac{RT}{c}\frac{d(\gamma c)}{dx} = \frac{RT}{c}\frac{dc}{dx}$.

[9]The keen reader will notice that the entire term, $H_i RT/c_i$, is treated as a constant during the integration, although it would seem obvious that $1/c_i$ should be integrated with respect to c_i. However, in a more fundamental approach to diffusion it can be shown that in the constant H_i, c_i is already included, and that division with c_i therefore cancels, allowing for a treatment of the term as a constant.

$$J_i = D_i \frac{c_0 - c_l}{l}, \quad D = \frac{H_i RT}{c_i} \tag{6.23}$$

The chemical potential procedure may actually also be used to describe the solvent flux predicted by the pore flow model. In this approach, there is assumed to be a pressure gradient in the membrane, while concentration is constant, and this will lead to an expression for solvent flux as described by Darcy's law (Wijmans and Baker, 1995).

The solution diffusion model may be applied to describe RO as illustrated in Figure 6.7. If a semi-permeable membrane is used to separate two solutions of different solute concentration (concentrate and permeate), the solvent activity will initially be highest in the permeate, which holds the highest solvent 'concentration', and the pressure will be equal on both sides of the membrane. This creates a continuous gradient in chemical potential from permeate to concentrate, and as a result solvent molecules will diffuse from permeate to concentrate to increase the solvent activity here. As this happens, pressure will increase, and when the pressure gradient balances the solvent activity, the chemical potential is equal across the membrane, and the system is at equilibrium. Because pressure is assumed to be constant in the membrane, solvent activity also becomes constant in the membrane at equilibrium. The pressure difference is the osmotic pressure.

If pressure in the concentrate is increased further, the process may be reversed, and solvent will flow from concentrate to permeate. In Figure 6.7, RO for a continuous system is shown. Here the system never reaches equilibrium, and as a result we see a difference in chemical potential across the membrane. It can be seen that this variation in chemical potential is caused by the increase in pressure, which reverses the gradient for solvent flow through the membrane.

RO may be described by applying the solution diffusion model on both interfaces of the membrane. Here 'c' is used as a subscript for the concentrate side, p as a subscript for the permeate side, and m as a subscript for the membrane. 'i' will denote solvent and 'j' solute.

At the membrane/concentrate interface the equation becomes

$$\mu_{i,c} = \mu_{i,c(m)} \tag{6.24}$$

Inserting the expression for chemical potential from Eqn (6.21) gives

$$\mu_i^0 + RT\ln(\gamma_{i,c}c_{i,c}) + V_{m,i}(P^0 - P_i^0) = \mu_i^0 + RT\ln(\gamma_{i,c(m)}c_{i,c(m)}) + V_{m,i}(P^0 - P_i^0) \Updownarrow \tag{6.25}$$

$$c_{i,c(m)} = K_i \cdot c_{i,c}, \quad K_i = \frac{\gamma_{i,c}}{\gamma_{i,c(m)}} \tag{6.26}$$

where K_i is the sorption coefficient.

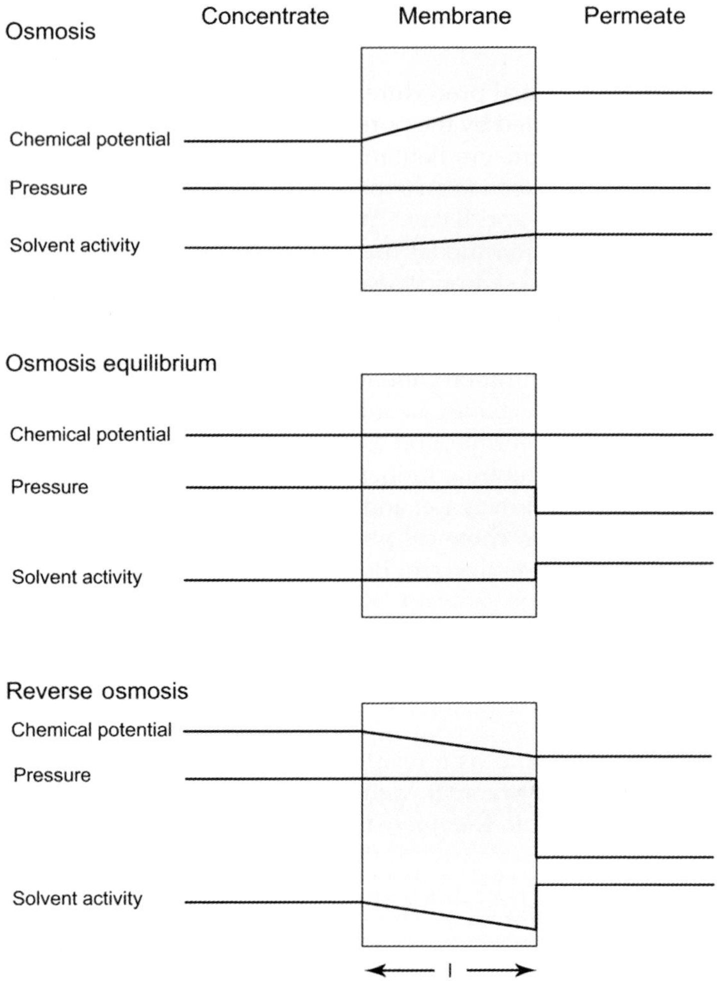

FIGURE 6.7 Application of the solution diffusion model on a semi-permeable membrane to describe conditions at osmosis, start and at equilibrium, and RO (Wijmans and Baker, 1995). (For colour version of this figure, the reader is referred to the online version of this book.)

The simple equation is the result of pressures being the same on both sides of the membrane interface for this side of the membrane. For the membrane/permeate interface, the same procedure may be applied, but according to the solution diffusion model, the pressures on either side of the interface will not be equal here.

$$\mu_{i,p} = \mu_{i,p(m)} \tag{6.27}$$

Inserting the expression for chemical potential from Eqn (6.21) gives

$$\mu_i^0 + RT\ln(\gamma_{i,p}c_{i,p}) + V_{m,i}(P^p - P_i^0) = \mu_i^0 + RT\ln(\gamma_{i,p(m)}c_{i,p(m)}) + V_{m,i}(P^c - P_i^0) \Updownarrow \tag{6.28}$$

$$c_{i,p(m)} = K_i \cdot c_{i,p} \cdot \exp\left(\frac{-V_{m,i}(P^c - P^p)}{RT}\right) \tag{6.29}$$

To describe the flux of solvent across the membranes, Eqns (6.26) and (6.29) are inserted into the Fickian diffusion model predicted by the solution diffusion model in Eqn (6.23).

$$J_i = \frac{D_i K_i}{l}\left(c_{i,c} - c_{i,p} \cdot \exp\left(\frac{-V_{m,i}(P^c - P^p)}{RT}\right)\right) \tag{6.30}$$

Equation (6.30) is rather complex, but through careful observations it may be simplified considerably. When the applied pressure balances the water activity gradient, $J_i = 0$, the pressure difference is equal to the osmotic pressure, $\Delta\pi$, and Eqn (6.30) simplifies to

$$c_{i,p} = c_{i,c} \cdot \exp\left(\frac{V_{m,i}\Delta\pi}{RT}\right) \tag{6.31}$$

Equations (6.30) and (6.31) may then be combined to simplify Eqn (6.30) further.

$$J_i = \frac{D_i K_i c_{i,c}}{l}\left(1 - \exp\left(\frac{-V_{m,i}(\Delta P - \Delta\pi)}{RT}\right)\right) \tag{6.32}$$

The term $-V_{m,i}(\Delta P - \Delta\pi)/(RT)$ is relatively small, and it may be assumed that it is close to zero (Wijmans and Baker, 1995). If this assumption is made, $1 - \exp(x) \rightarrow -x$ for $x \rightarrow 0$, and Eqn (6.32) may be rewritten as

$$J_i = A(\Delta P - \Delta\pi), \quad A = \frac{D_i K_i c_{i,c} V_{m,i}}{lRT} \tag{6.33}$$

If a solute, j, is considered instead of the solvent, the simplification made at Eqn (6.30) cannot be made since this was based on the flux of solvent being dependent on the pressure. However, as before $-V_{m,i}(\Delta P - \Delta\pi)/(RT)$ is small and since $\exp(x) \rightarrow 1$ for $x \rightarrow 0$, the solute flux becomes

$$J_i = B(c_{j,c} - c_{j,p}), \quad B = \frac{D_i K_j}{l} \tag{6.34}$$

The two resulting Eqns (6.33) and (6.34) represent an important prediction of filtration behaviour for RO membranes. The solvent flux is dependent on the pressure gradient, whereas the solute flux depends on the concentration gradient. Improved filtration may as such be achieved by increasing the pressure to obtain a higher TMP. This conclusion is of course dependent on the assumption of independence between the pressure and concentration gradients, and as we shall see this is not completely true. Another important fact that can be read from the two equations is that flux through a membrane can be increased by making the membrane thinner, a fact that has been very important in the development of membrane technology.

There is good experimental evidence for both the pore flow and the solution diffusion models, and it is clear that they cover two different regions of the pore range shown in Figure 6.1 (Wijmans and Baker, 1995). The solution diffusion model can be used to describe the region with very small or no pores and the pore flow model can be used to describe the region with larger pores. As already described in the introduction to the solution diffusion model, solutes moving through a pore in an RO membrane may be regarded as diffusing through the membrane because the size of the pore only allows one solute to pass into the pore at the time. It is also possible that the difference lies in the relative permanence of the pores. For membranes described by the solution diffusion model, the pores may only be present as statistical fluctuations. The polymer chains may be visualised as constantly moving adjacent to each other, and in this way creating pores that then disappears again in about the same timescale as the solute moves across the membrane. This would ensure that pore flow characteristics will not have time to develop. For membranes that can be described with a pore flow model, the pores could be thought of as more permanent, hereby allowing pore flow conditions to develop. The transition zone is estimated to be around 0.5–1 nm (Wijmans and Baker, 1995), which puts it right in the NF zone. This is especially interesting for the removal of micropollutants since they are often close to 1 nm in size. Membranes to filtrate micropollutants can as such be expected to fall into the solution diffusion/pore flow transition zone.

6.4.4. Prediction of Effect of Operational Variables on Filtration

From the derived model equations, the effect of some key operational variables can be deduced. Here we will focus on the effect on NF/RO membranes since these are most important for the removal of micropollutants.

The two most important parameters for NF/RO membranes are the water flux and solute rejection.

Water flux will increase linearly with applied pressure, as seen in Eqn (6.33). Increasing concentration in the feed will on the other hand lower the water flux, since the higher concentration of solutes results in a higher osmotic pressure, which will act as a resistance towards water flux. The increased osmotic pressure may also be viewed as a lowering of the effective pressure ($\Delta P - \Delta\pi$). If the temperature is increased it could seem from Eqn (6.33) that flux should decrease. However, because the diffusivity increases and viscosity of water decreases with temperature, the net effect of increasing temperature is to increase water flux. Because several parameters contribute to the effect of temperature, the correlation is not necessarily linear.

Solute flux is independent of pressure, but because the water flux increases with pressure, the net effect is a decrease in permeate concentration and as such an increase in rejection. The decreasing solute concentration in the permeate will of course create a larger concentration gradient, and in this way increase the solute flux. The net effect is however still an increase in rejection. Because solute concentration in the permeate is inversely correlated to water flow, the rejection will increase towards an asymptotic value. Through the same logic it holds that if rejection follows the solution diffusion model, an increase in feed concentration will lead to reduced rejection. Increasing feed concentrations is equivalent to increasing recovery. As the recovery is increased the solute concentration in the feed increases, and this leads to lower rejection.

The effect of temperature on solute rejection may vary. For an NF membrane (Desal 5 DK of GE Osmonics) with a pore size around 0.58 nm, it has been found that for dilute NaCl and Na_2SO_4 solutions (10^{-3} M), rejection was relatively stable in the investigated temperature range (12–50 °C) (Amar et al., 2009). However, as the concentration was increased (10^{-1} M) a decrease of around 10% in rejection was observed for the NaCl solution with increasing temperature. For the Na_2SO_4 solution the rejection did not decrease significantly with temperature. Overall, rejection was found to decrease with temperature, and the effect was largest for monovalent salts. The decrease in rejection may be explained by an increase in diffusivity of the solute, but it has also been found that an increase in temperature will increase the pore radius, which will lower the steric hindrance. In the same study the pore radius was found to increase from 0.58 to 0.67 nm for a temperature increase from 22 to 50 °C. The effect of the increased pore radius was to decrease the rejection of organic solutes such as glucose and arabinose from 65% to 46% for glucose and 50% to 25% for arabinose (Amar et al., 2009).

6.5. ISSUES FOR MEMBRANES FILTRATION – FOULING AND SCALING

The two main challenges for operation of membrane filtration processes are fouling and scaling. Fouling covers the deposition of organics, particles/colloids and microorganisms (biofouling) on the membrane while scaling covers the precipitation of ionic species that exceed their aqueous solubility on the membrane.

6.5.1. Fouling

The effect of fouling is to increase the required TMP to maintain a given flux because the fouling acts as an additional resistance. To understand the effect we may use the pore flow model, where it was shown that the correlation between flux and applied pressure can be described by Darcy's law. If we account for the osmotic pressure, which may be significant for NF/RO membranes, and express the hydraulic constant as the inverse product of the viscosity and membrane resistance, we get the following expression for flux (Peinemann and Nunes, 2010).

$$J = \frac{\Delta P - \Delta \pi}{\mu R_{\mathrm{m}}} \tag{6.35}$$

Equation (6.35) accounts for the flux of water through an unfouled membrane. In the case of fouling, the total resistance increases and may be accounted for by considering the resistances to be in series and therefore adding the resistance created by the fouling to the membrane resistance.

$$J = \frac{\Delta P - \Delta \pi}{\mu \left(R_{\mathrm{m}} + R_{\mathrm{fouling}} \right)} \tag{6.36}$$

From Eqn (6.36), it can be seen that the effect of fouling on the filtration process is to either increase TMP or decrease flux. If no measures are taken both may occur, but since the object for industry is a constant production, control systems will respond to fouling by increasing TMP. Industrial filtration processes can therefore be considered to be constant flux processes. This is in contrast to much laboratory work and models of the fouling process where pressure is held constant and the decline in flux is measured. We will return to this difference during the discussion of fouling models.

The fouling resistance can be divided into three separate contributions: adsorption, reversible and irreversible, which may be considered being in series as before.

$$J = \frac{\Delta P - \Delta \pi}{\mu (R_{\mathrm{m}} + R_{\mathrm{ads}} + R_{\mathrm{rev}} + R_{\mathrm{irrev}})} \tag{6.37}$$

For practical reasons it is beneficial to distinguish between the three types of fouling. Adsorption is unavoidable and will depend on the interplay between the specific membrane and water matrix. During continuous filtration, solutes will reach adsorption equilibrium with the membrane that upon reaching equilibrium will be constant during the lifetime of the membrane. If a switch in water matrix occurs, the equilibrium may shift and the adsorption may increase or decrease. Examples of such a switch could be a water treatment plant operating with multiple water sources that it can change between, backwash operations and chemical cleaning of the membranes. Because the shift in water type may lead to desorption of the adsorbed solutes, it is of critical importance to monitor and control these adsorption/desorption processes. A specific example of such an adsorption/desorption system has been observed for the filtration of the endocrine disrupting compound estrone with MF hollow fibre membranes (Chang et al., 2003).

The two other fouling terms, reversible and irreversible, represent fouling that occurs during operation. Fouling that may be removed by either a backwash or a chemical cleaning is termed reversible. This kind of fouling affects operation in that it determines the length of the intervals between successive backwashes/cleanings, which are important for the overall efficiency of the filtration process. Reversible fouling does however not affect the lifetime expectancy of the membranes because it can be removed. Fouling that is not removed during backwash/cleaning is called irreversible. It can be seen as a loss of flux or increase in TMP between successive backwashes/cleanings, which means that the membrane production efficiency will drop over time. Irreversible fouling will as such affect the lifetime expectancy of the membranes.

Fouling is also distinguished by the mechanism by which it occurs. In general there are four different types as shown in Figure 6.8 (Huang et al., 2008; Peinemann and Nunes, 2010):

1. Complete pore blocking
2. Partial pore blocking
3. Internal pore blocking
4. Cake layer

It is important to differentiate between the four fouling modes because they affect the filtration differently, and are affected differently by the operation of the filtration process. When foulants with a size larger than the pore diameter are present they may lead to the complete blocking of individual pores. Since the pores become completely blocked, no flow occurs through the affected pores, and this is seen as a reduction in the active membrane area. Instead of completely blocking a pore, foulants may deposit on the membrane in such a way that they only partially cover the pores. This also results in a reduction in the active membrane area, but

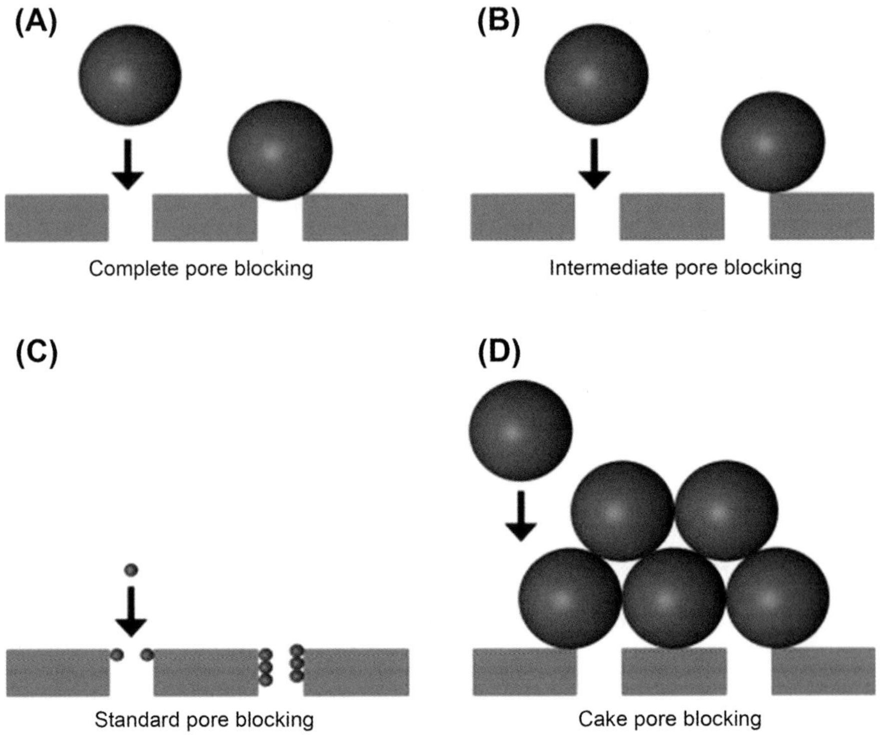

FIGURE 6.8 **Fouling mechanisms.** The four subplots show the four main fouling modes that fouling is usually divided into: (A) Complete pore blocking, (B) Intermediate pore blocking, (C) Standard pore blocking and (D) Cake filtration. (For colour version of this figure, the reader is referred to the online version of this book.)

since the pores are not completely blocked, the effect is not as severe as for complete pore blocking. If the foulants form bridges across the pores, the partial pore blocking may turn into a complete pore blocking. Importantly, the foulants need not be larger than the pores for partial pore blocking to occur. After the initial layer of fouling is formed, foulants may deposit on foulants rather than the membrane. When this happens, a cake is said to be formed, and this affects the membrane by adding and additional layer of resistance. Since the cake completely covers the membrane, the solution must first pass through the cake before being filtered by the membrane. All three of these fouling modes occur on the membrane surface. As such, they can be reduced by running the filtration in cross-flow mode. The shear and turbulence created by the feed flow will help to scourge the membrane surface and can alleviate fouling to some extent. The fourth mode of fouling, internal pore blocking, occurs when foulants smaller than the pore radius adsorb to the pore wall, and hereby decreases the effective pore radius. Because the fouling occurs in the pores, it is not affected by cross-flow filtration.

To describe the fouling process, a model for dead-end filtration at constant pressure was developed by Hermia (Hermia, 1982).

$$\frac{\mathrm{d}J}{\mathrm{d}t} = -kJ^{3-n} \tag{6.38}$$

In this equation, the numerator 'n' is used to distinguish the four different fouling modes. For complete pore blocking $n = 2$, for internal pore blocking $n = 1.5$, for intermediate pore blocking $n = 1$ and for cake filtration $n = 0$. In this way the severity of the different fouling modes can be seen. The smaller the value of n, the larger is the flux decline.

Later a model for the more complicated cross-filtration was also developed by Field et al. (Field et al., 1995).

$$\frac{\mathrm{d}J}{\mathrm{d}t} = -k(J - J_{ss})J^{2-n} \tag{6.39}$$

Here the system will approach a steady state situation in which the amount of new fouling is balanced by fouling removed by the cross-flow.

Depending on whether the system is set to dead end or cross-flow, Eqns (6.38) and (6.39) can be solved to give the correlations between flux and time seen in Table 6.4.

By measuring the flux during operation and fitting a plot of flux over time with the equations in Table 6.4, it may theoretically be investigated which type of fouling is responsible for the observed flux decline. However, the equations in Table 6.4 represent a fundamental approach to describe fouling, and in reality the equations are often difficult to use for an accurate description of the filtration process. In deriving the model, pores have been assumed to be vertical cylindrical, which may not be the case for real industrial membranes. Also, the models treat each type of fouling separately, whereas in reality, all fouling types may be occurring simultaneously. In real life, the observed fouling will often be a mixture of

TABLE 6.4 Equations for Fouling Modes in Dead-End and Cross-flow Filtration at Constant Pressure

	Dead End	**Cross-flow**
Complete pore blocking	$J = J_0 \cdot K_b A \cdot t$	$J = (J_0 - J_c) \cdot \exp\left(K_b A \cdot t \frac{\sigma J_0}{\varepsilon_0} t\right) + J_c$
Intermediate pore blocking	$J = J_0(1 + K_i \cdot (A \cdot J_0) \cdot t)^{-1}$	$\sigma t = \frac{1}{J_i}\left(\ln \frac{J_0 - J_i}{J_0} \cdot \frac{J}{J - J_i}\right)$
Internal pore blocking	$J = J_0\left(1 + \frac{1}{2} \cdot K_s \cdot (A \cdot J_0)^{0.5} \cdot t\right)^{-2}$	$\frac{1}{J^{0.5}} = \frac{1}{J_0^{0.5}} + \left(\frac{K_s'}{2}\right) A^{0.5} t$
Cake filtration	$J = J_0(1 + 2 \cdot K_s \cdot (A \cdot J_0)^2 \cdot t)^{-0.5}$	$Gt = \frac{1}{J_s^2}\left(\ln\left(\frac{J}{J_0}\frac{J_0 - J_s}{J - J_s}\right) - J_s\left(\frac{1}{J} - \frac{1}{J_0}\right)\right)$

the four fouling modes described above. In a typical scenario, the clean membrane will start by experiencing pore blocking. If the solution to be filtrated contains a variety of different particle sizes, this initial fouling can be both pore blocking caused by the larger particles as well as internal pore blocking. As an increasing number of pores become blocked and the surface is covered in a monolayer of fouling particles, cake filtration will start to occur. In dead-end filtration the cake layer will continue to build as more particles are led to the membrane, while in cross-flow filtration, the cake layer may reach an equilibrium size, where the amount of new fouling is equal to the amount removed by the cross-flow. As such, more practical tools for description of both the specific fouling processes, but also the fouling potential of a given water source, have been developed. One such practical tool is the modified fouling index (MFI).

The MFI is a simplified approach to determine the fouling potential and characteristics of a given water source. In the classical MFI, cake filtration is assumed to be the dominant fouling mechanism, and we may therefore use Eqn (6.36) to describe the filtration process. Furthermore, in determination of the MFI, the usual procedure is to measure the amount of permeate collected over a specific period of time, and it may as such be an advantage to rewrite Eqn (6.36) to depend on filtrate volume (V) and filtration time (t).

$$\frac{1}{A}\frac{\mathrm{d}V}{\mathrm{d}t} = \frac{\Delta P}{\eta\left(R_\mathrm{m} + R_\mathrm{f}\right)} \tag{6.40}$$

The fouling resistance may be calculated as the product of the amount of foulants brought into contact with the membrane and their ability to increase the resistance as seen in Eqn (6.41). Here V/A is the specific throughput (V_sp), C_b the foulant concentration and α is the specific resistance. Because the specific resistance and concentration of foulants can be difficult to determine, they are often expressed as the constant I, called the fouling index.

$$R_\mathrm{f} = \frac{V}{A} \cdot \alpha C_\mathrm{b} = V_\mathrm{sp} \cdot I \tag{6.41}$$

By integrating Eqn (6.40), a linear correlation between the filtration rate (t/V) and the filtrated volume can be obtained.

$$\frac{t}{V_\mathrm{sp}} = \frac{\eta R_\mathrm{m}}{\Delta P A} + \frac{\eta I}{2\Delta P A^2} V_\mathrm{sp} \tag{6.42}$$

The MFI is then defined as the slope of the fit.

$$\mathrm{MFI} = \frac{\eta \alpha C_\mathrm{b}}{2\Delta P A^2} \tag{6.43}$$

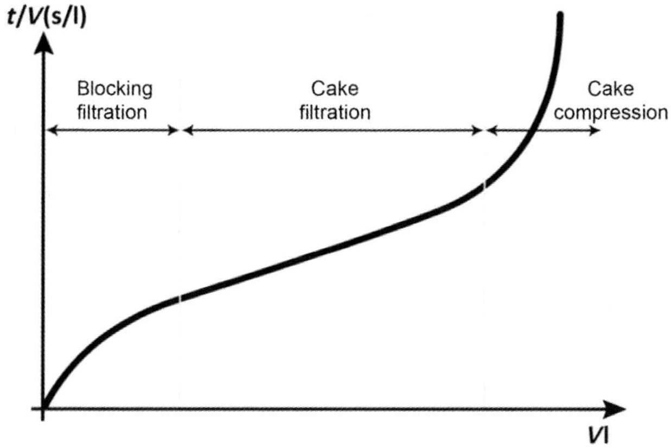

FIGURE 6.9 **Filtration time per filtrate volume plotted against collected filtrate volume.**

A typical plot of the filtration process is shown in Figure 6.9, where we can see how the MFI determination may be used to characterise the fouling process. The first part of the curve is nonlinear and corresponds to blocking filtration. After this initial pore blocking, cake filtration dominates, and the MFI may be determined as the slope of the curve. Depending on the operating pressure, the cake may start to compress, which will compact it and increase the resistance. This can be seen as the final region in the plot where curve increases nonlinearly.

The procedure for determining MFI has been standardised to enable comparison of MFI between different water sources. In the standardised procedure, a 0.45 μm MF membrane is used and filtration is conducted at 20 °C and a constant TMP of 2 bar (Schippers and Verdouw, 1980). The process can also be run at constant flux, as described by Boerlage et al. (Boerlage et al., 2004).

From Eqns (6.42) and (6.43), it can be seen that as MFI increases the time to collect a specific volume of permeate decreases. A high MFI will therefore indicate that the water has a high fouling potential. However, the stress is on potential. For RO and NF processes, the interesting characteristic of the water is its ability to foul the RO/NF membrane, but since the pore size of the MF membrane is 0.45 μm, the MF membrane is limited to investigating the fouling potential of foulants at or larger than 0.45 μm. Foulants smaller than this, such as colloidal particles, are not accounted for unless they adsorb to the membrane, but they may still significantly foul the RO/NF membrane. To handle smaller foulants, variations of the classical MFI have been proposed where UF (Boerlage et al., 2002; Roorda and van der Graaf, 2001) and NF (Khirani et al., 2006) membranes have also been used. In these cases the process is called MFI-UF and MFI-NF.

Furthermore, by determining the MFI of several different membranes, a Combined Fouling Index may be constructed that takes into account the effect of different foulants. In this way an improved prediction of the fouling potential on a subsequent NF/RO membrane can be obtained, and the effect of specific types of foulants may be investigated. In an investigation of the effect of different pretreatment techniques on the fouling potential of river water on an NF membrane, Choi et al. determined the MFI of the pretreated river waters with three membranes (Choi et al., 2009): a hydrophilic 0.45 μm membrane, a hydrophobic 0.45 μm membrane and a UF membrane with a molecular cutoff at 100,000 Da. In this way it was possible to investigate how the different pretreatment techniques affected hydrophilic compounds, hydrophobic compound and dissolved matter. Also, by combining the three MFIs, a significant better correlation was found between the fouling index and the observed rate of flux decline for the NF process.

However, these MFIs rely on the use of proxy membranes to determine the fouling potential, and they assume only cake filtration to be occurring. Also, in their derivation it is commonly assumed that pressure is constant, which is in stark contrast to how membranes are operated in real life. Here a constant production is desired, and it is therefore permeate flux and not TMP that is held constant. To overcome these shortcomings a unified membrane fouling index (UMFI) has been developed by Huang et al. (Huang et al., 2008). In this approach to the MFI, specific flux, $J' = J/TMP$, is used instead of the ratio t/V. The big advantage of using specific flux is that the UMFI then can be determined for both constant flux and constant pressure filtration. UMFI may be determined by plotting the inverse normalised specific flux against the specific throughput, as given by Eqn (6.44).

$$\frac{1}{J'_{sp}} = 1 + \text{UMFI} \cdot V_{sp}, \quad J'_{sp} = \frac{J_{sp}}{J_{sp0}} = \frac{\frac{J_s}{\Delta P}}{\frac{J_{s0}}{\Delta P}} \tag{6.44}$$

Equation (6.44) is based on the UMFI for cake filtration. This is done since the equations for internal and pore blocking can be shown to approximate cake filtration when the concentration of foulants in the water is sufficiently low, which may be a good assumption for most natural waters (Huang et al., 2008). Specific UMFI equations for each of the four fouling modes have been developed, and can be applied if necessary (Huang et al., 2008).

As such it can be seen that typical approaches for MFI and UMFI are very similar, and both can be used to estimate the fouling potential of a given water source, but they cannot be directly compared. MFI has a unit of s/l^2, while UMFI has a unit of m^2/l^2 or $1/m$. The difference stems from the use of normalised specific flux for the UMFI. As mentioned before, the use of specific flux makes it possible to determine the UMFI for both constant pressure and constant flux. Furthermore, the normalisation eliminates the impact of the total membrane area, which is important for the scaleup of UMFI from laboratory to industrial scale.

EXAMPLE 2: EXPERIMENTAL DETERMINATION OF UMFI FOR MF AND UF PRETREATMENT OF GROUND WATER

One of the major foulants in ground water is iron colloids, resulting from the oxidation of iron(II) to the low soluble iron(III). It is therefore necessary to remove these colloidal particles before an RO/NF treatment of the ground water, and this may be accomplished by using MF/UF processes as a pretreatment. Two ceramic membranes, 3 μm MF with a surface area of 0.09 m^2 and 0.04 μm UF with a surface area of 0.05 m^2, were therefore investigated for this purpose, and to evaluate their applicability, the UMFI was measured. The experiment was conducted at a constant flux of 3000 $l/(m^2\,h)$.

Selected data collected from the experiment with the 3 μm MF membrane are shown in Table 6.5.

By plotting the data from the two experiments and making a linear fit, the UMFI for the two membranes can be determined.

$$\mathrm{UMFI}_{0.04\ \mu\mathrm{m}} = 0.0129\ \frac{\mathrm{m}^2}{\mathrm{l}} = 12.9\ \frac{1}{\mathrm{m}}$$

$$\mathrm{UMFI}_{3.0\ \mu\mathrm{m}} = 0.0037\ \frac{\mathrm{m}^2}{\mathrm{l}} = 3.7\ \frac{1}{\mathrm{m}}$$

As could be expected, the UF membrane has the highest UMFI index due to its smaller pore size. This is also an indication that the UF membrane is capable of removing a larger degree of the iron colloids, since if both removed the iron colloids completely or to the same degree, identical cake layer should build up resulting in equal increase in the fouling resistance and therefore UMFI. As a pretreatment for RO/NF treatment of the ground water, the UF membrane does therefore show the highest potential.

Another observation that can be made is that for the 3 μm membrane an initial pore blocking seems to take place. The linear region is first properly established after the first 150 l/m^2. This can also explain why the intersection with the ordinate is not 1 as expected from Eqn (6.44). The constant value of 1 is obtained by the normalisation of the membrane resistance, and if an initial pore blocking takes places, this may have the effect of creating an apparent higher membrane resistance, R^2 (Figure 6.10).

(Continued)

EXAMPLE 2: EXPERIMENTAL DETERMINATION OF UMFI FOR MF AND UF PRETREATMENT OF GROUND WATER (*cont'd*)

TABLE 6.5 Data from Filtration of Aerated Ground Water with 3 μm MF Membrane

Filtration Time (s)	TMP (bar)	Volume Filtered	Specific Throughput (l/m^2)	Specific Flux (m^3/(m^2 h bar)	Normalised Specific Flux	Inverse Normalised Specific Flux
0	0.25	–	–	11.5	–	–
68	0.38	5.40	60.0	8.44	0.73	1.37
121	0.45	9.18	102	6.70	0.58	1.72
190	0.54	15.3	170	5.98	0.52	1.93
244	0.59	19.1	212	5.34	0.46	2.16
296	0.64	23.5	261	4.98	0.43	2.32
365	0.69	27.3	303	4.34	0.38	2.66
418	0.75	31.8	354	4.05	0.35	2.85
487	0.80	37.1	412	3.81	0.33	3.03
540	0.84	41.1	456	3.61	0.31	3.19
609	0.91	48.4	537	3.49	0.30	3.30
662	0.94	50.0	555	3.21	0.28	3.59
718	0.98	54.5	605	3.09	0.27	3.73
783	1.03	59.7	663	2.97	0.26	3.89

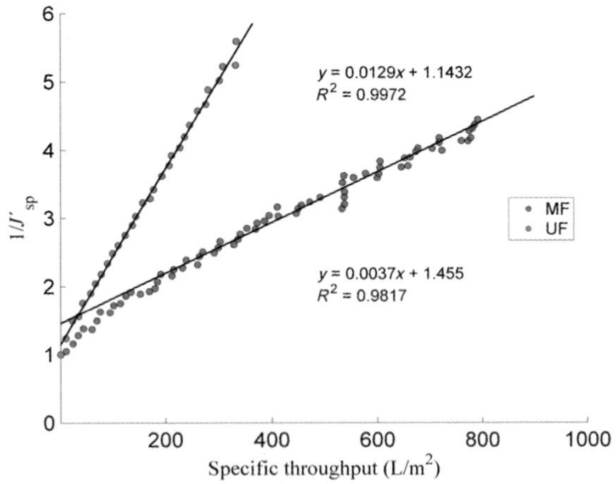

FIGURE 6.10 UMFI analysis of filtration of aerated ground water. (For colour version of this figure, the reader is referred to the online version of this book.)

When using NF and RO membranes to remove micropollutants, the MFI or UMFI can be used to determine the potential of mainly organic or particular matter in the water source to foul the membranes, or to evaluate the effect of different pretreatment strategies on the fouling potential. The fouling indexes can however not be used to evaluate the severity of the other issue for NF/RO membranes, scaling.

6.5.2. Scaling

As previously discussed, RO and NF membranes are capable of retaining dissolved ionic species. When water is forced through an RO/NF membrane, these species will be retained to some degree resulting in an increase in concentration. If the ionic product of one or more of these species on the feed side of the membrane exceeds the saturation index, precipitation may occur on the membrane.

$$A_xB_{y(s)} \rightleftharpoons xA^{y+}_{(aq)} + yB^{x-}_{(aq)} \tag{6.45}$$

$$\mathrm{IAP} = \gamma^x_{A^{y+}}\left[A^{y+}\right]^x \cdot \gamma^y_{B^{x-}}\left[B^{x-}\right]^y \quad \text{at equilibrium} \quad K_{sp} = \mathrm{IAP} \tag{6.46}$$

A common way of indicating the degree of saturation for a given species is through the saturation index, SI.

$$\mathrm{SI} = \frac{\mathrm{IAP}}{K_{sp}} \tag{6.47}$$

The value of SI will then show whether a species is undersaturated, saturated or supersaturated.

- $\mathrm{SI} < 1 \rightarrow$ Undersaturated $\rightarrow$ dissolution
- $\mathrm{SI} = 1 \rightarrow$ Saturated
- $\mathrm{SI} > 1 \rightarrow$ Supersaturated $\rightarrow$ precipitation

Scaling may generally affect membrane efficiency in three ways:

1. Increase of pressure (at constant flux)
2. Increasing pressure drop along membrane module
3. Decreasing retention

As for fouling, the scale layer acts as an additional layer of resistance, and if a system operates at constant production, a higher TMP is required to overcome the increase in resistance. Most NF/RO membranes are operated in spiral modules, and scale may therefore take up a significant part of the flow path created by the spacer. In this way the deposited scale will disrupt cross-flow conditions across the membrane by creating turbulence, and this will lead to a larger pressure drop across the membrane module. Pressure losses are unwanted because they increase the cost of the filtration since less energy may be recovered from the pressurised concentrate, but also because it may physically damage the membrane. The lower pressure at the concentrate end of the membrane module may also lead to a decreased retention because a lower TMP at the concentrate end reduces the water flux, whereas the solute transport is unaffected. The scale itself may also reduce retention. As can be seen in the solution diffusion model, the concentration difference that influences the solute transport is the difference between the permeate concentration and the feed concentration directly adjacent to the membrane surface. This can be understood as the feed that the membrane actually 'sees'. Deposition of scale increases the concentration as seen by the membrane and will therefore create a larger concentration gradient across the membrane.

To determine whether a given water source may result in scale formation on a membrane, the different ion products may be calculated and compared with solubility products. However, as real waters consist of a range of different ions, there are many possible ionic compositions, each of which may exist in different crystal structures. Add to this that the calculations must be performed at different levels of recovery to determine if and when a water source will form scales. If the water source has a high ionic strength, the calculations may be further complicated. The higher ionic strength will affect the activity constant of the ionic species, and at very high concentrations salting out may occur. Increasing ionic strength will reduce the electrical double layer (DLL) of particles in solution to a degree where the repulsion of the DLL is weaker than the attractive forces between the particles. In such cases precipitation may occur earlier than predicted by calculations of the saturation index. To overcome this considerable work load, software with prebuild models may be used to predict, which species will become supersaturated and when this happens.

EXAMPLE 3: USE OF SOFTWARE PHREEQC INTERACTIVE TO DETERMINE MAIN SCALANTS AND DEGREE OF POSSIBLE RECOVERY

We wish to investigate the possibility of applying RO to treat ground water, and are interested in determining what the main scalants will be and when they can thermodynamically be expected to precipitate. In the investigation we look at two different ground waters, one with a low salinity (Astrup water) and another with high salinity (Hvidovre water). The composition is given in Table 6.6.

An initial speciation calculation allowed to obtain equilibrium, and determination of saturation indexes for species in each ground water type shows that even without any pre concentration sulphur minerals

EXAMPLE 3: USE OF SOFTWARE PHREEQC INTERACTIVE TO DETERMINE MAIN SCALANTS AND DEGREE OF POSSIBLE RECOVERY *(cont'd)*

may tend to precipitate if the conditions are reducing (pe ≈ −2). This happens because SO_4^{2-} is reduced to H_2S, which then precipitate with iron. Some iron(III) oxides may also precipitate.

As such any NF/RO membranes should first be applied after a pretreatment for removal of iron species. If an aeration/sand filtration system is applied before any RO/NF processes an oxic environment (pe ≈ 15) is created. This prevents H_2S formation and will lead to precipitation of iron and manganese oxides that may subsequently be removed in a sand filter. From the saturation indexes for iron and manganese under oxidising conditions, in Table 6.7, a complete removal of these may be assumed. Although manganese is not predicted to precipitate from the Astrup water, it is known to co-precipitate with iron. In the simulation iron and manganese concentration may as such be set to zero, and the simulation can be run again to identify other minerals that may precipitate. An evaporation of water is used to simulate the complete retention of minerals. The results can be seen in Figure 6.11.

We find that it is the calcium species that become supersaturated as the recovery is increased. It can be seen that especially calcium fluoride (CaF_2) will become supersaturated relatively quickly, and this may be important when the general alkalinity is low since calcium fluoride then may be dominating as seen for Astrup. However, because fluoride is only present in low concentrations, the precipitated amount will be small relative to some of the other supersaturated species. The same can be said for the phosphate species, hydroxyapatite. The main challenge will be calcium carbonate species (calcite, aragonite and dolomite[10]) and calcium sulphate species (anhydrite and gypsum). From the SI plots it can be seen that carbonate species will be more important.

If we neglect fluorite and hydroxyapatite, a recovery of 93% may be obtained for the ground water from Astrup before the solubility for calcite is exceeded. For Hvidovre the higher calcium concentration and alkalinity mean that a recovery of only 8% will be possible before the solubility for calcite will be exceeded.

Another interesting parameter to investigate is the amount of scale formation. Figure 6.11 shows that calcite is the most important scale of the calcium carbonate and sulphate species. Above a recovery of 70% dolomite has the highest SI for Hvidovre water, but often dolomitisation is not observed to occur, and we may therefore assume calcite precipitation to dominate. If we do this we can use PHREEQCI to estimate how much calcite will be precipitated from 1 l of ground water (Figure 6.12)

(Continued)

EXAMPLE 3: USE OF SOFTWARE PHREEQC INTERACTIVE TO DETERMINE MAIN SCALANTS AND DEGREE OF POSSIBLE RECOVERY (*cont'd*)

TABLE 6.6 Composition of Ground Waters

Component	Astrup DGU-NR 121.1081 (mg/l)	Hvidovre DGU-NR 208.2616 (mg/l)
Ca^{2+}	24	220
Mg^{2+}	6.30	33
Na^{+}	21	100
K^{+}	1.8	8
Fe^{2+}	12	0.66
Mn^{2+}	0.43	0.04
Cl^{-}	34	290
SO_4^{2-}	54	180
NH_4^{+}	0.027	0.31
NO_3^{-}	0	4.4
F^{-}	1.5	0.49
P	0.19	0.011
Alkalinity as HCO_3^{-}	47	400
CO_2	25	0
pH	6.7	6.88

TABLE 6.7 Iron and Sulphur Precipitation under Reducing and Oxidising Conditions

Mineral Phase	Astrup DGU-NR 121.1081 (mg/l)	Hvidovre DGU-NR 208.2616 (mg/l)
REDUCED CONDITIONS		
Pyrite (FeS_2)	8.11	11.18
Mackinawite (FeS)	−0.21	0.09
Hematite (Fe_2O_3)	3.53	2.38
Goethite (FeOOH)	0.80	0.23
OXIDISED CONDITIONS		
$Fe(OH)_3$	2.00	3.27
Goethite (FeOOH)	7.25	8.52
Hematite (Fe_2O_3)	16.42	18.97
Manganite (MnOOH)	−6.51	3.99
Pyrolusite ($MnO_2 \cdot H_2O$)	−14.72	7.23

EXAMPLE 3: USE OF SOFTWARE PHREEQC INTERACTIVE TO DETERMINE MAIN SCALANTS AND DEGREE OF POSSIBLE RECOVERY (*cont'd*)

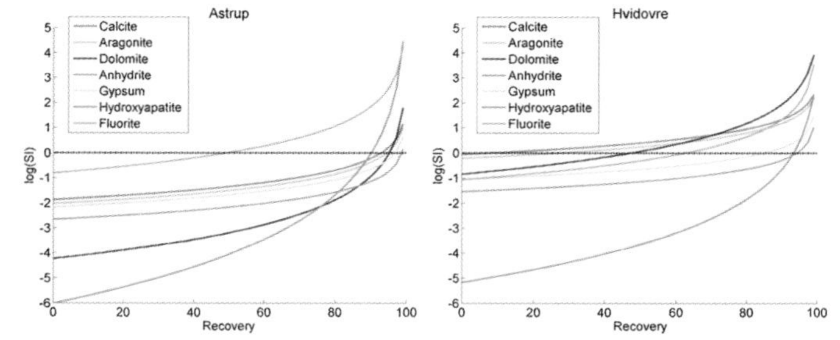

FIGURE 6.11 **Impact of degree of recovery on the saturation indexes of common ground water species.** (For colour version of this figure, the reader is referred to the online version of this book.)

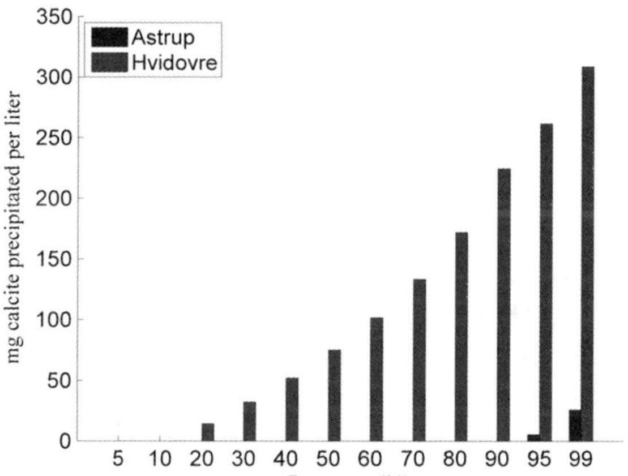

FIGURE 6.12 **Precipitation of calcite.** (For colour version of this figure, the reader is referred to the online version of this book.)

[10]Dolomite consist of equimolar amounts of calcium and magnesium.

As seen from the simulation in Example 3 iron, manganese, calcium and at reducing conditions sulphide species will be important for scaling because of their low solubility. Other important species that were not part of the example are silica and sulphate-based scales of trace metals such as barium and strontium. What type of scaling agent will occur depends on the specific water type, and an initial investigation of the water composition should always be made before filtration is undertaken. In the investigation software such as PHREEQCI or relevant scaling indexes such as Langelier Saturation Index, but even with these tools it can be difficult to get precise results. Both techniques are based on the use of saturation indexes, which are defined from thermodynamic equilibriums, and do not take precipitation kinetics into consideration. It is as such possible for a solution to be kinetically stable even though it is supersaturated. This was used in the example when calcite was assumed to precipitate even though dolomite had a higher SI value. PHREEQCI can use kinetics, but to do this, rate constants must be determined by experiments or found in literature.

To increase the recovery, scale formation can be reduced by use of techniques such as softening or scale inhibitors. The literature on these is extensive and will not be covered here.

6.5.3. Concentration Polarisation

Most RO membranes today are used for desalination of sea water. The osmotic pressure of sea water can be calculated with the Van Hoff equation, where deviations from ideality are accounted for by the unit less osmotic coefficient, ϕ_i. The Van Hoff equation can be derived from the solution diffusion model (Wijmans and Baker, 1995).

$$\Delta\pi = \phi_i \Delta C_i RT \tag{6.48}$$

Average sea water can be assumed to have a salinity of 3.5 wt%, which can be assumed to be NaCl. Average temperature can be estimated to 15 °C and the osmotic coefficient can be set to 0.95. If these values are used the osmotic pressure becomes

$$\Delta\pi = 0.95 \cdot 2 \left(\frac{\frac{3.3\%}{100\%} \cdot 10^6 \ \frac{\mathrm{g}}{\mathrm{m}^3}}{58.5 \ \frac{\mathrm{g}}{\mathrm{mol}}} \right) \cdot 8.314 \ \frac{\mathrm{J}}{\mathrm{mol\ K}} \cdot (15 + 273.15)K = 27.2 \ \mathrm{bar} \tag{6.49}$$

This is the theoretical pressure that must be overcome to have a net production. If we compare this value with actual operational pressures at large-scale RO installations, see Table 6.8, it is seen that the actual values are much higher.

Why do we see this difference?

The main contributor is of course the recovery. The membrane filtration increases the concentration hereby increasing the osmotic pressure. This can be approximated as a doubling of the osmotic pressure. Furthermore, as we have seen in the discussion about fouling, the membrane in itself has a certain resistance that must be overcome. Depending on the

TABLE 6.8 Operational Pressures at RO Plants

RO Installation	Operating Pressure (bar)
Ashkelon (Israel)	60
Tuas (Singapore)	59
La Rosiére (Jersey)	65

membrane, the membrane resistance may add 5–10 bar. Pressure losses also add 1–2 bar. Still these contributions cannot completely explain the observed pressure difference, and they cannot explain why fouling and scaling occur earlier than predicted. The explanation is the phenomenon known as concentration polarisation.

When solutes are transported to a membrane through which they cannot penetrate, their direction of diffusion will be against the flow towards the membrane surface. If the transport of solutes to the surface is larger than the diffusion away, there will be a net build-up of solutes near the surface. This phenomenon is called concentration polarisation (Figure 6.13).

Concentration polarisation can be described by setting up a mass balance, where we assume the process to be at steady state, meaning that the transports of solutes into and away from the boundary layer are equal.

$$\frac{dM_i}{dt} = 0 = J \cdot C_i - J \cdot C_{i,p} - D_{ij}\frac{dC_i}{dx} \tag{6.50}$$

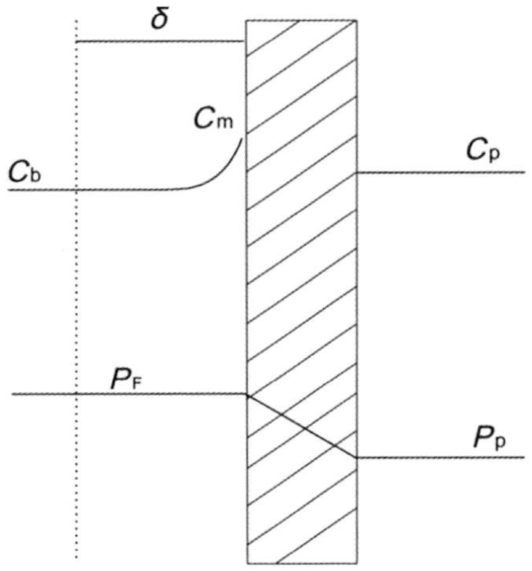

FIGURE 6.13 **Conceptual model of concentration polarisation.**

If we integrate across the boundary layer we get a correlation of concentration and flux.

$$C_m = C_p + (C_b - C_p)e^{\left(\frac{J}{k}\right)} \tag{6.51}$$

where k is the mass transfer coefficient, which can be estimated from the diffusion coefficient, Reynolds number and Sherwood number (Crittenden et al., 2012). Equation (6.51) shows us that by increasing flux the concentration at the membrane surface increases exponentially. This is important because the osmotic pressure is dependent on the concentration that the membrane 'sees', and the consequence is that we cannot keep increasing production without obtaining significant pressure losses through concentration polarisation. Also, because the concentration of solutes is higher at the surface compared with the bulk solution, fouling and scaling will occur sooner than expected from the bulk concentrations.

The existence of concentration polarisation is also one of the reasons why RO/NF membranes are operated in cross-flow mode. The shear and turbulence created by the cross-flow lower the thickness of the concentration boundary layer at the membrane surface.

6.6. VERSATILITY OF MEMBRANES – EXAMPLES OF APPLICATIONS

With the decreasing price of membranes the number of real world applications is increasing, especially in two sectors: desalination and waste water treatment. The increasing population and water consumption have led to the development of water-poor regions where the price of water is sufficiently high to justify the use of advanced techniques such as membranes. For now the places experiencing water scarcity are mainly found in dry and highly populated areas in the Middle East and North Africa, but also in more developed countries such as Australia are membranes being used for desalination, and even in Denmark is desalination being considered as a possible source of drinking water to the Copenhagen area. The prime example today on the use of membranes for production and reuse of drinking water is however none of the before mentioned places, but Singapore.

6.6.1. Singapore – NEWater

With an area of 720 km^2, close to the area of the Danish island of Bornholm or the German city of Hamburg, and a population of 5.3 million, Singapore is densely populated. Singapore has only relatively few natural drinking water resources, and it has therefore been necessary to apply a different strategy than seen elsewhere in the World. This strategy consists of four sources of water, or 'taps' as they are popularly called:

1. Imported water from Malaysia
2. Catchment water
3. NEWater
4. Desalination

Imported water accounts for around 40% of the total water supply, but the aim is to reduce this to zero by 2060. This increases pressure of the other three taps. Catchment water is the collection of rain water in large reservoirs, which is then treated to reach drinking water quality. This is itself an unusual method for producing drinking water, but it is the NEWater and desalination that have made Singapore one of the World membrane hubs.

The water supply of Singapore is illustrated in Figure 6.14, where the blue line is NEWater and the dark red line is the desalination. NEWater is as such the recovery of waste water. In traditional waste water treatment, the treated water is disposed of, but in Singapore it undergoes further treatment to turn it into drinking water. The steps are MF, RO and UV disinfection.

The MF step removes particles and bacteria, while the RO removes dissolved organics and viruses. The UV disinfection acts as an additional safety barrier.

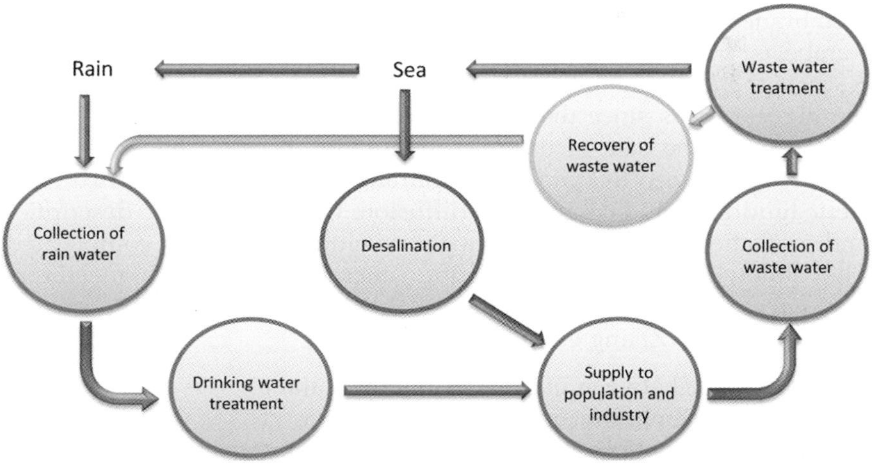

FIGURE 6.14 **Schematic representation of Singapore's water supply.** Blue line is the NEWater. (For interpretation of the references to colour in this figure legend, the reader is referred to the online version of this book.)

6.6.2. Méry-sur-Oise – NF of Surface Water

Application of membranes in not restricted to desalination and waste water treatment; they are also used in cases where traditional drinking water treatment techniques do not suffice. One example of this is found in Europe north of Paris at the Méry-sur-Oise drinking water treatment plant.

The original water works used coagulation, flocculation, sand filtration, ozone and granular activated carbon to treat the water from the Oise river, but because of the high content of organics it could not meet the required drinking water standards, and even had to stop production in 1998 when November rains increased the total organic carbon (TOC) content of the river water above 10 mg/l. Therefore, in 1999, NF was added at the end of the process train, with a capacity of 140,000 m^3/day. Besides achieving lower TOC content, the NF is also capable of lowering the content of pesticides and increases the softness of the water (Cyna et al., 2002; Ventresque et al., 2000).

6.7. MEMBRANES FOR MICROPOLLUTANT REMOVAL

As seen with the Méry-sur-Oise plant, one of the applications of membranes is the removal of micropollutants such as pesticides. The current state of the art in removal of micropollutants is the use of activated carbon, but activated carbon suffers from a number of drawbacks mainly related to saturation, bacterial growth, price and size, and therefore membranes have been investigated as a possible alternative (Plakas and Karabelas, 2012). Most micropollutants are around 1 nm in size and this restricts the use of membranes to NF and RO. NF membranes have had the highest focus since they operate at lower pressures and have pore sizes around the size of the micropollutants. Because NF membranes operate in between RO and UF membranes, the rejection is a mixture of steric hindrance and competitive diffusion, which makes the description and prediction of the rejection of a given solute increasingly complicated. The following parameters of the micropollutant and membrane may affect the rejection (Kiso et al., 2011; Košutić et al., 2005; Plakas and Karabelas, 2012; Zhang et al., 2004):

1. Relative size between pores and micropollutant
2. Charge of membrane and micropollutant
3. Relative hydrophobicity/hydrophilicity of micropollutant and membrane
4. Dipole moment of micropollutant

Furthermore the feed water composition and the induced changes of the membrane during filtration may also be important for the resulting rejection. Parameters of this category are (Plakas and Karabelas, 2008, 2009):

5. Ionic content and composition of feed water
6. Fouling of membrane
7. Presence of other micropollutants

In general the steric hindrance induced by the ratio between size of pores and micropollutant is accepted as the most important parameter, but this is case dependent, and the other parameters cannot be ignored. It has been found that for uncharged compounds that do not interact with the surface of the membrane, the steric model presented in Section 6.4.2 by Kiso et al. can give satisfying results (Kiso et al., 1992, 2010, 2011), but whether this applies to micropollutants has not been extensively investigated. Another size parameter that has been used extensively is the molecular weight cut-off (MWCO). MWCO is defined as the molecular weight at which a rejection of 90% of a given compound is obtained, and because of the readily availability of the molecular weight this approach has been popular. Many studies have found correlation between the MWCO of a membrane and its rejection capability of micropollutants (Boussahel et al., 2000, 2002; Zhang et al., 2004), underlining the importance of the size parameter. However, the use of MWCO is not perfect since it does not account for the spatial geometry of the micropollutants, and deviations from the molecular weight/MWCO correlation have been observed (Mohammad and Ali, 2002; Van der Bruggen et al., 1998). For best results a steric model taking length and width of the micropollutants into consideration should be used, but the MWCO can be used as an approximated indicator for the expected applicability of a given membrane.

Since NF membranes are usually characterised by their surface charge, a presence of chargeable groups on micropollutants can be very important for the observed rejection. The majority of NF membranes are negatively charged because of the presence of carboxylic groups in the active surface layer, and pesticides with a negative charge are therefore rejected to a higher degree than would otherwise have been expected if the rejection was purely steric (Berg et al., 1997). This also highlights the importance of the pH of the feed water. At pH values lower than the pKa value of a given micropollutant, the compound will be negatively charged and may therefore experience a changed rejection. The charged membrane surface may also affect uncharged micropollutants by orienting along the axis of their dipole moment. This was shown by Van der Bruggen et al. through a comparison of the rejection of pesticides and saccharides. Here it was found that a high dipole moment leads to a lower rejection (Van der Bruggen et al., 1998).

Since compounds with comparable polarity are better able to mix, the relative hydrophobicity/hydrophilicity between membrane and micropollutant may also be important for the efficiency of the filtration process. In a study of aromatic and non-phenylic pesticides, a clear correlation was identified between the log K_{ow} of the pesticides and the degree of adsorption to the membrane (Kiso et al., 2000, 2001). Adsorption can play an important part in the rejection of a micropollutant. In a diffusive transport mechanism the first step will be adsorption to the membrane surface, and the existence of a diffusive flux is part of the reason why complete rejection is not seen. Furthermore if micropollutants adsorb to the membranes, there will be a build up of micropollutants which may desorb as a result of changing feed water conditions or during a cleaning of the membrane.

Feed water composition may affect the rejection in several ways depending on which of the feed water components that are investigated. In general, rejection of micropollutants is found to be higher in real waters compared with distilled or demineralised water used in laboratory studies, and this is ascribed to adsorption of ions to the surface of the membrane. In studies of the tangential streaming potential of membranes in solutions with varying ionic composition and concentration it has been shown that the zeta potential increases with especially anion concentration. Anions have a lower hydrated radius compared with cations, and adsorption of anions will therefore tend to prevail over cations (Elimelech et al., 1994; Szymczyk et al., 2010). The adsorption of ions to the membrane surface reduces the pore radius either by direct adsorption in the pores or by partly covering the entrance of the pore, and hereby increases the rejection. It has also been suggested that the higher ionic concentration leads to a contraction of the electrical double layer in the pores, which hereby reduces the repulsion between the pore walls and allows the pores to contract (Nghiem and Shäfer, 2005).

Presence of organic compounds in the feed water may also affect the rejection performance, but the underlying mechanisms are complex and may lead to both higher and lower rejections. It has been found that for triazine pesticides, the presence of low molecular weight fractions of humic compounds may lead to higher rejection values through complexation between the humic compounds and the triazines. However, in the same study a lower rejection was also observed for one of the investigated membrane/triazine pair, illustrating the complex effect of the humic acids. The presence of divalent ions such as calcium and magnesium may further affect the rejection because of the ability of these ions to participate in complex formation. In the same study, increased rejection was seen even at low concentrations of calcium (4 mg/l) (Plakas and Karabelas, 2009). Not only humic compounds but also other micropollutants may affect rejection. It has been found that for tight NF membranes multi-solute feed water may decrease the rejection of individual micropollutants, while for NF membranes with larger pores rejection can be seen to increase. The decrease has been explained by competitive adsorption of the micropollutants, which may increase the flux of micropollutants through the membrane because they experience less adsorption through the membrane. For the loose NF membrane the presence of a larger micropollutant may increase the rejection of the smaller micropollutants because the larger compound adsorbs to the pore wall and hereby decreases the effective pore radius (Plakas and Karabelas, 2008). Finally, organics in the feed water may foul the membrane surface and hereby change the surface chemistry of the membrane, which may again both decrease and increase rejection. If the micropollutants are preferential mixed with foulants in the fouling layer the concentration of micropollutants at the surface will increase because of diffusion from the feed water to the fouling layer, and this may lead to an increased flux of micropollutants through the membrane and hereby a decreased rejection. If on the other hand the fouling acts as an additional barrier for the micropollutants, the rejection may increase (Plakas and Karabelas, 2011).

Overall membranes have proven to be very effective at removing micropollutants, but there are several challenges left to investigate. Some of the most important are an improved understanding of the physicochemical interactions between membrane and micropollutant and the role of fouling layers, development of membranes that can achieve complete rejections to ensure micropollutant-free permeates and improved treatment techniques for the concentrate.

For the last point AOPs can possibly come to play an important role and if they can be successfully integrated with the membrane filtration process, both processes may benefit from it.

6.8. MEMBRANES IN AOPS

Two of the disadvantages of the use of AOPs are the high energy consumption that often lies around several kWh/m^3, and the long reaction times. Most often degradation of micropollutants is found to follow pseudo first-order kinetics, as described by Eqn (6.52).

$$-r = kC \tag{6.52}$$

From this it follows that the rate of degradation increases with concentration, and because it is a first-order process, the half-life is a constant given by Eqn (6.53)

$$t_{1/2} = \frac{\ln(2)}{k} \tag{6.53}$$

which may be generalised to Eqn (6.54), where it is seen that for any given percentage degradation, the time is constant. This means that if we have a micropollutant, then it takes the same amount of time to go from an initial concentration of 1 mg/l to 0.1 mg/l as it does going from 10 mg/l to 1 mg/l.

$$t = \frac{\ln\left(\frac{C_0}{C}\right)}{k} \tag{6.54}$$

However, as can be seen from the two sets of numbers a larger mass of micropollutant is removed when we have a higher initial concentration. Since time is linearly correlated to energy consumption in an AOP process, the conclusion is that the energy to mass of degraded pollutant ratio can be improved by increasing the concentration.

$$E_{\text{mass}} = \frac{P \cdot t}{\Delta m} = P \cdot \frac{\ln\left(\frac{C_0}{C}\right)}{k} \cdot \frac{1}{\Delta C \cdot V}, \quad \left[\frac{kJ}{kg}\right] \tag{6.55}$$

where P is the electrical power in watt used by AOP and V the treated volume. For AOPs in which other chemicals like H_2O_2 and/or O_3 are consumed, these can be considered constant and included into the pseudo first-order rate constant, if they are assumed to be added after filtration to obtain the same concentration as without membrane filtration. In Example 4, a concrete example of the effect of concentration on the efficiency of an AOP is given. In practice such a beneficial effect has been demonstrated by Miralles-Cuevas et al. (Miralles-Cuevas et al., 2013) for a photo-Fenton process combined with NF.

EXAMPLE 4: THEORETICAL EFFECT OF CONCENTRATION ON THE REMOVAL OF A MICROPOLLUTANT

Case story: Endocrine disruptors are becoming a major concern in the environment. In Denmark they have been found to cause sexual deformities in fish communities in several local fjords. The endocrine disruptors are discharged to the fjords with waste water, which means that by treating the waste water for these compounds, the environmental condition of the fjords could be improved.

In a case study at the waste water treatment plant in Vejle, Denmark, the effect of introducing immobilised photocatalytic surfaces to an existing UV disinfection unit was evaluated by investigating the degradation of the endocrine disruptor 17α-ethinylestradiol (EE2) (Madsen and Søgaard, 2012). EE2 originates from the use of birth control pills and is one of the most potent endocrine disruptors.

EXAMPLE 4: THEORETICAL EFFECT OF CONCENTRATION ON THE REMOVAL OF A MICROPOLLUTANT *(cont'd)*

The study found that the introduction of photocatalytic surfaces would have a positive effect on the removal of EE2 by increasing the rate of degradation with 66%. However, because of the high flow and the relative low intensity of the UV system only around 3% of the EE2 could be expected to be degraded, and it was clear that a different approach was necessary. One such could be the implementation of a membrane filtration unit. Under assumption of a flow of 1000 m^3/h, 108 UV lamps with a UVC output of 175 W, a rate constant of 1.14 $(kWh\ UVC/m^3)^{-1}$,[11] a starting concentration of 1 μg/l and a 99% degradation, the effect of concentrating the water stream can be calculated and plotted in Figure 6.15.

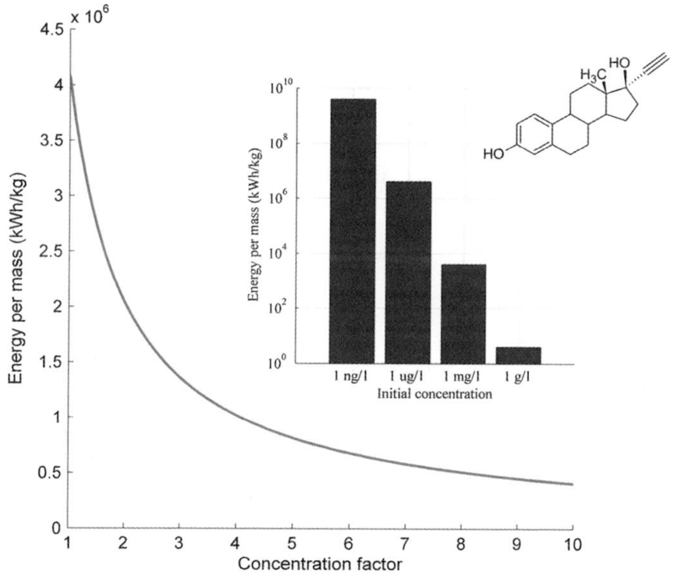

FIGURE 6.15 Energy required for degradation of 1 kg of EE2 with photocatalysis under assumption of 99% degradation at different starting concentrations. (For colour version of this figure, the reader is referred to the online version of this book.)

[11]You may note the rather unusual unit for the rate constant $(kWh\ UVC/m^3)^{-1}$. This is however a convenient way of going from laboratory data to real-life application. Laboratory-sized UV lamps are often less efficient in converting electrical energy to UV light, but by converting from time to energy per volume, results from the laboratory can easily be converted to real life. Furthermore, in practice it will also often be the amount of energy used that is of interest rather than the time.

Besides the effect of increasing the concentration, membranes may benefit the use of AOPs in other ways.

A membrane process can be designed to allow the micropollutants to permeate, while removing particular and colloidal matter that may otherwise reduce the efficiency of the AOP. This may be particularly useful for waters with a high content of NOM and where UV is used in the AOP process. By reducing the turbidity of the water, UV light will penetrate farther into the water matrix, and shadow zones on or behind particles are avoided.

Another benefit of using a membrane filtration process before the AOP is that the problem of formation of degradation intermediates can be controlled. In AOP literature, there is a tendency of focusing on the initial degradation step of a specific compound without considering what it is transformed into. However, the complete oxidation of a compound usually involves several oxidation steps, and it is possible that during some of these steps, the toxicity of the degradation intermediate will be larger than the original compound. In this case, the degradation intermediates must be taken of. Furthermore, it is often not one degradation intermediate that is formed, but several, and this may complicate the process control. If membranes are used to remove the micropollutants, then degradation intermediates formation during the following AOP will only take place in the concentrate. Degradation intermediates are as such completely removed from the treated water.

References

Amar, N.B., Saidani, H., Palmeri, J., Deratani, A., 2009. Effect of temperature on the rejection of neutral and charged solutes by Desal 5 DK nanofiltration membrane. Desalination 246, 294–303.

Berg, P., Hagmeyer, G., Gimbel, R., 1997. Removal of pesticides and other micropollutants by nanofiltration. Desalination 113, 205–208.

Boerlage, S.F.E., Kennedy, M., Tarawneh, Z., De Faber, R., Schippers, J.C., 2004. Development of the MFI-UF in constant flux filtration. Desalination 1, 103–113.

Boerlage, S.F.E., Kennedy, M.D., Dickson, M.R., El-Hodali, D.E.Y., Schippers, J.C., 2002. The modified fouling index using ultrafiltration membranes (MFI-UF): characterisation, filtration mechanisms and proposed reference membrane. J. Memb. Sci. 197, 1–21.

Boussahel, R., Bouland, S., Moussaoui, K.M., Montiel, A., 2000. Removal of pesticide residues in water using the nanofiltration process. Desalination 132, 205–209.

Boussahel, R., Montiel, A., Baudu, M., 2002. Effects of organic and inorganic matter on pesticide rejection by nanofiltration. Desalination 145, 109–114.

Bowen, W.R., Mohammad, A.W., Hilal, N., 1997. Characterisation of nanofiltration membranes for predictive purposes – use of salts, uncharged solutes and atomic force microscopy. J. Memb. Sci. 126, 91–105.

Bowen, W.R., Sharif, A.O., 1994. Transport through microfiltration membranes – particle hydrodynamics and flux reduction. J. Colloid Interface Sci. 168, 414–421.

Chang, S., Waite, T.D., Schäfer, A.I., Fane, A.G., 2003. Adsorption of the endocrine-active compound estrone on microfiltration hollow fiber membranes. Environ. Sci. Technol. 37, 3158–3163.

Choi, J.-S., Hwang, T.-M., Lee, S., Hong, S., 2009. A systematic approach to determine the fouling index for a RO/NF membrane process. Desalination 238, 117–127.

Crittenden, J.C., Trussell, R.R., Hand, D.W., Howe, K.J., Tchobanoglous, G., 2012. Water Treatment – Principles and Design, third ed. John Wiley & Sons, Inc.

Cyna, B., Chagneau, G., Bablon, G., Tanghe, N., 2002. Two years of nanofiltration at the Méry-sur-Oise plant, France. Desalination 147, 6975.

De Pinho, M.N., Semião, V., Geraldes, V., 2002. Integrated modeling of transport processes in fluid/nanofiltration membrane systems. J. Memb. Sci. 206, 189–200.

Deen, W.M., 1987. Hindered transport of large molecules in liquid-filled pores. AIChE J. 33, 1409–1425.

Elimelech, M., Chen, W.H., Waypa, J.J., 1994. Measuring the zeta (electrokinetic) potential of reverse osmosis membranes by a streaming potential analyzer. Desalination 95, 269–286.

Field, R.W., Wu, D., Howell, J.A., Gupta, B.B., 1995. Critical flux concept for microfiltration fouling. J. Memb. Sci. 100, 259–272.

Hermia, J., 1982. Constant pressure blocking filtration laws – application to power-law non-Newtonian fluids. Trans. Inst. Chem. Eng. 60, 183–187.

Huang, H., Young, T.A., Jacangelo, J.G., 2008. Unified membrane fouling index for low pressure membrane filtration of natural waters: principles and methodology. Environ. Sci. Technol. 42, 714–720.

Khirani, S., Aim, R.B., Manero, M.H., 2006. Improving the measurement of the Modified Fouling Index using nanofiltration membranes. Desalination 191, 1–7.

Kiso, Y., Kitao, T., Jinno, K., Miyagi, M., 1992. The effects of molecular width on permeation of organic solute through cellulose acetate reverse osmosis membranes. J. Memb. Sci. 74, 95–103.

Kiso, Y., Muroshige, K., Oguchi, T., Hirose, M., Ohara, T., Shintani, T., 2011. Pore radius estimation based on organic solute molecular shape and effects of pressure on pore radius for a reverse osmosis membrane. J. Memb. Sci. 369, 290–298.

Kiso, Y., Muroshige, K., Oguchi, T., Yamada, T., Hhirose, M., Ohara, T., Shintani, T., 2010. Effect of molecular shape on rejection of uncharged organic compounds by nanofiltration membranes and on calculated pore radii. J. Memb. Sci. 358, 101–113.

Kiso, Y., Nishimura, Y., Kitao, T., Nishimura, K., 2000. Rejection properties of non-phenylic pesticides with nanofiltration membranes. J. Memb. Sci. 171, 229–237.

Kiso, Y., Sugiura, Y., Kitao, T., Nishimura, K., 2001. Effects of hydrophobicity and molecular size on rejection of aromatic pesticides with nanofiltration membranes. J. Memb. Sci. 192, 1–10.

Košutić, K., Furač, L., Sipos, L., Kunst, B., 2005. Removal of arsenic and pesticides from drinking water by nanofiltration membranes. Sep. Purif. Technol. 42, 137–144.

Kucera, J., 2010. Reverse Osmosis – Industrial Applications and Processes. Schriverner.

Madsen, H.T., Søgaard, E.G., 2012. Case study of treatment of waste water for 17α-ethinylestradiol and microorganisms with UV and photocatalysis in an on-going process of introducing AOP techniques in the Danish water sector. Water Pract. Technol. 7.

Miralles-Cuevas, S., Arques, A., Maldonado, M.I., Sánchez-Pérez, J.A., Rodriguez, S.M., 2013. Combined nanofiltration and photo-Fenton treatment of water containing micropollutants. Chem. Eng. J. 224, 89–95.

Mohammad, A.W., Ali, N., 2002. Understanding the steric and charge contributions in NF membranes using increasing MWCO polyamide membranes. Desalination 147, 205–212.

Nakao, S., Kimura, S., 1981. Analysis of solutes rejection in ultrafiltration. J. Chem. Eng. Jpn. 14, 32.

Nghiem, L.D., Shäfer, A.I., 2005. Trace contaminant removal with nanofiltration. In: Shäfer, A.I., Fane, A.G., Waite, T.D. (Eds.), Nanofiltration, Principles and Applications. Elsevier Advanced Technology, Oxford, UK, pp. 479–520.

Otero, J.A., Mazarrasa, O., Villasante, J., Silva, V., Prádanos, P., Calvo, J.I., Hernández, A., 2008. Three independent ways to obtain information on pore size distributions of nanofiltration membranes. J. Memb. Sci. 309, 17–27.

Peinemann, K.-V., Nunes, S.P. (Eds.), 2010. Membranes for Water Treatment, fourth ed. WILEY-VCH.

Plakas, K.V., Karabelas, A.J., 2008. Membrane retention of herbicides from single and multi-solute media: the effect of ionic environment. J. Memb. Sci. 320, 325–334.

Plakas, K.V., Karabelas, A.J., 2009. Triazine retention by nanofiltration in the presence of organic matter: the role of humic substance characteristics. J. Memb. Sci. 336, 86–100.

Plakas, K.V., Karabelas, A.J., 2011. A systematic study on triazine retention by fouled with humic substances NF/ULPRO membranes. Sep. Purif. Technol. 80, 246–261.

Plakas, K.V., Karabelas, A.J., 2012. Removal of pesticides from water by NF and RO membranes – a review. Desalination 287, 255–265.

Roorda, J.H., van der Graaf, J.H.J.M., 2001. New parameter for monitoring fouling during ultrafiltration of WWTP effluent. Water Sci. Technol. 43, 241–248.

Schippers, J.C., Verdouw, J., 1980. The modified fouling index, a method of determining the fouling characteristics of water. Desalination 32, 137–148.

Szymczyk, A., Fievet, P., Bandini, S., 2010. On the amphoteric behavior of Desal DK nanofiltration membranes at low salt concentrations. J. Memb. Sci. 355, 60–68.

Van der Bruggen, B., Schaep, J., Maes, W., Wilms, D., Vandercasteele, C., 1998. Nano-filtration as a treatment method for the removal of pesticides from ground water. Desalination 117, 139–147.

Ventresque, C., Gisclon, V., Bablon, G., Chagneau, G., 2000. An outstanding feat of modern technology: the Mery-sur-Oise Nanofiltration Treatment Plant (340,000 m^3/day). Desalination 131, 1–16.

Wijmans, J.G., Baker, R.W., 1995. The solution-diffusion model: a review. J. Memb. Sci. 107, 1–21.

Zhang, Y., Van der Bruggen, B., Chen, G., Braeken, L., Vandecasteele, C., 2004. Removal of pesticides by nanofiltration: effect of the water matrix. Sep. Purif. Technol. 38, 163–172.

CHAPTER

7

Advanced Iron Oxidation at Drinking Water Treatment Plants

*Professor MSO Erik G. Søgaard**, *Project leader Daniel Anobaah Ankrah*†

* Section of Chemical Engineering, Department of Biotechnology, Chemistry and Environmental Engineering, Aalborg University Esbjerg, Niels Bohrs Vej 8, DK 6700 Esbjerg, Denmark

† Dansk GeoservEx A/S, Aarhus, Denmark

Ground water is considered to be the cleanest source of drinking water. This is because of the properties of the soil to be able to purify the water as it moves in the saturated zone of the soil. Actually, the soil acts as a huge filter similar to the sand filters that are used by waterworks to remove precipitated colloidal particles from the raw water on its way to drinking water (Figure 7.1).

Implementation of the drinking water directives states a series of maximum admissible concentration values for many chemical elements occurring in drinking water. However, as ground water is used for drinking purposes with only simple treatment, many of these maximum admissible concentration values are highly relevant for ground water protection.

One of the Danish government's policies to protect ground water in Denmark was to classify ground water abstraction areas, to protect valuable areas for drinking water interest.

Similarly, they mapped out areas are used to establish site-specific ground water protection zones and associated regulations of land use to prevent ground water pollution.

7.1. DRINKING WATER TREATMENT IN DENMARK

In Denmark, ground water is considered as the favoured source of drinking water and it accounts for about 100% of the drinking water

Chemistry of Advanced Environmental Purification Processes of Water
http://dx.doi.org/10.1016/B978-0-444-53178-0.00007-9

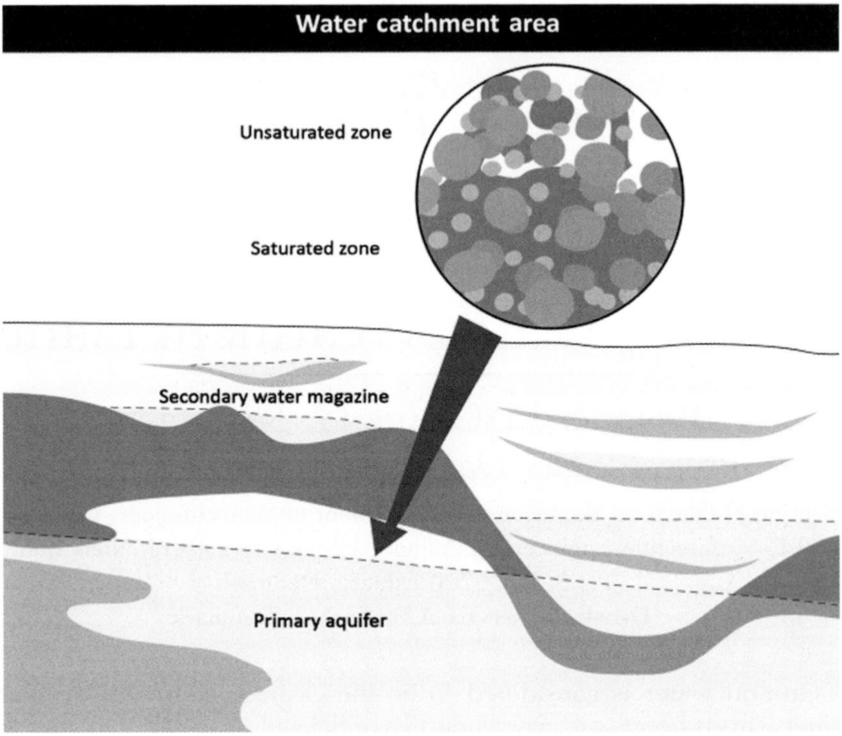

FIGURE 7.1 **Sketch of ground water zone with primary and secondary aquifers.** The water table between saturated and unsaturated primary aquifer is highlighted. *Reproduced with permission from Forsyningen Esbjerg.* (For colour version of this figure, the reader is referred to the online version of this book.)

supply sources. Therefore, most research and development and technological improvements on water treatment are geared towards ground water purification.

Historically, civilisation started and concentrated along areas where there was ample source of water. At that time, people knew relatively nothing about the connection between disease and drinking water. If there was water treatment at all, it was performed only for melioration of the cosmetic value and taste. In recent years, however, increasing population and improvement of scientific knowledge have resulted in many questions being raised about potability or safety of drinking water. The drinking water standards and regulations in Denmark were developed out of a growing concern of the need to protect people from illnesses caused by contaminated drinking water. The standards have therefore developed and expanded over a period of time as knowledge of the health effect of contaminants increases and the treatment technology to remove contaminants improved (Pontius, 1990) (Figure 7.2).

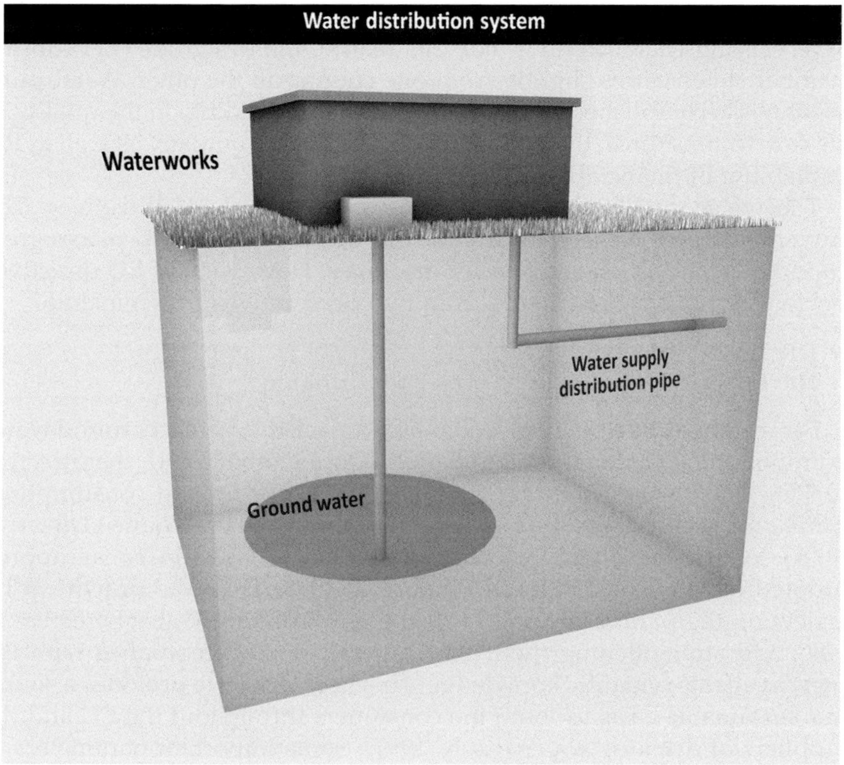

FIGURE 7.2 **Sketch of a smaller waterwork in Denmark with water boring, housing for simple drinking water treatment and pipe for water supply system.** *Reproduced with permission from Forsyningen Esbjerg.* (For colour version of this figure, the reader is referred to the online version of this book.)

An overview of legislation protecting drinking water standards in the European Union (EU) and Denmark as well as the design of the water treatment plants (WTPs) basically used in Denmark to attain the required standards is given below.

7.2. DRINKING WATER STANDARDS IN THE EU AND DENMARK

The drinking water standards in Denmark were either adopted directly from the EU directives on drinking water or developed out of the framework of the EU directives and improved to either prevent or remove contaminants from raw water for human consumption. The drinking water quality standards may be the same from one country to the other among the EU states but the raw water taken from the water sources may have varying characteristics. For this reason, the design of WTPs in the member states varies slightly from one country to the other. Water purification is accomplished by several methods, depending on the quality of the raw water source, the philosophy of the water company as well as the availability of financial resources.

The water quality in EU is governed by several EU directives. The implementation of the water framework directive by the EU places great importance on the usage of water resources. However, the EU directives tackle water quality standards from two perspectives which include

1. The raw water quality and
2. The quality of treated fresh water for drinking.

For instance, the directive 98/83/EC is enacted to protect ground water from pollution by dangerous materials. On the other hand, the directive 80/778/EEC focuses on treated water quality for human consumption. Similar to the directive 80/778/EEC, the EU enacted the Council Directive 98/83/EC on the quality of water intended for human consumption, adopted by the Council on November 3, 1998. This was drawn up by reviewing the parametric values of the old Drinking Water Directive of 1980, and strengthening them where necessary in accordance with the latest available scientific knowledge. This new Directive provides a sound and sustainable basis for both the consumers throughout the EU and the suppliers of drinking water (Kiely, 1997). Some important parameters in the EU drinking water directive and their limit values are as shown in Table 7.1.

It is worth noticing that the advent of the Danish legislation places much importance on drastic reduction of contaminants to levels stipulated by the legislation. In view of this, water treatment companies are increasingly looking for ways to improve the process efficiency for the removal of unwanted chemical constituents, including compounds of Fe and Mn.

The limit value for Fe set by the World Health Organization (WHO) is 0.3 mg/l (WHO, 1996). Unlike the WHO limit values, it could be seen from Table 7.1 that EU has set a parametric value of 0.2 mg/l for Fe. However, in Denmark, several water treatment companies are targeting Fe concentration in drinking water to about 0.1 mg/l. Fulfilling these rigorous standards of Fe in water supply systems demands a more efficient technology which will reduce operational cost and make judicious use of the raw water available (Kiely, 1997; Winther et al., 2003; Drinking water, 2010).

Fe and Mn are usually concentrated in the percolating water, which is in contact with soil particles, minerals and occasionally man-made materials like steel and polymer pipes used for the transportation of water to the waterworks for further purification. This makes the

TABLE 7.1 EU Drinking Water Directive Parameters (Kiely, 1997)

Group Description	Parameter	Maximum Admissible Concentration
Organoleptic parameters	Colour	20 mg/l Pt/Co scale
	Turbidity	10 mg/l SiO_2
	Odour	Dilution of 2 at 12 °C
	Taste	Dilution of 2 at 12 °C
Physicochemical parameters	Temperature	25 °C
	pH	$6.5 < pH < 8.5$
	Conductivity	400 μS/cm
	Chlorides	250 mg/l Cl
	Sulphates	250 mg/l SO_4
	Calcium	100 mg/l Ca
	Magnesium	50 mg/l Mg
	Sodium	150 mg/l Na
	Potassium	12 mg/l K
	Aluminium	0.2 mg/l Al
	Total dry residues	1500 mg/l
Parameters concerning substances undesirable in excessive amounts	Nitrates	50 mg/l NO_3
	Nitrites	0.1 mg/l NO_2
	Ammonium	0.5 mg/l NH_4
	Kjeldahl N	1 mg/l N
	Oxidisability	5 mg/l O_2
	Hydrogen sulphide	Undetectable μg/l
	Substances extractable in Chloroform	No increase above Background
	Hydrocarbons	10 μg/l
	Phenols	0.5 μg/l C_6H_5OH
	Boron	1000 μg/l B
	Surfactants	200 μg/l (lauryl sulphate)
	Organochlorines	1 μg/l
	Iron	200 μg/l Fe
	Manganese	50 μg/l Mn
	Copper	100 μg/l Cu
	Zinc	100 μg/l Zn
	Phosphorus	5000 μg/l P_2O_5/l
	Fluoride	1000 μg/l F
	Suspended solids	0
	Barium	100 μg/l Ba
Parameter concerning toxic substances	Arsenic	10 μg/l As
	Cadmium	5 μg/l Cd
	Cyanides	50 μg/l Cn
	Chromium	50 μg/l Cr
	Mercury	1 μg/l Hg
	Nickel	50 μg/l Ni
	Lead	50 μg/l Pb
	Antimony	10 mg/l Sb
	Selenium	10 mg/l Se
	Pesticides	0.1 μg/l
	PAHs	0.2 μg/l

(*Continued*)

Table 7.1 EU Drinking Water Directive Parameters (Kiely, 1997)—cont'd

Group Description	Parameter	Maximum Admissible Concentration
Microbiological parameters	Total coliforms	PMN $\leq$ 1/100 ml
	Faecal coliforms	PMN $\leq$ 1/100 ml
	Faecal streptococci	PMN $\leq$ 1/100 ml
	Sulphite-reducing clostridia	PMN $\leq$ 1/20 ml
Minimum required for softened water	Total hardness	>60 mg/l $CaCO_3$
	Alkalinity	>30 mg/l HCO_3
	NVOC	4 mg/l

quality of water differ from one location to the other. There are localised areas in Denmark and many other countries where huge amounts of Fe have degraded the water quality seriously enough to prevent usage of the water for human consumption (Ankrah and Søgaard, 2009b).

Today, Fe removal processes are known worldwide. However, there are still wide gaps in our knowledge of the ostensibly simple processes of aeration, precipitation and filtration used to remove Fe from ground water. A special advanced biological method for Fe removal of is now used in many waterworks worldwide either because the right conditions for it have been arranged or because the present conditions after some time turned into the right ones for biological Fe removal.

Fe in ground water should not present health hazards to consumers supplied with fresh water from ground water for drinking purposes. However, for a longer period it can become poisonous particularly to men. The presence of Fe in ground water is objectionable to the consumers as those contaminants could precipitate in fresh water distribution systems. When precipitated, Fe may cause problems of unpleasant odour, taste in drinking water and beverages and yellow to brown stains in plumbing and laundered clothing. Additionally, Fe in distribution systems may promote growth of microorganisms, thereby forming slimy tissues several millimetres in thickness, which could accumulate and eventually block the water distribution systems. Microorganisms could also reduce oxygen levels in the water, and where chemicals are used for disinfection they can reduce the effectiveness of those chemicals.

7.2.1. Occurrence of Fe

Fe is the sixth most abundant element in the universe. It constitutes about 34.8% of the Earth, 5% of the crust and 34 ppb in the ocean. Fe is

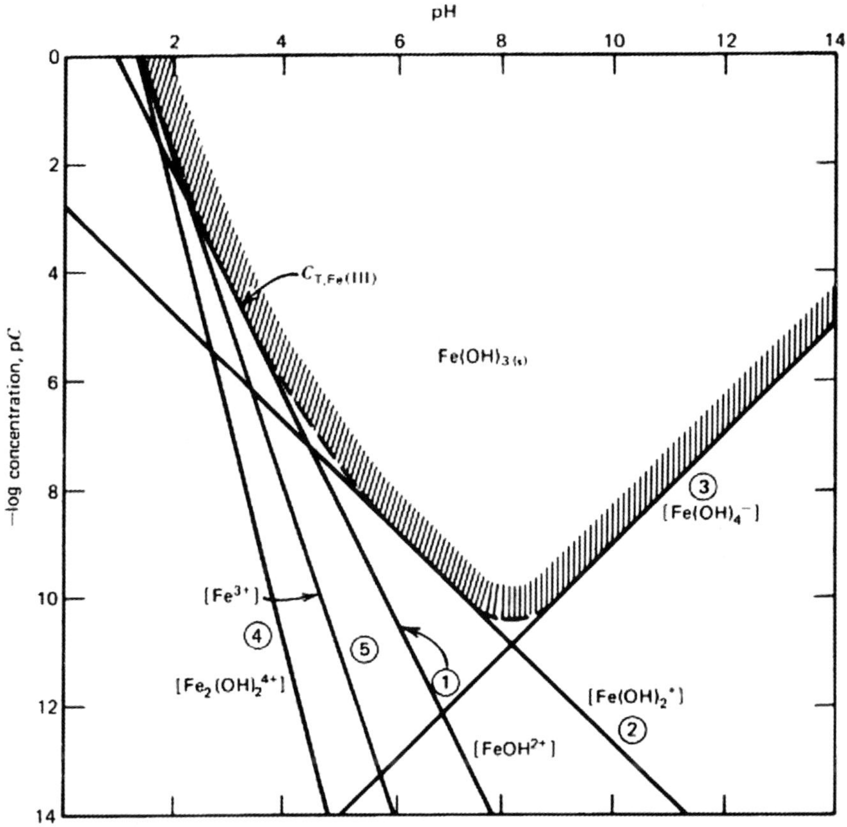

FIGURE 7.3 **Occurrence of Fe compounds in water as function of pH.** The dimer (4) and Fe^{3+} (5) are only present at low pH, hydrolysed Fe(III) compounds (1) and (2) at higher pH and tetrahydroxyferrate (3) at much higher pH. The precipitate (ferrihydrate) has its maximum between pH 7 and 8. *From Snoyink and Jenkins (1980). This material is reproduced with permission from John Wiley & Sons, Inc.*

mostly found in igneous rocks and in partially water-soluble siderite (Fe(II)carbonate). Fe is omnipresent in soils, sediments and ground water, and have profound influence on the chemistry of water in aquifers and subsurface (O'Neill, 1995) (Figure 7.3).

As mentioned, WTPs are designed to produce water of the desired quality by employing appropriate technology, which is cost-effective and also based on the raw water quality. Almost 100% of the raw water source for drinking water supply is abstracted from ground water in Denmark. Therefore, with the enactment of legislation to map out and protect the ground water sources, makes the treatment process quite simpler than in most other EU countries.

7.3. A TYPICAL DRINKING WTP FOR SIMPLE WATER TREATMENT IN DENMARK (VR. GJESING WTP)

Vester Gjesing WTP treats water extracted from six wells placed alongside the eastern bank of Fovrfeld Bæk in the city of Esbjerg, Denmark. Furthermore, there are other drillings in the water field, and water from these drillings is pumped for treatment at a neighbour waterwork which is connected to Vester Gjesing treatment plant by the water distribution system. Approximately, 1.2 million m^3 of water is abstracted from the catchment area per year, of which 30% is treated on the neighbour waterwork. The drillings are conducted through a clay layer and all drillings are equipped with submerged pumps, with frequency converters given it an active field. This means that extraction is more sustainable because the water field is put under so less strain as possible and large depression cone around the single drilling that otherwise could bring water from unwanted zones into the well is avoided. All drillings and their pumping systems are equipped with above-ground lockable raw water site containers. About 10-m protection zones around the wells are established.

7.3.1. Geology

Water has been drawn from this catchment area for about 70 years and it is located on Esbjerg Hill Island. The drillings are screened with filters from 48 to 138 m below ground (bgs). One of the southern wells has only filters at more than 100 m in depth. The north and southern wells have dense clay layers between 10 and 30 bgs, including fluctuating clay and a silt layer down to approximately 70 bgs. Also, the north and southern wells have records of mica down more than 130 bgs. There are no geological models present which can describe the vulnerability and potential areas with exceptional ground water production within the specific extraction area.

7.3.2. Water Quality

The ground water is slightly alkaline with pH around 8. There is a fairly high iron content, which varies between 4 and 12 mg/l. The ground water is free of nitrate with low sulphate content and there have never been traces of pollution with pesticides or any other environmentally pollutants found in the water field. (See also Table 7.2.)

TABLE 7.2 Raw Water Quality and Drinking Water Quality of the Two WTPs

	Astrup		Vr. Gjesing	
Contents	**Raw Water**	**Treated Water**	**Raw Water**	**Treated Water**
pH	6.8	7.9	7.7	7.8
Conductivity mS/cm	28	32	37	37
H grades	4.8	6.5	7.9	7.8
Ca (mg/l)	26	39	50	49
Mg (mg/l)	3.6	4.4	3.9	4.1
K (mg/l)	1.6	1.7	1.8	1.8
Na (mg/l)	19	20	21	21
NH_4 (mg/l)	0.032	<0.006	0.04	0.017
Mn (mg/l)	0.47	<0.005	0.41	<0.005
Fe (mg/l)	13	0.031	5.1	0.013
Cl (mg/l)	33	32	33	34
F (mg/l)	0.22	0.16	0.15	0.16
SO_4 (mg/l)	43	44	35	35
Aggr. CO_2 (mg/l)	16	9.2	4	4
HCO_3 (mg/l)	50.6	76.5	123	121
Total P (mg/l)	0.27	0.008	0.32	0.012
Oxygen (mg/l)	0.4	10.6	0.2	11.5
CH_4 (mg/l)	<0.005	<0.005	0.008	<0.005
As (μg/l)	0.60	<0.030	<0.030	<0.030
Ba (μg/l)	45	31	37	25
B (μg/l)	17	17	18	22
Pb (μg/l)	<0.025	0.24	<0.025	0.031
Cu (μg/l)	<0.040	0.18	<0.040	0.13
Zn (μg/l)	9.6	4.7	13	8.2
Ni (μg/l)	0.30	8.0	0.10	0.16
NVOC (mg/l)	0.51	0.42	0.88	0.64
Temperature (°C)	8.4	8.3	8.2	8.5
Turbidity (FTU)	46	0.53	42	0.13
Inorg. ash (mg/l)	200	210	230	240
Colour	Green brown	Colourless	Yellow	Colourless
Smell	Unpleasant	Normal	Unpleasant	Normal

7.3.3. Treatment Processes

Figure 7.4 shows the water treatment process train at Vr. Gjesing WTP.

The screened raw water is pumped from the wells and fully physically aerated to oxidise elements such as iron and manganese and to strip off volatile compounds like small amounts of methane and hydrogen sulphide. Oxidation processes play a pivotal role in converting Fe from their soluble forms into insoluble ones. It is therefore only when the elements, initially present in their reduced forms, are successfully oxidised that they can be precipitated to enable them to be removed by sand filters. Then it is channelled into the first filtration unit, where iron and eventual organic substances are filtered out. The pH of the water is about 7.5, which is appropriate for the precipitation of Fe, Mn and humic substances. Some other natural impurities present in the water can be co-precipitated with the precipitated iron oxyhydroxides as well. At the second filtration unit, the water, which now have contents of mainly Mn and some residual Fe, is filtered. The water is then considered to be clean; so it can be released into a fresh water tank and brought into the water distribution system for the consumers. The backwash water produced as a result of washing the filter materials (beds to prevent them from being choked up) is channelled into a wastewater tank where after precipitation of particles the clear surface water is redirected into the main treatment line after treating it with UV to avoid development of

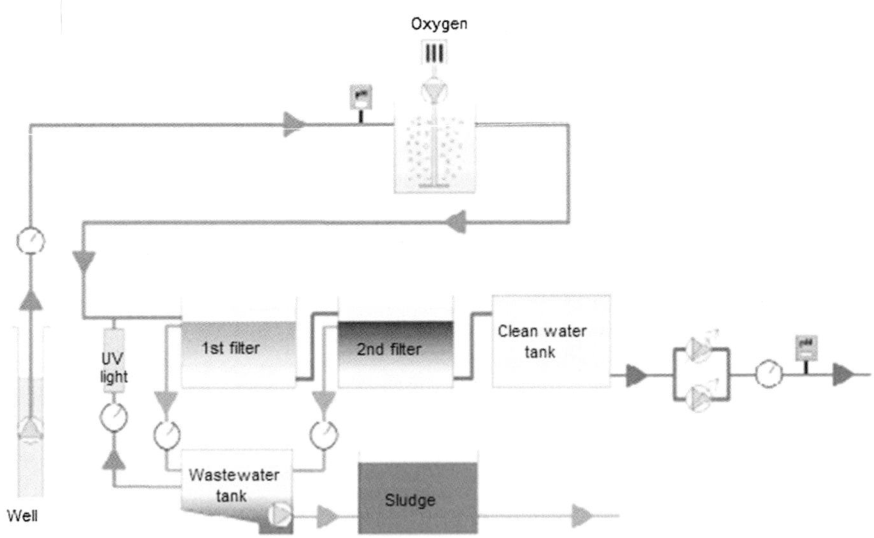

FIGURE 7.4 **Water treatment process at Vr. Gjesing WTP.** *Reprinted with permission from Esbjerg Water Supply; reproduced with permission from Forsyningen Esbjerg.* (For colour version of this figure, the reader is referred to the online version of this book.)

pathogens in the water. The sludge is disposed of initially in a local sink but later it will be stored at a landfill.

Ground water treatment in Denmark, to a larger extent, does not differ from catchment area to catchment area. However, water treatment engineers design the water treatment process based on the ground water quality. As can be seen from the description of the drinking WTP of Astrup below a very advanced purification methodology was necessary to apply to remove huge amounts of Fe from the raw water with the purpose to use it for drinking water when a scarcity of pure ground water near Esbjerg arose in 1993.

7.4. ASTRUP DRINKING WTP USING ADVANCED BIOLOGICAL FILTRATION

7.4.1. Catchment's Area

Astrup drinking WTP treats water extracted from seven drilled wells in the small city of Forum north of Esbjerg. All seven drillings are equipped with submersible pumps and above-ground lockable covers. The drillings are screened in intervals 75—125 bgs, and all drillings are sealed with clay pluggings by penetration of the clay layer. A strict law has been made to protect all the drillings (about 200 m in radius) from effluents of wastewater or any polluted liquid that can drain into the ground water table. In addition to this, the drillings are protected against pollutants such as fertiliser, by help of plants planted around them in a density of about $30 \times 30\ m^2$. The pumps are equipped with frequency converters and there is active field site control in the field, which implies that the extraction like in Vr. Gjesing is as low as possible and large depression cone around the single drillings therefore is avoided.

7.4.2. Geology at Astrup WTP

The water catchment area is also placed on Esbjerg Hill Island and is stretched according to a performed hydraulic modelling southeast. The quaternary sediment is at least 150 m thick in the catchment area and consists of an upper sediment of sand between 20 to approximately 50 bgs, varying with low permeable layers consisting of mainly silt. Beneath the first level to an approximately depth of about 120 m is again sand, in which all wells are filter screened. Most of the surrounding area is used for conventional farming.

7.4.3. Water Quality

The raw water at Astrup WTP is slightly acidic with pH value of about 6.5. It has a very high Fe concentration between 5.5 and 19 mg/l and is

TABLE 7.3 Velocity Constant for the Precipitation of Fe(II) to Fe(III) by Purely Physicochemical Fe Removal (from Søgaard et al., 2001)

Examples of k_o Values ($\times 10^{13}$ M^{-2} atm^{-1} min^{-1}) from the Expression of Purely Physicochemical Fe^{2+} Oxidation/Precipitation Mentioned in the Text

k_O *VALUE*	*T/°C*	*pH*	*FE[(II)]$_O$/μM*	*MEDIUM*	*REFERENCE*
0.5–2.4	10.5–14.5	7.5–7.8	30–90	Ground water	Ghosh et al., (1966)
2.9	25	6.8	30–90	Well water	Ghosh (1974)
60	?	6.25	?	Tap water	Stumm and Lee (1961)
1.9 ± 0.7	10	7.0	6	Oxic lake	Davison and seed (1983)
2.2 ± 0.8	10	6.8–7.2	7–400	Anoxic lake	Davison and seed (1983)
8 ± 2.7	20.5	6.4–7.5	<50	Na_2CO_3/CO_2	Stumm and Lee (1961)
16–21	25	6.3	<100	Na_2CO_3/CO_2	Theis and Singer (1974)

free of nitrate. The ground water is also free of pesticides and other anthropogenic applied substances. (See also Table 7.3)

7.4.4. Treatment Processes

Figure 7.4 shows the treatment process train at Astrup WTP. Screened raw water is pumped from the wells directly into the primary sand filter, where Fe and some organic substances are co-precipitated and filtered off by the sand filter system. The oxygen content in that filter is only slightly increased from almost 0 to about 4 mg/l by help of a small fall (30 cm) of the water before it hits a plate and the water table of the filter (Figure 7.5).

No pumping of air into the water is taking place at this part of the treatment. After this filtration step the pH of the filtrate is regulated above the initial 6.5 to about 7.3 by the addition of $Ca(OH)_2$. After this, the water is now vigorously aerated and channelled to a second filtration tank where mostly Mn, residual Fe and organic substances as well as other impurities are filtered off from the water (Hem and Lind, 1983). The water is then released into the fresh water tank. The unique characteristic at Astrup WTP is that the iron is removed by a biological method in the first filtration step involving iron-oxidising bacteria. The reason to let the raw water in the first filtration step be exposed slightly with atmospheric air and create a small gradient of oxygen to only about 4 mg/l is that in this way it is possible to promote the growth of the Fe bacteria in the filters (Figure 7.6).

The filter material are backwashed to clean it and prevent it from being choked up with Fe/clayish precipitates and pieces of organic materials.

FIGURE 7.5 **Raw water is slightly aerated to about 4 mg/l of oxygen by help of a 30 cm fall at top of the primary filter by hitting a plate placed above the filter sand.** *Reproduced with permission from Forsyningen Esbjerg.* (For colour version of this figure, the reader is referred to the online version of this book.)

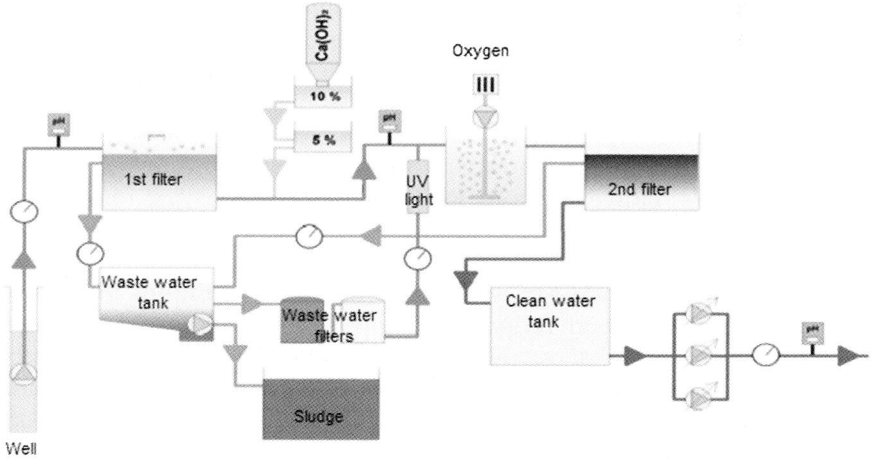

FIGURE 7.6 **Water treatment process at Astrup WTP.** *Reproduced with permission from Forsyningen Esbjerg.* (For colour version of this figure, the reader is referred to the online version of this book.)

The backwash water with precipitated sludge is channelled into the wastewater sink. In the sink, the wastewater is permitted to settle and the upper part of the water is filtered in two closed filters, which differ in filter materials from the original primary filter to retain small colloidal particles that still could be present in the backwash water. The water is treated with UV light to control the development of pathogens and channelled back into the main treatment stream exactly before the fully aeration step with pressurised air and the second filtration step. If some particles should still be left in the water they are supposed to precipitate in the secondary filter together with Mn and residual Fe from the first filtration step. (See also Yang et al., 2004; Zhang et al., 2005.)

7.4.4.1. Comparison of the Two Selected Drinking WTPs

Table 7.2 compares the water quality parameters for the two WTPs, Astrup and Vr. Gjesing. The two raw waters have a very similar contents of Mg, K, Na, Cl, F and sulphate and these compounds are largely unaffected by their transportation in the WTP and have almost remained their original concentrations. Only very small amounts of the heavy metals Pb, Cu and Ni are present but their contents are increasing from raw water to drinking water probably because of their escape from metallic transportation pipes of the water conduction systems. Small amounts of NH_3, P, As and CH_4 present in the raw water are almost eliminated by the water treatment system. P and As removal is connected to the removal of Fe (Galal-Gorchev and Stumm, 1963). Contents of Ba and Zn are decreased but not eliminated; however, their contents are small. Also small contents of B are present and seems to be unaffected by the treatment methodology. The Ca content and Hardness (H) grades of Vr. Gjesing are constant, contrary to the values from Astrup where an addition of $Ca(OH)_2$ increases pH, Ca content and H grades. Because of the smaller amount of Ca in the raw water of Astrup the amount of aggressive carbon dioxide (see later) is a little higher than that at Vr. Gjesing.

Different levels of concentrations exist in the raw water at the two drinking WTPs. However, both WTPs are able to treat or remove the contents to the allowable limits as indicated in Table 7.2 compared with Table 7.1. The efficiency of the two WTPs can be seen especially from their ability to remove Fe and Mn, which had high concentrations in the raw water. But, they were both reduced to below 0.1 mg/l for Fe and 0.01 mg/l for Mn, which are the allowable limit values in Denmark. It can also be seen that the treatment increases the levels of heavy metals as Pb, Cu and Ni probably from transportation pipes inside the WTP whereas the concentration levels Ba and Zn decrease. Turbidity, colour and smells all decrease. The levels of macro ions Na, K, Mg and Ca seem not to be influenced by the simple treatment.

7.4.4.2. Fe Removal at Vr. Gjesing Drinking WTP

Vr. Gjesing WTP operates by the normal aeration–filtration method of removing Fe from the raw water which is a purely physicochemical method. Its raw water is first aerated to oxidise and precipitate Fe and subsequently let it become filtered by the sand filters. Actually, no precipitation step by help of a precipitation tank is present at this drinking WTP and the oxidation precipitation process is going on in the sand filter catalysed by the sand grains that has become coated with the precipitated iron hydroxide in form of amorphous ferrihydrate. This iron oxide is a nanosized crystalline form with lots of defects because of foreign compounds present and its very fast precipitation reaction (Cornell and Schwertman, 2003). At Vr. Gjesing the simple purely physicochemical method is enough to purify the ground water, probably because of the optimal pH (7.5–8.0), which favours the purely physicochemical method of Fe removal.

The chemical process that is taking place can be presented by the following reaction scheme:

$$4Fe^{2+}{}_{(aq)} + O_{2(aq)} + 10H_2O_{(l)} \rightarrow 4Fe(OH)_{3(s)} + 8H^+{}_{(aq)} \tag{7.1}$$

As can be seen from the reaction above oxygen picks up an electron from $Fe^{2+}{}_{(aq)}$ which becomes oxidised and precipitated as $Fe(OH)_{3(s)}$. The reaction is expected to take place at the interface between the ferrihydrate-coated sand grains and the water. The hydrated metal ion $Fe(H_2O)_6^{3+}{}_{(aq)}$ can act as an acid with pK_a 2.2 and by giving of $H^+{}_{(aq)}$ from the inner sphere of Fe^{3+} that is strongly complexing the six water molecules. By this it gets more neutralised and becomes an integrated part of the precipitated ferrihydrate. The process can be considered as a chemical catalytic Fe removal reaction in which the coated sand grains act as the catalyst. Without the sand grains in an open precipitation tank the reaction will need production of over critical sized stable colloidal ferrihydrate nuclei to let the oxidation precipitation reaction takes place and remove Fe. Many drinking WTPs were originally built with precipitation tanks for that purpose followed by a sand filter to remove the colloids. Nowadays, these tanks are often skipped if the water does not need any pretreatment before filtering.

For Fe in ground water, the kinetic behaviour of the oxidation/precipitation process may be affected by conditions in addition to pH also the presence of carbon dioxide, silicates and organic material. These conditions can influence the speed of a chemical reaction and also affect the reaction's mechanism and transition states. Mathematical models that can describe the characteristics of the Fe oxidation in the ground water have been constructed since the early 1960s (Stumm and Lee, 1961).

$Fe(OH)_3$ in Eqn (7.1) also can be considered as an oxidised hydrolysis product of $Fe^{2+}{}_{(aq)}$. In 1964 the following expression for the oxidation/precipitation process was proposed (Morgan and Stumm, 1964):

$$-\frac{d[Fe^{2+}]}{dt} = k[Fe^{2+}][OH^-]^2 pO_2 \tag{7.2}$$

Integrating the equation keeping the partial pressure of dioxygen and pH constant results in

$$-\ln\frac{[Fe^{2+}]_t}{[Fe^{2+}]_0} = k[OH^-]^2 pO_2 t \tag{7.3}$$

The rate constant was found to be $k = 2.6 \times 10^{13}$/min/atm/mole^2 at 20.5 °C

As discussed earlier the above shown kinetics of the homogenous oxidation of Fe^{2+} is derived from the fundamental oxidation process that takes place in the often obsolete precipitation tanks at drinking WTPs. As mentioned Fe oxygenation is greatly accelerated in the presence of surfaces and this is an important activity of the coatings of the sand grains in the sand filter oxidation/precipitation process. To further elaborate a simultaneous oxidation of iron occurs where the poorly ordered ferrihydrate acts as highly reactive sorbent of dissolved Fe^{2+} and other minerals in the ground water. The Fe(III) solids thereby is considered to form a strong inner sphere complex with Fe^{2+}. The present Fe(III) therefore catalyses Fe^{2+} oxidation (e.g. Barry et al., 1994). At constant pH and O_2 concentration, the rate equation can be written as

$$r_1 = -\frac{d[Fe(II)]}{dt} = (k + k'\,[Fe(III)])\,[Fe^{2+}] \tag{7.4}$$

where

k = rate constant for the homogeneous reaction = $k_o pO_2[OH^-]^2$

k' = rate constant for the heterogeneous reaction = $k_{so} K pO_2[OH^-]^2$

k_o and k_{so} are the real rate constants for the oxidation reactions and K is the adsorption constant for the adsorption of Fe^{2+} on solid Fe(III) (ferrihydrite). [Fe(III)] is normally given in m^2/l.

The effect of Fe(III) autocatalysis is becoming more pronounced when Fe(III) precipitates are exceeding 5–10 mg/l (Tamura et al., 1976).

We can consider the two processes going on in (1) a precipitation tank and (2) a sand filter as related processes where in the first case critical nuclei of ferrihydrite need to be present to accelerate the process whereas in the second case they are already present in form of the coated sand grains. In the last case the coating mainly contains ferrihydrite but some part of it can have aged into more stable iron oxides as for example goethite and lepidocrosite.

Table 7.3 below shows results of some velocity constants k_o from investigations of purely physicochemical precipitation of Fe(II) at different conditions in open containers.

7.4.4.3. *Fe Removal at Astrup Drinking WTP*

The problem with huge amounts of Fe(II) in ground water that shall be converted to drinking water is the oxidation and precipitation of Fe(II) in the sand filters resulting in large amounts of precipitated ferrihydrite. Normally, backwash should be performed a couple of times every day to avoid clogging of the filters. This would mean that much less drinking water could be produced compared with the normal capacity of the WTP. When the backwash process is going on the production of drinking water is suspended. To increase the time water can be produced an advanced Fe(II) precipitation technique was planned to be a part of the new built WTP in Astrup in 1995.

Besides purely physicochemical and chemical catalytic processes of iron precipitation, bacterial mediation in Fe removal is also known actively to increase the velocity of the oxidation and precipitation of Fe(II) in the sand filters (Søgaard et al., 2000). Therefore, biological involvement in the general Fe^{2+} oxidation could become a possibility by the original planning of the conditions in the filters at Astrup WTP. Compared with the normal purely physicochemical iron oxidation/precipitation methodology the oxidation kinetics is further enhanced when Fe(III) oxides in association with a bacteria intermix to form the precipitated bacteriogenic Fe oxides. These consist partly of bacteria exopolymers (see further details below) and partly of hydrous ferric hydroxides as for example ferrihydrite.

In biological oxidation/precipitation of Fe(II) like this, the oxidation is often attributed to one or more particular iron-precipitating bacteria species. The kinetic model of the influence of bacteria can as a function of the bacterial exopolymer density and total dissolved Fe^{2+} concentration be written as

$$r = -k\,[\text{bacterial exopolymer sites}][\text{Fe(II)}]\left[\text{OH}^-\right]^2 \text{pO}_2 \qquad (7.5)$$

where [bacterial exopolymer sites] means the amount of extra surfaces in the sand filter that can accelerate the adsorption, precipitation and oxidation of Fe(II) compared with the chemical catalytic expression in Eqn (7.4).

Therefore, a summary of the general rate expression also taking into account the difference in reaction kinetics that was found to be pH dependent and involving homogenous, heterogeneous and bacterial exopolymers sites can be written as

$$
\begin{aligned}
r_{\text{tot}} &= -\frac{d\left[Fe^{2+}\right]}{dt} = k_0\left[Fe^{2+}\right]\left[OH^-\right]^2 pO_2 \\
&+ k_{\text{so}}[Fe(III)]\left[Fe^{2+}\right]\left[OH^-\right]^2 pO_2 \\
&+ k_b[\text{bacterial exopolymer sites}]\left[Fe^{2+}\right]\left[OH^-\right]^2 pO_2 \\
&+ r_{\text{other}}
\end{aligned}
\tag{7.6}
$$

where

k_o = the homogeneous rate constants in water $1/(M^2$ s atm). The expression is dependent of pH but at pH between 5 and 8 the present expression fits to experimental investigations;
k_{so} = overall rate constant for heterogeneous, abiotic processes $1/(M^2$ s atm $m^2)$;
[Fe(III)] = heterogeneous, abiotic surface area (m^2/l). It can be calculated from BET surface measurements and amounts of solid ferrihydrate or coated sand grains present in the active areas of the sand filter;
k_b = rate constant for heterogeneous catalytic Fe(II) oxidation involving iron-precipitating bacteria. The term will be repeated for various strains and corrected for specific conditions (units are $m^2/(l\ M^2$ s atm));
r_{other} = the oxidation rate attributable to processes not considered explicitly, such as the presence of other reactive species in homogeneous solution or as heterogeneous particles.
(By inspiration from Barry et al., 1994.)

Table 7.2 shows velocity constants from Fe(II) oxidation and precipitation where the constant k is a combination of different parts of the expression above: $k = k_o + k_{so}[Fe(III)] + k_b$ [bacteria exopolymer sites].

By comparing Tables 7.3 and 7.4 we can see that involving sand filters with heterogeneous surfaces in form of coated sand grains, colloidal precipitated particles and the exopolymers from iron-precipitating bacteria can increase the velocity of the process by a factor of about 1000 from $10^{13}/M^2$/atm/min which can be considered as the velocity constant for the purely physicochemical oxidation (homogenous oxidation), where the raw water is normally oxidised by only aeration in a precipitation tank.

The chemical catalytic process itself which means the heterogeneous precipitation process going on in the sand filters by help of the coated sand grains is able to increase the velocity by about 10–100 times.

TABLE 7.4 Velocity Constant for the Oxidation/Precipitation of Fe(II) by Biological Iron Removal (Søgaard et al., 2001)

k Values ($\times 10^{16}/M^2$/min/atm) for Biological Fe^{2+} Oxidation/Precipitation in a Sand Filter with Strains of the Fe-Precipitating Bacteria *Gallionella ferruginea* Using the Kinetic Expression for Fe^{2+} Oxidation/Precipitation Mentioned in Eqn (7.3)

k *VALUE*	*T/°C*	*pH*	*FE[(II)]$_O$/μM*	*MEDIUM*	*REFERENCE*
3.3–4.4	10.2–10.7	5.7–5.8	2.3	Polluted ground water	Søgaard et al., (2001)
10–12	10.2–10.6	5.7–5.8	21	Polluted ground water	Søgaard et al., (2001)
19–21	10.2–10.7	5.6	52	Polluted ground water	Søgaard et al., (2001)
7.7–13	10.3–10.9	5.7	126	Polluted ground water	Søgaard et al., (2001)

A 1000 times increase in the k value when the Fe(II) were precipitated from the filters by biological means can be explained by the presence of the bacteria exopolymers. Its name comes from that they are present outside the cell membrane of the bacterium. These exopolymers which also in general is called biopolymers form a mesh network in between the pore spaces of the sand filter material, when the bacteria get in contact with the Fe(II) and only a slightly aeration of the water. As a result, the Fe^{++} in solution either directly adsorbs onto the surfaces of the exopolymers and becomes oxidised or is attached to the exopolymers in a colloidal microcrystalline form (Ankrah and Søgaard, 2009b).

7.4.4.4. *Other Factors Affecting the Rate of Oxidation*

Several other factors affects the rate of Fe oxidation. The kinetic Eqn (7.3) depicts that the rate of Fe oxidation is dependent on the concentration of Fe^{2+}, dissolved oxygen and hydroxide ions or, equally, the pH of the ground water. Therefore, at pH of about 6–8.5, which is the typical pH area of ground water, has a remarkable influence on the reaction. For every unit increase in pH, the rate will be increased by 100-fold (see Figure 7.7).

Similarly, a correlation exists between alkalinity and the rate of oxidation (Jobin and Ghosh, 1972).

The reaction proceeds at a higher rate in a solution of higher alkalinity. This is as a consequence of the CO_2 and HCO_3^- buffer system which resists the changes in acidity that occur during the oxidation reaction (Stumm and Singer, 1966).

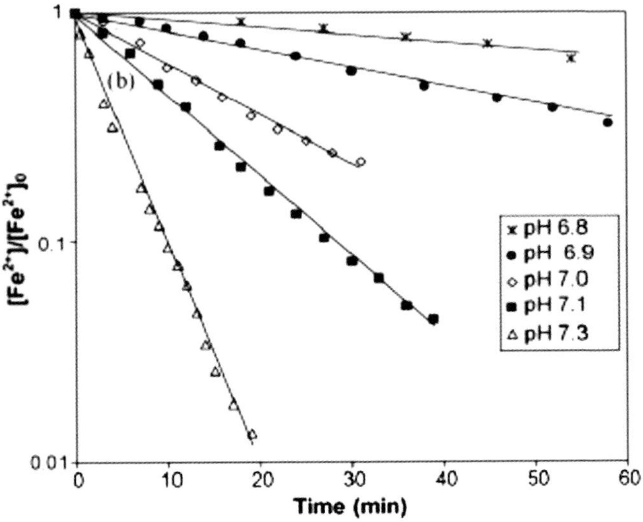

FIGURE 7.7 **Evolution of Fe^{2+} concentration versus time as a function of pH for $[Fe]_0 \simeq 10$ mg/l.** *From El Azher et al. (2008); reproduced with permission from Elsevier.*

The alkalinity of ground water provides in that way the buffer capacity to prevent pH to drop during the oxidation of Fe(II) in the WTP as could be expected from reaction scheme (1). In most cases, alkalinity also influences the characteristics of the Fe precipitate. Normally, at high alkalinity (above 250 mg/l $CaCO_3$), carbonate Fe precipitates are formed rather than Fe hydroxides and a larger part of the precipitate is in the Fe(II) form rather than in the Fe(III) form. Contrary, to the high alkalinity, at lower alkalinity values, slower oxidation and poorer floc formation of Fe precipitate occur (Snoeyink and Jenkins, 1980).

7.4.4.5. Possible Catalytic Effect of Some Elements

Oxidation is accelerated by anions such as F^-, $H_2PO_4^-$, and HPO_4^{2-} and lowered by others in the order $ClO_4^- > NO_3^- > Cl^- > H_3SiO_4 > Br^- > I^- > SO_4^{2-}$. For instance, silicate is an integral part of the sand grains and also clay minerals. Silicate can be thought of as salts of quarts, SiO_2, which forms silicic acid, H_4SiO_4. The presence of silicate can interfere with Fe oxidation and its presence is one of the reasons for that mainly ferrihydrate and not the more stable iron oxides goethite and lepidicrosite are formed (Weber and Stumm, 1965; Cornell and Schwertmann, 2003). The other anions may compete with oxygen for the positions on the surfaces of the precipitated iron oxides and in that way delay the uptake of an electron from Fe(II) resulting in a slower oxidation/precipitation reaction.

Robinson et al., in 1975, demonstrated that silicates can catalyse the rate of Fe(II) oxidation. In the presence of silicates at pH from 6.5 to 7.2 the following expression was obtained:

$$-\frac{d[Fe^{2+}]}{dt} = \left\{k(OH^-)^2 + k_{si}(H_4SiO_4)^{1/2}(OH^-)^{1/2}\right\}pO_2[Fe^{2+}] \qquad (7.7)$$

The above equation shows that the presence of silicates interferes with the rate of Fe oxidation as compared with the rate of Fe oxidation without silicates. Silicate is also known to interfere with hydrolysis thereby increases nucleation and finally prevents the growth of the particles of hydrous Fe(III) oxides. This is as a result of silicic acid competition with water and hydroxides for the coordination sites on the Fe. At silicate levels of 7 mg/l the Fe(III) precipitates are generally amorphous (Olsen and O'Melia, 1973).

7.4.4.6. Oxidation in the Presence of Fe(II)carbonate

The mechanism underlining the oxidation of Fe^{2+} in the presence of carbonate is that the introduction of oxygen allows dissolved CO_2 over saturating level of the ground water to escape, causing an increase in pH, thereby leading to oversaturation of Fe(II)carbonate. Crystallisation of the Fe(II)carbonate finally can occur:

$$Fe^{2+} + HCO_3^- \rightleftarrows FeCO_3(s) + H^+ \qquad (7.8)$$

The solubility of the total carbonic species in water can therefore be dominated by the formation of Fe carbonate when it becomes the most stable Fe species (Ghosh et al. 1966).

Also, from Figure 7.8 it can be extrapolated that if $Fe(OH)_2$ should be present at all, it will not be precipitated below pH values of about 9.5. This hypotheses was inferred by Olsen and Twardowski who used thermodynamic calculations and a floc filterability approach to investigate the precipitation of Fe(II)hydroxide, Fe(II)carbonate and Fe(III)hydroxide. They indicated that it was logical to anticipate Fe(II)carbonate (siderite) to precipitate in WTPs treating hard water (high levels of $Ca^{++} + Mg^{++}$). If that will be the case, then the rate of oxidation of Fe(II) will be very slow because it is going to happen from a solid substance (Just, 1908; Stumm and Singer, 1966; Olsen and Twardowski, 1975). Neither at Vr. Gjesing WTP nor at Astrup WTP the water is hard but below 60 mg/l which is the condition for soft water.

7.4.4.7. Oxidation in the Presence of Organic Matter (Humic Substances)

Humic materials in form of humic or fulvic acids being it in the dissolved or in a colloidal solid phase in water may carry functional group

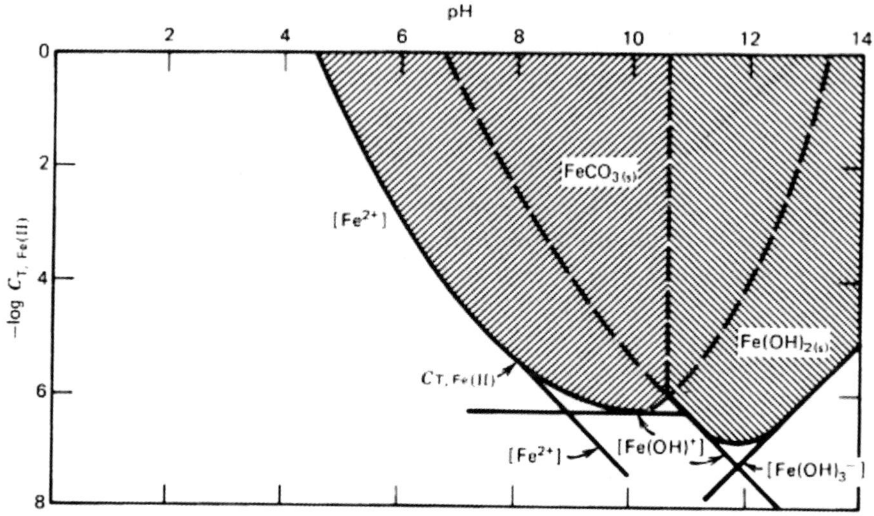

FIGURE 7.8 Stability of solid Fe(II)compounds in water as a function of pH originally containing 10^{-5} M of total Fe species. *From Snoeyink and Jenskins (1980). This material is reproduced with permission of John Wiley & Sons, Inc.*

such as carboxyl (COOH) and hydroxyl (OH). Such groups can act as ligands to form complexes or coordinates and form chelates with metals such as Fe(III) (Vanloon and Duffy, 2005; Faust and Aly, 1998).

Therefore organic humic materials can act as inhibitors such that the ferrous–ferric system which is formed in the process of oxidation could act as electron scavenging system for the oxidation of the organic material by oxygen. As a consequence, the entire rate of oxidation of Fe(II) in the water and its precipitation as ferrihydrite may be slowed down. If this is a problem sometimes it is necessary to add a coagulation compound in form of commercial Al polymers to co-precipitate the humic substances and Fe or $KMnO_4$ to increase the velocity of oxidation of Fe(II). Often we meet these application at WTP near the coastal regions of Denmark where the depths of water wells are rather short (Figure 7.9).

The reaction scheme for the oxidation of Fe(II) with $KMnO_4$ can be seen below.

$$Fe^{++}(aq) + MnO_4^-(aq) + H_2O \rightarrow Fe(OH)_3(s) + MnO_2(s) + H^+(aq)$$

This oxidation reaction that also precipitates Fe and Mn will co-precipitate the humic substances to form a common sludge in the sand filter systems. At other WTPs where the problems with humic substances are smaller it is not necessary to add Al coagulants or $KMnO_4$ because the normal precipitation of ferrihydrite and also manganese oxides will be able to co-precipitate smaller amounts of humic substances.

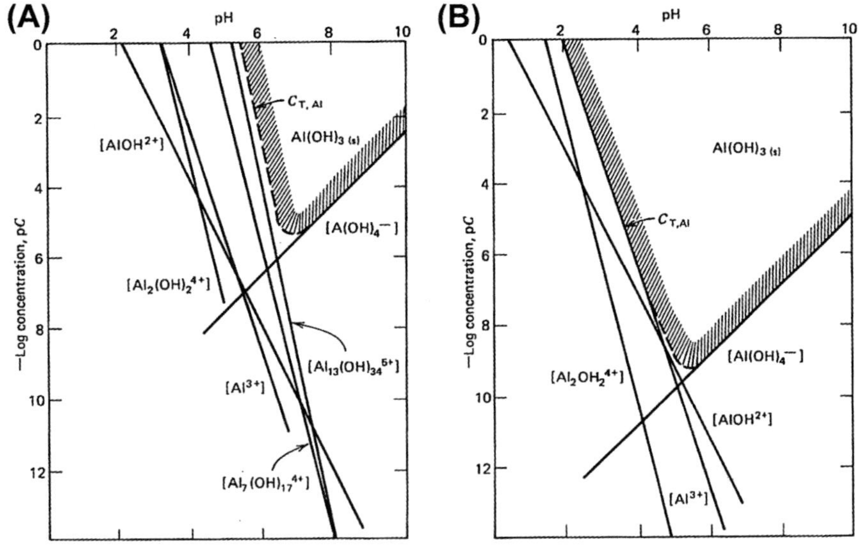

FIGURE 7.9 **The species of Aluminium compounds in water as a function of pH.** At pH 6–7 where Al's principally form is an inorganic polymer $Al_{13}O_4(OH)_{24}(H_2O)_{12}$ the co-precipitation with the humic substances and Fe will take place (Sogaard, 2002). *From Snoeyink and Jenkins (1980). This material is reproduced with permission of John Wiley & Sons, Inc.*

7.4.4.8. Surface Properties of Precipitates in Sand Filters

Surface speciation of Fe oxides highly affects the adsorbing capacity of anions or cations to their surfaces and it is invariably dependent on the pH of the solution. The pH at which there is no net charge on the surface of Fe oxides is the point of zero net charge (pzc). Usually, for ferrihydrite pzc is found for pH about 7.8–7.9 (Davis and Leckie, 1979; Manceau et al., 1992). The pzc of ferrihydrite is in this way dependent on the absorption and desorption of protons in the water to oxygen of ferrihydrite. If the raw water contains anions as phosphate, sulphate, silicate, carbonate or fluoride they can become adsorbed at inner or outer sphere of iron by replacing hydroxyl groups and move pzc to lower pH. Alternatively, contrary to this adsorption of anions as sodium, potassium, magnesium and calcium as ion pairs with the oxide surface can shift pzc to higher pH values. However, the adsorption constants of anions to iron oxides are about 1000 higher than the corresponding absorption constants of cations (Smith and Jenne, 1991).

One way of measuring the surface speciation and charge of colloidal ferrihydrite particles which are the results of the oxidation and precipitation of Fe(II) in the drinking water treatment system is by measuring the zeta potential of the particles. We can consider the water with its precipitated colloid particles as a very thin dispersion in a weak electrolyte solution comprising its macro ions. The zeta potential measurements of the particles are based on electrophoresis where the movement of the particles is measured relative to the electro osmotic movement of the electrolyte. Zeta potential is a very good index of the charge and therefore the magnitude of interaction between colloidal particles and their relative stability in the water without tendency to aggregation or flocculation. However, as the idea with the filtering part of the WTP mainly is to avoid that colloid particles move on to the drinking water tank it is of interest to neutralise the particles so they aggregate and will be easier filtered and left in the sand filter. The magnitude of the zeta potential in this way gives an indication of the potential stability of the colloidal systems. If all the particles have large negative or positive zeta potential, then they will repel each other and there is dispersion stability. On the other hand, if the particle have low zeta potential value, then there is no force to prevent the particle being attracted by long-range van der Waals forces, the surface charge of the colloidal particles are said to be destabilised. In this way, the particles will coagulate and flocculate. If pzc of the particles is not present at the actual conditions of the water another possibility is to let the charged particles become attracted to the already precipitated sludge in the filter. Therefore, it is also of interest to measure the zeta potential of already present sludge that partly make up the coating of the sand grains in the filter. This measurement can be done by measuring the streaming potential of the iron oxide-coated sand grains. Table 7.5 shows the results of measurements of zeta potential of colloidal particles and streaming potential of the filter sand grains from a WTP where the penetration of colloid-sized particles through all filter steps installed in series and their presence in the drinking water tank was a problem.

As it can be seen from Table 7.5 the present colloidal ferrihydrite particles have negative zeta potential whereas the filter media shows a slightly positive streaming potential.

7.4.4.9. Biological Fe Removal from Ground Water for Drinking Purposes

As mentioned above unlike at Vr. Gjesing WTP, Astrup WTP removes Fe by a biological method. The process design at Astrup WTP benefited from the relatively low pH and low redox potential E_h of the raw water. Only a slightly aeration before the first filtration stage was arranged (Figure 7.5). The slight aeration adds a minimal amount of oxygen up to about 4 mg/l to the raw water and keeps the amount of CO_2 in the raw water mainly in form of HCO_3^-, which is required for the Fe bacterial metabolism. The redox potential typically increases from −60 to 100 mV during the first filtration step, for example from purely anaerobically to aerobically (Figure 7.5). Raw water pH is typically a little below 7.

TABLE 7.5 Zeta Potential of Particles in Water from Five Water Wells, after Aeration Tank and after Each of Four Filters in Series and Streaming Potential of Filter Medium in Filtration Step 1

Sample Identity	Zeta Potential (mV)
Water sample from Well 2	−18
Water sample from Well 3	−10
Water sample from Well 4	−24
Water sample from Well 5	−23
Water sample from Well 6	−21
Water sample from Aeration stage	−16
Water sample from Filter 1	−13
Water sample from Filter 2	−14
Water sample from Filter 3	−14
Water sample from Filter 4	−14
Sample Identity	Streaming Potential (mV)
Streaming potential of filter media	+2.8

One of the most common iron-oxidising bacteria found in the sand filtration units of drinking WTPs is *Gallionella ferruginea* and it is also the bacterium present and active in Astrup WTP. The optimal conditions for *G. ferruginea* is a pH slightly below 7 and an E_h above +200 mV (Ehrlich, 1996). Hanert reports E_h values between +200 mV and +320 mV with an oxygen constant between 0.1 and 1 mg O_2/l (Hanert, 1992). If O_2/H_2O was the only redox couple present an oxygen content of 0.1 mg/l will correspond to an E_h of 370 mV at pH 7. Therefore, the much lower values of E_h found in the (Degremont, 1991) raw water must be a result of a change in redox conditions originally being determined by couples as SO_4^{2-}/HS^- and to a smaller extent $Fe^{++}/Fe(OH)_3$ and probably others like Mn^{++}. When aeration is taking place and level of O_2 increases the O_2/H_2O couple becomes more important. If, for example, the couple SO_4^{2-}/HS^- was considered to be the only present couple in the raw water E_h would have been about −35 mV with a typical content of sulphate in the water. As it can be seen from Table 7.2 the sulphate concentration seems to decrease a little by the treatment of raw water in the WTPs. Sulphur was also found in minor amounts in the sludge from Astrup WTP (Søgaard et al., 2000). Generally, the best conditions for biological iron removal is with an rH_2 greater than 14 but the boundary between purely physico-chemical and biological iron precipitation is not well defined. $rH_2 = -\log(pH_2) = E_{h(measured)}/29.6\ \text{mV} + 2\ pH_{(measured)}$ corresponding to a rewriting of Nernst equation for the reduction potential of H^+ (see, e.g. Erhlich, 1996).

After the first filtration stage at Astrup WTP, the water is the heavily aerated by the injection of compressed air and dosed with $Ca(OH)_2$ to increase pH and give it the optimal pH required for Mn removal (Søgaard et al., 2000). Removal of aggressive carbon dioxide which is measure of an excess of the amount required to precipitate present concentrations of calcium ions as calcium carbonate is also important; aggressive carbon dioxide can be considered as a measure of the corrosivity in the water supply system and it can also lead to scaling.

7.4.5. The Effect of Fe-Oxidising Bacteria in Filter Media

Since Fe-oxidising bacteria were discovered, they are increasingly recognised as a principal part of a wide range of geochemical reactions in aqueous environments (Ferris, 2005). The prospects of using it to remove Fe from drinking water has been vigorously exploited by water treatment units and the mechanism underlining the iron removal process by the iron bacteria is considered to be because of the presence of a huge amount of templates that the exopolymers from the Fe-oxidising bacteria produce between the sand grains of the filter. These exopolymers result in a much faster rate of Fe precipitation than found in normal filters. The Fe-oxidising bacteria in that way play a supplementary role to the physicochemical iron removal process under certain conditions such as pH slightly below 7, initially low E_h and positive gradients of oxygen content. Moreover, the presence of biological phenomenon in the filters reduces the iron concentration in the treated water also by breaking the problem of the difficulty to remove Fe—organic or Fe—silicate complexes and as earlier mentioned also by increasing the rate of oxidation of Fe(II) to Fe(III) (Ridgway et al., 1981, Søgaard et al., 2001). The precipitated sludge has a density seven to nine times higher than the density of similar sludge from conventional sand filters and its residence time in the filter can be longer than for the normal colloidal purely physicochemical precipitated sludge. The reason is its high content of biopolymer that is also the reason for the increase of velocity of oxidation/precipitation of Fe(II) (Søgaard et al., 2000). It also means that backwash does not need to go on so often as in conventional filters.

7.4.6. Metabolism of Iron Removal Bacteria

The oxidation reaction of Fe(II) to Fe(III) in the form of ferrihydrite is exothermic with a Gibbs free energy of −64 kJ/mol.

$$2Fe^{2+}(aq) + \tfrac{1}{2}O_2(aq) + 5H_2O \rightarrow 2Fe(OH)_3(s) + 4H^+(aq) + \text{Energy} \quad (7.9)$$

The iron-oxidising bacteria derive their essential energy requirements through a strictly chemolithotrophic process (Stumm, 1992). Near-neutral pH the iron-precipitating bacteria are supposed to obtain energy by help of the following reaction chain involving the electrochemical half reactions of (4.1):

$$2Fe^{2+}(aq) \rightarrow 2Fe^{3+}(aq) + 2e^{-} \tag{7.10}$$

$$2H^{+} + \tfrac{1}{2}O_2 + 2e^{-} \rightarrow H_2O(l) \tag{7.11}$$

$$2Fe^{3+}(aq) + 6H_2O(l) \rightarrow 2Fe(OH)_3(s) + 6H^{+}(aq) \tag{7.12}$$

In the model of metabolism reaction (7.9) is taking place at the surface of the cell membrane of the bacteria. The electron is transported through the membrane by help of the electron transfer chain in the membrane that consist of the cytochromes c and a and some other similar co-enzymes. In the cell the electron meets oxygen and a proton and reduces oxygen to water as showed in reaction (7.10). The proton from (7.10) is a part of the proton motive force $\Delta Pmf = \Delta\psi + \Delta pH$ which is a combination of the membrane potential and the gradient in pH. This proton motive force is used to generate ATP in the bacterium cell.

Reaction (7.11) takes place outside the cell membrane where ferrihydrate is precipitated (Bird et al., 2011; Banfield and Welch, 2000; Erlich, 1996).

The stalks of the iron-oxidising bacteria are principally made of carbohydrates in form of polysaccharides rich in carboxyl groups (Emerson and Ghiorse, 1993; Chan et al., 2009, 2011). The stalks are grown together with the cells and twisted because of the rotation of the cell during its growth. Besides the iron-precipitating bacterium *G. ferruginea* which is the most often found bacterium in the sand filters of drinking water treatments plants also *Leptothrix* species can be found. Details of the growths of another iron-oxidising bacterium *Mariprofundus ferrooxydans* show that the formation of the stalks can go on with a velocity of in average of about 2 μm/h. Between each cell division each cell was found to produce a stalk with an average length of 31 μm. The stalk begins with adhesion to a solid material and remains stationary while the cells can leave the stalks. The stalks bifurcate during the cell division (Chan et al., 2011). The metabolic reason for creation of the stalks is probably the cells method to avoid encrustation. The oxidised Fe(II) in form of Fe(III) establishes a complex with the acidic carboxyl groups of the polysaccharides which have lost a proton at the neutral-near pH. That complex is rather stable and the Fe(III) precipitates present directly in the exopolymers shows no sign of crystallinity as found by selected area electron diffraction (Søgaard et al., 2000).

The process is an enzyme-mediated oxidation of Fe^{2+} with a concomitant fixation of carbon dioxide into an assimilable nutrient for the iron-oxidising bacteria. As a result, precipitation of Fe^{3+} salts occurs either by enzymatic action of autotrophic bacteria (intracellular) or by catalytic action of polymers excreted by the bacteria sheath (extracellular). The original source of carbon dioxide is principally from the ground water where organic compounds in form of humic substances are transformed by anaerobic fermentation in ground water and by gravity transferred by rainwater and ground water to, for example, 100 m deep below the ground surface. Humic substances are present in ground water either chelated with Fe^{2+} or existing solely in the anoxic zone (Bromfield, 1956; Søgaard et al., in press).

Several strains of iron-oxidising bacteria may be involved in water treatment systems but in all cases, the biological oxidation process is seen as catalytic in nature, causing a rapid oxidation of soluble ferrous Fe to insoluble ferric hydroxides found to be more compact or superior to the precipitates formed from other treatment processes (Colmer et al., 1950, Søgaard et al., 2000).

7.5. BIOTIC CONDITIONS FOR IRON PRECIPITATION

A shift from abiotic to biotic precipitation is governed by the chemical and the physical properties of the water (Boettcher and Viswanathan, 1991; Mouchet, 1992). The most important conditions for the biotic precipitation are enumerated at neutral or slightly acidic pH, a change from negative redox potential to redox potentials up to about 200–320 mV and oxygen levels changing from 0 to 2–3 mg/l, together with considerable amount of CO_2. However, redox potential and pH are the main factors that determined whether biotic precipitation will take place or not. The stringent pH and E_h limits required by the iron bacteria describe them as very good example of organism that develops under neither strongly reducing conditions nor in a fully oxidised milieu (Colmer et al., 1950).

Table 7.6 below compares three of the most important iron bacteria in terms of their characteristic features as well as conditions required for their growth and development.

Mouchet, 1992, Søgaard et al., in press and Sharma et al., 2005, described the iron-oxidising bacteria as gradient organisms. This means that they possess a unique property of causing oxidation and precipitation of soluble Fe^{2+} at conditions which straddles the theoretical boundary between the field of Fe^{2+} solubility and the formation of Fe^{3+} precipitates, as defined by the thermodynamic analysis of the electrochemical equilibrium. See Figure 4.1. They therefore thrive well under conditions that are neither completely reduced nor completely oxidised. In WTPs, the conditions could be manipulated and optimised by aerating or stripping off CO_2 from the ground water.

TABLE 7.6 Comparison of Conditions and Features of *Gallionella*, *Leptothrix* and *Sphaerotilus* (Bergey and Holt, 1994; Balows et al., 1992)

	Gallionella	*Leptothrix*	*Sphaerotilus*
	Chemolithoautotrophic	Chemoheterotrophic	Chemoorganotroughs
Cell shape and size	Kidney shape Width: 0.5–0.7 μm Length: 0.8–1.8 μm Gram-negative cell walls	Rod shape Width: 0.6–1.4 μm Length: 1.0–12.0 μm Gram-negative cell walls	Rod shape Width: 1.2–2.4 μm Length: 2.0–10 μm Gram-negative cell walls
Sheaths/Stalk	Width: 2 nm Shape: Twisted Surface structure: Dissolves in reducing agents and weak acids	Width: 1.5–3 μm Shape: Hollow sometimes and straight Surface structure: Rough Detaches from the body intermittently	Width: Not specified Shape: Hollow sometimes and straight Surface structure: Smooth
Energy/carbon source	Carbon dioxide in form of principally HCO_3^-	Sugars, organic acids and glycerol	Carbon sources like sugars, organic acid. They also metabolise nitrogen compounds
Type of iron produced	Ferrihydrite (Ferric hydroxide)	Ferrihydrite (Ferric hydroxide)	Ferrihydrite (Ferric hydroxide
Living conditions	Anaerobic and microaerophilic Oxygen content: 0.1–3 mg/l increasing Temp: About 8–16 °C Oxidation/reduction potential: Negative −300 mV increasing Fe^{2+} content: 2–25 mg/l pH: 6–7.6	Slightly aerobic to aerobic Oxygen content: 1 mg/l Temp: 10–40 °C Oxidation/reduction potential: Not specified Fe^{2+} content: Not specified pH: 7.5	Anoxic and aerobic Oxygen content: <0.1 mg/l – full saturation oxygen saturation Temp: About: 10–40 °C Oxidation/reduction potential: Not specified $Fe2^{+}$ content: Not necessary pH: 5.4–9

In general, several bacteria genera and species are found in a variety of soils and aquatic habitats associated with Fe. They require special conditions for their growth and development and, as a consequence, produce extracellular polymeric substances. The required condition are mostly water, temperature, nutrient source, carbon dioxide and electron acceptor source, which in mostly ferric iron in the case of the iron-oxidising bacteria. These bacteria usually form a community encapsulated within the self-developed polymeric matrix, which they produce and adhere to surfaces such as flocs of Fe particles produced in WTPs or in nature (Bromfield, 1974).

The physiology and the ultrastructure of some of these microorganisms have been investigated more detailed (e.g. Van veen et al., 1978; Dondora, 1975; Andersen and Pedersen, 2003). Many iron bacteria could easily be identified directly by observation under the microscope because of their distinct characteristic sheath secretion. Examples of such bacteria are of the genus *Sphaerotilus*, for example *Sphaerotilus natans*, *S. dichotoma*, and *Leptothrix* group, for example *Leptothrix ochracea*, *Leptothrix lopholea*, *Leptothrix discophora*, *Leptothrix cholodnii* (see Figure 7.10(B)). Others like *G. ferruginea* are recognised by their elongated helical or twisted stalks, composed of numerous intertwine micro fibrils but a non-stalk-forming strain has also been reported to exist (Figure 7.10(A)). Without the stalks, *G. ferruginea* consist of a

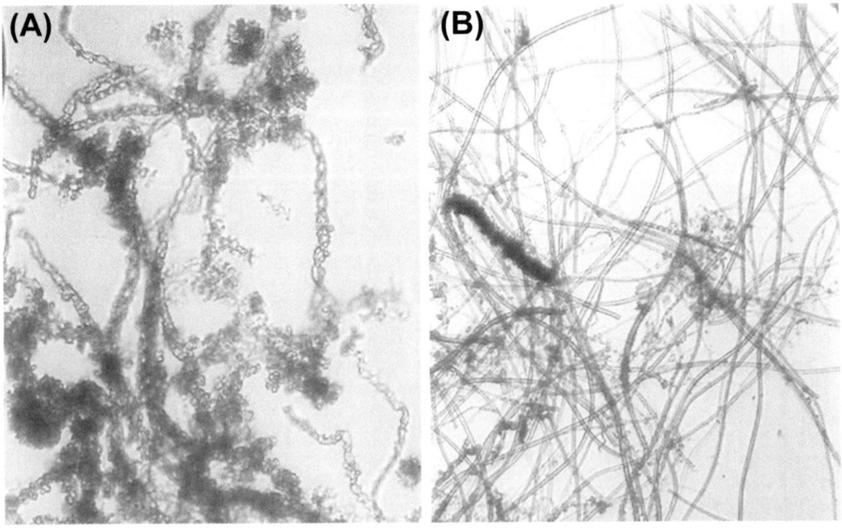

FIGURE 7.10 (A) Light microscope image of stalks of *Gallionella* sp; (B) light microscope image of sheaths of *Leptothrix* sp. *From Ankrah and Søgaard (2009b); Nanotechnology Perceptions www.unibas.ch/colbas/ntpr.* (For colour version of this figure, the reader is referred to the online version of this book.)

kidney-shaped mycoplasmodial cell body, which gives rigidity to the cell wall (Hallbeck and Pedersen, 1995). On the other hand, the Eubacteria in the family Siderocapsacea, such as *Siderocapsa treubii, Siderocapsa Major, Siderocapsa treubii*, and so on, might not have stalks or sheaths, making them extremely difficult to identify.

It is believed that all the above-mentioned Fe bacteria can also cause the oxidation of manganese if the concentration of manganese exceeds that of iron in the ground water (Katsoyiannis and Zouboulis, 2004). However, there are other strains of bacteria that have a more specific manganese oxidation activity, namely *Pseudomonas* (e.g. *Ps. manganoxidans*); Metallogenium (e.g. *M. personatum, M. symbioticum*); and Siderobacteriales such as *Leptothrix* (e.g. *Leptothrix echinata, Leptothrix lopholes*) (Bromfield, 1974).

Methods like the16S ribosomal DNA sequence analyses and Generalised transducing phage (SN-T) are known to be useful tools used extensively to build up phylogenic trees, assist in the analysis of microbial communities and also identify gene sequence within single bacteria species. However, report(s) of the phylogenetic and evolutionary relationship investigations have not yet been published on sludge from an operational drinking WTP (Beumer and Robinson, 2005).

7.6. FUNCTIONS OF EXOPOLYMERS IN SAND FILTERS

Today, the exact function of the sheath does not remain obscure. Investigations of sludge samples from filters have revealed that massive growth of bacteria sheath have profound catalytic influence on the oxidation of iron. Researchers have also argued that the sheath has nutritional and ecological consequences for the iron bacteria. This is evident in their ability to grow in slow running water low in nutrients, where the presence of sheath allows the bacteria to attach themselves to solid surfaces. Furthermore, the sheath defends the organism from predators and parasites. Work done by Hallbeck and Pedersen in 1995 demonstrated that *G. ferruginea* stalks give them the unique possibility to colonise and survive in habitats with high nutrients of iron, which produces toxic oxygen radicals (e.g. O_2^-) during the chemical oxidation process. Under such conditions, the iron is normally inaccessible to the bacteria without defence system against the oxidation of iron.

7.7. CONTENT OF THE Fe BACTERIAL STALK

Research done by Balashov (1970) to digest the stalk of *G. ferruginea* by proteolytic enzymes gave the strongest conviction that the sheaths are protein in nature. Till date, although it is known that the encrustation of the sheaths of iron-precipitating bacteria could be protein in nature, it has however been found that besides proteins, the sheath of *S. natans* contains larger amounts of carbohydrates and smaller amounts of lipids (Takeda et al., 2003). Similarly, investigation conducted in 2004 and repeated recently indicated that the exopolymers contain carbohydrates, proteins and lipids (Søgaard and Simonsen, 2004). A more detailed analysis of the chemical composition done by Takeda et al. in 2005 revealed that the sheath of *S. natans* is a complex of a polysaccharide, glycine and cystein-rich peptides. It was also indicated that the sheath polysaccharide is composed of glucose and (*N*-acetyl)galactosamine in an approx molecular ratios of 1:4, constituting about 70% of the sheath (Takeda et al., 2005). Other inorganic substances than Fe, for example Na, Si, P, S, Cl, Ca, Cu, were found as well (Ridgway et al., 1981, Søgaard et al., 2001). If these elements possess a function in the exopolymers or are present mainly because of the drying of samples before energy dispersion spectroscopy is not known.

7.8. CONCLUSION

The purely physicochemical method for Fe oxidation and precipitation or the chemical catalytic process in sand filters may not always be effective in removing Fe contaminants from ground water for drinking purposes. Therefore, chemical oxidants are sometimes needed to enhance the oxidation process. Biological oxidation/precipitation provides an alternative to the conventional, purely physicochemical, routes, with numerous advantages. The iron-precipitating bacteria is now well known to be able to increase rates of precipitation of iron in drinking water treatment systems. A clear metabolic reason for the huge amount of work in producing iron-precipitating sheaths or stalks that these bacteria perform is still uncertain. However, the bacteria could play a catalytic role by adsorbing the amorphous ferrihydrites thereby increasing the rate of iron oxidation and subsequent precipitation. Because of the numerous advantages that biological iron removal offers over the purely physico-chemical method it has the potential to become the most favoured method of iron removal from drinking water where the conventional method has reached its limits (Ankrah and Søgaard, 2009a; Dimitrakos Michalakos et al., 1997).

References

Andersen, C.R., Pedersen, K., 2003. *In situ* growth of *Gallionella* biofilms and partitioning of lanthanides and actinides between biological material and ferric oxyhydroxides. Geobiology 1, 169–178.

Ankrah, D.A., Søgaard, E.G., 2009a. The applicability of iron and manganese precipitating bacterial in drinking water systems. Nanotechnol. Perceptions 5, 209–216.

Ankrah, D.A., Søgaard, E.G., 2009b. A review of biological iron removal. In: The Thirteenth International Water Technology Conference (IWTC 13), pp. 999–1005. Hurghada, Egypt.

Balashova, V.V., 1970. The ultrastructure of *Gallionella filamenta*. Mikrobiologica 39, 1429–1437.

Balows, A., Truper, H., Dworkin, M., Harder, W., Schleifer, K., 1992. The Prokaryotes Second Edition: A Handbook on the Biology of Bacteria: Ecophysiology, Isolation, Identification, Applications. Springer-Verlag, New York.

Banfield, J., Welch, S.A., 2000. Microbial controls on the mineralogy of the environment. In: EMU Notes in Mineralogy, vol. 2. Eötvös University Press, Budapest, pp. 173–196 (Chapter 5).

Barry, R.C., Schnoor, J.L., Sulzberger, B., Sigg, L., Stumm, W., 1994. Iron oxidation kinetecs in an acidic alpine lake. Water Res. 28 (2), 323–333.

Bergey, D.H., Holt, J.G., 1994. Bergey's Manual of Determinative Bacteriology. Lippincott Williams & Wilkins, Philadelphia.

Beumer, A., Robinson, J.B., 2005. A broad-host-range, generalized transducing phage (SN-T) acquires 16S rRNA genes from different genera of bacteria. Appl. Environ. Microbiol. 71 (12), 8301–8304.

Bird, L.J., Bonnefoy, V., Newman, D.K., 2011. Bioenergetic challenges of microbial iron metabolisms. Trends Microbiol. 19 (7), 330–340.

Boettcher, B., Viswanathan, M.N., 1991. Biological removal of iron from ground water. Water Sci. Technol. 23 (7/9), 1437–1446.

Bromfield, S.M., 1956. Oxidation of manganese by soil microorganisms. Biol. Sci. 9, 238–252.

Bromfield, S.M., 1974. Bacteria oxydation of manganese ions as affected by organic substrate concentration and composition. Soil Biol. Biochem. 6, 383–392.

Chan, C.S., Fakra, S.C., Edwards, D.C., Emerson, D., Banfield, J.F., 2009. Iron oxyhydroxide mineralization on microbial extracellular polysaccharides. Geochim. Cosmochim. Acta 73, 3807–3818.

Chan, C.S., Fakra, S.C., Emerson, D., Fleming, E.J., Edwards, K.J., 2011. Lithotrophic iron-oxidizing bacteria produce organic stalks to control mineral growth: implications for biosignature formation. ISME J. 5, 717–727.

Colmer, A.R., Temple, K.L., Hinkle, M.E., 1950. An iron-oxidizing bacteria from an acid drainage of some butumen and coal mines. Bacteriology 59, 317–328.

Cornell, R.M., Schwertmann, U., 2003. The Iron Oxides. Wiley-VCH GmbH & Co. KGaA.

Davison, W., Seed, G., 1983. The kinetics of the oxidation of ferrous iron in synthetic and natural waters. Geochim. Cosmochimi. Acta 47, 67–79.

Davis, J.A., Leckie, J.O., 1979. Speciation of adsorbed ions at the oxide/water interface. In: Jenne, E.A. (Ed.), Chemical Modeling in Aqueous Systems, ACS Symposium Series 93. Amer. Chem. Soc., pp. 299–321.

Degremont, 1991. Water Treatment Handbook, sixth ed., vol. 1 ISBN: 2.9503984.1.3.

Dinitrakos Michalakos, G., Martineznieva, J., Vayenasd, V., Lyberatos, G., 1997. Removal of iron from potable water using a trickling filter. J. Water Res. 31, 991–996.

Dondoro, N.C., 1975. The Sphaerotilus Leptothrix group. Microbiology 29, 407–427.

Drinking Water http://www.mst.dk/udgiv/publications/2001/87-7944-519-5/html/kap07_eng.htm Date cited, July 17, 2010.

Ehrlich, H.L., 1996. In: Geomicrobiology, second ed. Marcel Dekker, New York.

El Azher, N., Gourich, B., Vial, C., Belhaj Soulami, M., Ziyad, M., 2008. Study of ferrous iron oxidation in Morocco drinking water in an airlift reactor. Chem. Eng. Process. Process Intensif. 47, 1877–1886.

Emerson, D., Ghiorse, W.C., 1993. Ultrastructure and chemical composition of the sheath of *Leptothrix discophora* SP-6. J. Bacteriol. 175, 7808–7818 (Cornell University Ithaca, New York).

Faust, S.D., Aly, O.M., 1998. Chemistry of Water Treatment, second ed. ISBN: 1-57504-011-5.
Ferris, F.G., 2005. Biogeochemical properties of bacteriogenic iron oxides. Geomicrobiol. J. 22 (3), 79–85.
Galal-Gorchev, H., Stum, W., 1963. The reaction of ferric iron with ortho-phosphate. J. Inorg. Nucl. Chem. 25.
Ghosh, M.M., O'Connor, J.T., Engelbrecht, R.S., 1966. Precipitation of iron in aerated ground water. J. San. Eng. Div. Proc. ASCE 90, 199.
Ghosh, M.M., 1974. Oxygenation of ferrous iron(II) in highly buffered waters. In: Rubia, A.J. (Ed.), Aqueous Environmantal Chemistry of Metals Chapter 5, Ann Arbor Science, pp. 193–217.
Hallbeck, L., Pedersen, K., 1995. Benefits associated with the stalk of Gallionella ferruginea, evaluated by comparison of stalk-forming and non-stalk forming strains and biofilm studies in situ. Microbial Ecol 30, 257–268.
Hanert, H.H., 1992. The genus *Gallionella*. In: Balows, et al. (Eds.), The Prokaryotes, second ed., vol. IV. Springer Verlag, New York, pp. 4082–4088.
Hem, J.D., Lind, C.J., 1983. Nonequilibrium models for predicting forms of precipitated manganese oxides. Geochim. Cosmochim. Acta 47, 2037–2046.
Jobin, R., Gosh, M.M., 1972. Effect of buffer intensity and organic matter on the oxygenation of ferrous iron. J. AWWA 64, 590–595.
Just, G., 1908. Kinetic investigation of the autoxidation of ferrous bicarbonate in aqueous solution. Z. Phys. Chem. 63.
Katsoyiannis, I.A., Zouboulis, A.I., 2004. Biological treatment of Mn(II) and Fe(II) containing groundwater: kinetic considerations and product characterization. Water Res. 38, 1922–1932.
Kiely, G., 1997. Environmental Engineering. International edition. ISBN: 0-07-709127-2.
Manceau, A., Charlet, L., Boisset, M.C., Didier, B., Spadini, L., 1992. Sorption and speciation of heavy metals on hydrous Fe and Mn oxides. From microscopic to macroscopic. Appl. Clay Sci. 7 (1–3), 201–203.
Morgan, J.J., Stumm, W., 1964. The Role of Multivalent Metal Oxides in Limnological Transformation, as Exemplified by Iron and Manganese. Paper Presented at Second International Conference, Water Poll. research, Tokyo.
Mouchet, P., 1992. From conventional to biological removal of iron and manganese in France. J. AWWA 84 (4), 158–167.
Olsen, L.L., O'Melia, C.R., 1973. The interaction of Fe(II) with Si(OH). J. Inorg. Nucl. Chem. 35, 1977–1985.
Olsen, L.L., Twardowski Jr, C.J., 1975. $FeCO_3$ vs. $Fe(OH)_3$ precipitation in water-treatment plants. J. Am. Water Works Assoc. 67 (3), 150–153.
O'Neill, P., 1995. Environmental Chemistry, Second ed. Chapman and Hall, Great Britain. ISBN 0 412 48490 0.
Pontius, F.W., 1990. Water Quality Treatment, A Hand Book of Community Water Supply, fourth ed. ISBN: 0-07-001540-6.
Ridgway, H.F., Olsen, B.H., Means, E.G., 1981. Iron bacteria in drinking water distribution systems: Elemental analysis of Gallionella stalks, using X-Ray Energy. Dispersitive microanalysis. Appl. Environ. Microbiol. 41 (1), 288–297.
Sharma, S.K., Petrusevski, B., Schippers, J.C., 2005. Biological iron removal from groundwater. Water Supply 54 (4), 239–247.
Smith, R.W., Jenne, E.A., 1991. Environ. Sci. Technol. 25 (3), 525–531.
Snoeyink, V.L., Jenkins, D., 1980. Water Chemistry. John Wiley. ISBN:0-471-05196-9.
Stumm, W., 1992. Chemistry of the Solid-Water Interface. J. Wiley & Sons.
Stumm, W., Lee, G.F., 1961. Oxygenation of ferrous iron. Ind. Eng. Chem. 53, 143–146.
Stumm, W., Singer, P.C., 1966. Precipitation of iron in aerated groundwater. discussion. J. San. Eng. Dev. Proc. ASCE 92, 120.

Søgaard, E.G., 2002. Production of Coagulation Agent PAX-14. Contents of Polyaluminium Chloride Compounds paper presented at chemical water and waste water treatment, Berlin, Germany.

Søgaard, E.G., Simonsen, C., 2004. Morphology and Content of Exopolymers. International Workshop on Geomicrobiology. Aarhus University, Denmark.

Søgaard, E.G., Medenwaldt, R., Abraham-Peskir, J., 2000. Conditions and rates of biotic and abiotic iron precipitation in selected Danish freshwater plants and microscopic analysis of precipitate morphology. Water Res. 34 (10), 2675–2682.

Søgaard, E.G., Aruna, R., Abraham-Peskir, J., Bender Koch, C., 2001. Conditions for iron precipitation by *Gallionella ferruginea* in a slightly polluted groundwater. Appl. Geochem. 16, 1129–1137.

Takeda, M., Nakamori, T., Hatta, M., Yamada, H., Koizumi, J., 2003. Structure of the polysaccharide isolated from the sheath of *Sphaerotilus natans*. Int. J. Biol. Macromol. 33, 245–250.

Takeda, M., Makita, H., Ohno, K., Nakahara, Y., Koizumi, J., 2005. Structural analysis of the sheath of a sheathed bacterium, *Leptothrix cholodnii*. J. Biol. Macromol. 37, 92–98.

Tamura, H., Goto, K., Nagayama, M., 1976. The effect of ferric hydroxide on the oxygenation of ferrous ions in neutral solutions. Corros. Sci. 16, 197–297.

Theis, T.L., Singer, P.C., 1974. Complexation of iron(II) by organic matter and its effect on iron(II) oxygenation. Environ. Sci. Technol. 8 (6), 569–573.

Vanloon, G.M., Duffy, S.J., 2005. Environmental Chemistry, a Global Perspective, second ed. Oxford University press, New York. 515 pp.

Van Veen, W.L., Mulder, E.G., Deinema, M.H., 1978. The Sphaerotilus-Leptothrix group of bacteria. Microbiol. Rev. 42 (2), 329–356.

Weber, W.J., Stumm, W., 1965. Formation of a silicato-Iron (III) complex in dilute aqueous solution. J. Inorg. Nucl. Chem. 27.

WHO, 1996. Guidelines for Drinking Water Quality: Health Criteria and Other Supporting Information, second ed., vol. 2. WHO, Geneva.

Winther, L., Linde, J.J., Winther, H., 2003. Vandforsyningsteknik. ISBN: 87-502-0945-0.

Yang, H., Li, D., Zhang, J., Hao, R., Li, B., 2004. Design of biological filter for iron and manganese removal from water. J. Environ. Sci. Health A Tox. Hazard. Subst. Environ. Eng. A39 (6), 1447–1454.

Zhang, J., Li, D., Wang, J.H., Yang, H., Wang, B., 2005. Operational performance of biological treatment plant for iron and manganese removal. Water Supply Res. Technol. 54 (1), 15–24.

CHAPTER

8

Advanced Arsenic Removal Technologies Review

Ph.D. fellow Krzysztof P. Kowalski

Section of Chemical Engineering, Department of Biotechnology, Chemistry and Environmental Engineering, Aalborg University Esbjerg, Niels Bohrs Vej 8, DK 6700 Esbjerg, Denmark

8.1. INTRODUCTION

Arsenic is a metalloid, which naturally occurs in an environment, but is proven to have negative influence on a human health, and even in higher doses was used as a poison. For example it was found in the body of King George III of the Great Britain and it is considered to be responsible for his madness (BBC NEWS Medical Mysteries). Nowadays the arsenic occurrence has been proven to be responsible for many diseases in Bangladesh and India, where it occurs in drinking water. Thus it has been considered as one of the six world's worst pollutions in 2010, in Blacksmith's Institute Report, and situation in Bangladesh was described as 'the largest mass poisoning of a population in history… beyond accidents at Bhopal, India, in 1984 and Chernobyl, Ukraine, in 1986' (Blacksmith Institute).

Therefore the arsenic problem has been considered as one of the main issues within researches and authorities. And for the past 20 years interest in this problem solution has increased significantly as a consequence of new findings about arsenic effects on human health resulting in implementation of stricter drinking water quality guidelines (Figure 8.1). Moreover the discussion about further changes in potable water quality guidelines is ongoing because of consumer expectance and available technologies, especially in Europe and North America (Van Halem et al., 2009).

However to understand the removal processes it is needed to know about better arsenic properties, especially the derivatives that occur in the

Chemistry of Advanced Environmental Purification Processes of Water
http://dx.doi.org/10.1016/B978-0-444-53178-0.00008-0

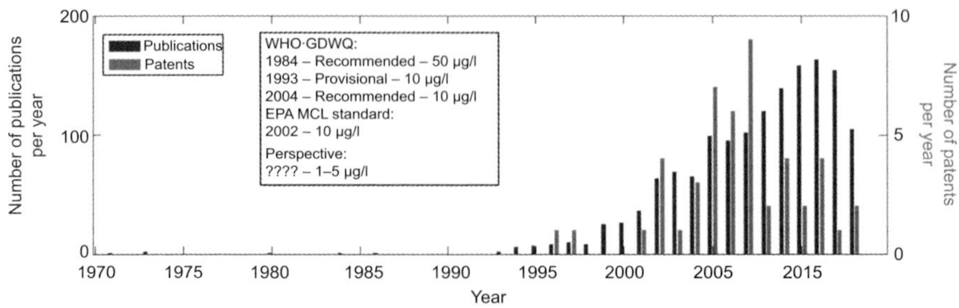

FIGURE 8.1 Number of publications and patents concerning arsenic removal for drinking water production as a response to decreasing allowance of As in drinking water by WHO and EPA (WHO-GDWQ – Guidelines for Drinking-water Quality, EPA-MCL – U.S. Environmental Protection Agency Maximum Contaminant Level). (For colour version of this figure, the reader is referred to the online version of this book.)

environment, because the technologies are based on shifting from one arsenic compound to another.

8.1.1. Main Arsenic Species

The most common arsenic compounds that naturally occur are arsenite and arsenate. The first one, with As oxidation state +3, can be found in a non-ionised form (H_3AsO_3) in most of the pH range (pH = 0–9). And only in strong basic conditions (pH > 9) it is changed to its ionised forms like $H_2AsO_3^-$, $HAsO_3^{2-}$ and AsO_3^{3-} (Figure 8.2). The arsenate, with As oxidation state +5, can be found in ionised forms in most of the pH range (pH = 2–14). The arsenate ions are $H_2AsO_4^-$, $HAsO_4^{2-}$ and AsO_4^{3-}, and occurrence in such forms is the basis of their removal technologies. The main difference between arsenite and arsenate is redox condition; the arsenite occurs in a reductive environment (low redox potential) and the arsenate in a higher redox potential environment. Moreover the species differs in solubility, that is As(III) and As(V) oxides solubility is, respectively, 10.1 and 66 g in 100 ml of water (Andersen et al.).

8.1.2. Source and Occurrence of the Arsenic Compounds

The sources of the arsenic compounds in the environment can be anthropogenic activities or natural processes. The anthropogenic sources are concerned as products, and its production facilities, where arsenic compounds are used, like wood preservatives, paints, dyes, pharmaceuticals, herbicides and semiconductors (Environmental Protection Agency of the United States, 2002). The main anthropogenic sources of the arsenic compounds are mining and smelting industries. Once released, arsenic species can undergo different processes resulting in distributing it in the

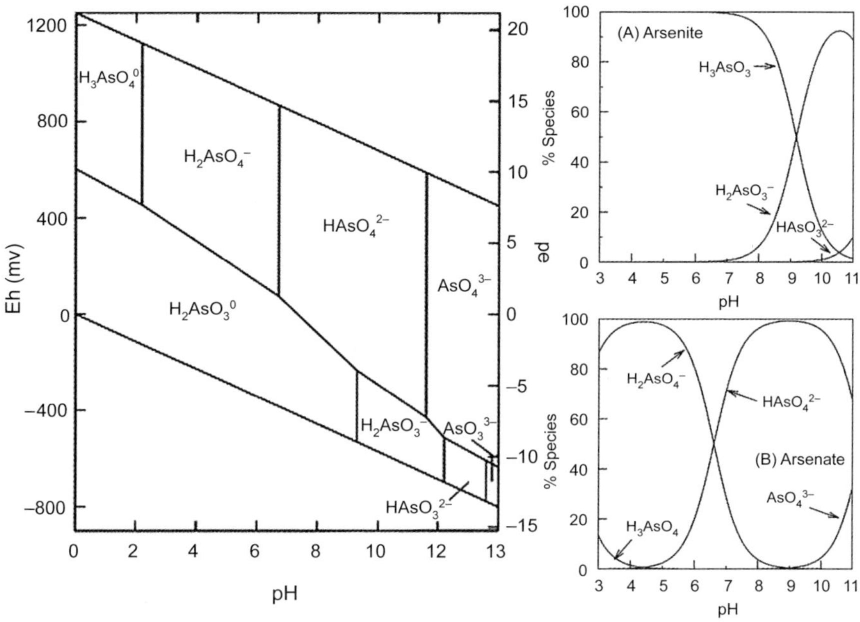

FIGURE 8.2 Eh-pH diagram for aqueous As species in the system $As-O_2-H_2O$ at 25 °C and 1 bar total pressure, on the left, and (A) arsenite and (B) arsenate speciation as a function of pH (ionic strength of about 0.01 M; redox conditions have been chosen such that the indicated oxidation state dominates the speciation in both cases), on the right. *Reproduced from Smedley and Kinniburgh (2002) with permission from Elsevier.*

environment. The general pathways of arsenic release are presented in Figure 8.3. Regardless the arsenic source and distribution pathway in nature the attention shall be made on possibility of involving living organisms (biota), which might have serious repercussions.

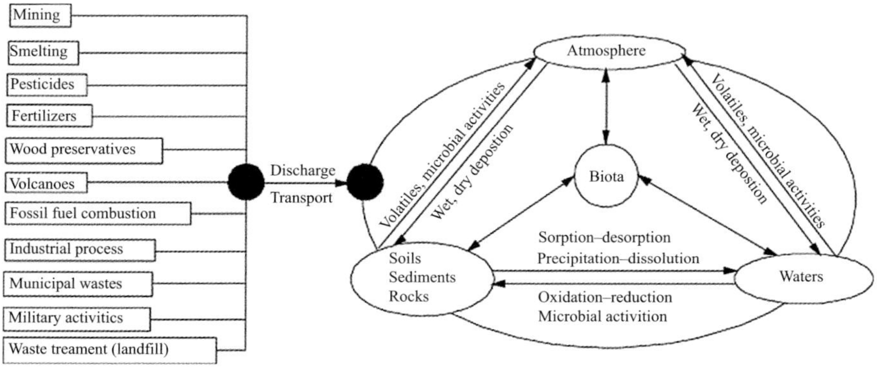

FIGURE 8.3 **A simplified diagram of the arsenic cycle.** *Reproduced from Wang and Mulligan (2006) with permission from Elsevier.*

Natural arsenic can be found in mineral and hydrothermal veins. The mineral composition depends on the physical and chemical conditions, especially other metals, resulting in diverse mineral compositions and properties (Table 8.1). Additionally, the minerals can be also found in smaller amounts associated to clays and rocks. Arsenic minerals occurrence in the ground is more problematic as it might influence the ground water quality, and in a consequence the drinking water. Depending on the conditions, reducing and high pH/oxidizing, the occurrence of ground water total arsenic concentrations was predicted for different regions around the world (Figure 8.4). Although contaminated aquifers may be found in each region it is important to note that the prediction only indicates probability areas, where clean aquifers may be also found (Amini et al., 2008). Generally, ground water with a high probability of arsenic contamination appears to be widespread across the world. However there is a region known from high level of arsenic occurrence in ground water. Therefore the worst

TABLE 8.1 Major Arsenic Minerals Occurring in Nature

Mineral	Composition	Occurrence
Native arsenic	As	Hydrothermal veins
Niccolite	$NiAs$	Vein deposits and norites
Realgar	AsS/As_4S_4	Vein deposits, often associated with orpiment, clays and limestones, also deposits from hot springs
Orpiment	As_2S_3	Hydrothermal veins, hot springs, volcanic sublimation product
Cobaltite	$CoAsS$	High-temperature deposits, metamorphic veins
Arsenopyrite	$FeAsS$	The most abundant As mineral, dominantly mineral veins
Arsenolite	As_2O_3	Secondary minerals formed by oxidation of arsenopyrite, native arsenic and other As minerals
Claudetite	As_2O_3	Secondary minerals formed by oxidation of realgar, arsenopyrite arsenic and other As minerals
Scorodite	$FeAsO_4*2H_2O$	Secondary mineral
Hoernesite	$Mg_3(AsO_4)_2*8H_2O$	Secondary mineral, smelter wastes
Conichalcite	$CaCu(AsO_4)(OH)$	Secondary mineral
Pharmacosiderite	$Fe_3(AsO_4)_2(OH)_5*5H_2O$	Oxidation product of arsenopyrite and other As minerals

Adapted from** Smedley and Kinniburgh (2002) **with permission from Elsevier.

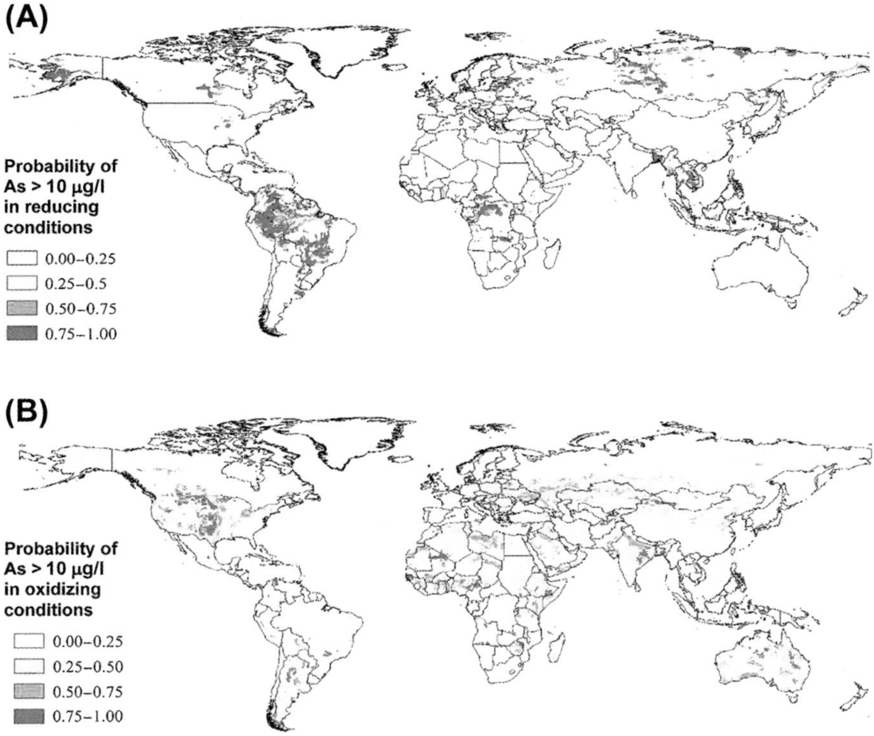

FIGURE 8.4 Modelled global probability of geogenic arsenic contamination in ground water for (A) reducing ground water conditions and (B) high-pH/oxidizing conditions where arsenic is soluble in its oxidized state. *Reprinted with permission from Amini et al. (2008). Copyright 2008 American Chemical Society.* (For colour version of this figure, the reader is referred to the online version of this book.)

situation, concerning arsenic presence in ground water, is in Bangladesh, where more than 50% of the wells exceed the concentration of 5 μg/l, and in some cases the amount of arsenic exceeds 1000 μg/l (Figure 8.5). Moreover, it has been estimated by WHO that 57 million of the people drink arsenic-contaminated water (>10 μg/l) and the situation is called the largest mass poisoning in the history (Steenhuysen).

Moreover, a consequence of arsenic release to environment can be spreading in other drinking water resources. For example the analysis of more than 2700 samples of river sediments collected in Europe showed that released arsenic compounds might be a major environmental problem. The median value of all analysis was 8 μg/l, but the mean value of 14.1 μg/l points out that in some areas arsenic might be an essential problem (Chon et al., 2010). Therefore the wide arsenic incidence in environment causes its presence in food, especially in fishes that might live in contaminated waters, which is the easiest way of arsenic spreading.

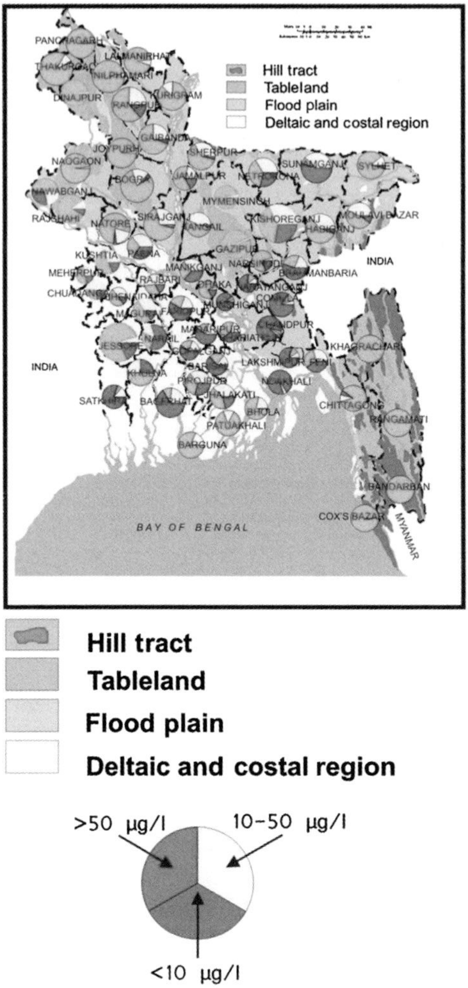

FIGURE 8.5 Four geomorphological regions of Bangladesh and the status of As contamination in ground water in 64 districts of Bangladesh. *Adapted from Chakraborti et al. (2010) with permission from Elsevier.* (For colour version of this figure, the reader is referred to the online version of this book.)

The examples of arsenic concentrations in various food groups are presented in Table 8.2. The high amounts in meats and fishes might be related with bioaccumulation in the animal tissues and bones, which shows the biggest tendency to absorb arsenic compounds.

8.1.3. Behaviour of Arsenic in Water Environment

When the arsenite and the arsenate will get to the surface waters they follow different reactions, which pathways depend on the water condition and composition. The arsenic compounds can be dissolved,

TABLE 8.2 Mean Concentrations (ng/g) of Total Arsenic in Food

Fats, Oils, Sweets		Vegetables	
Beet sugar	12.2	Beans (green)	2.1 J
Cane sugar	23.8	Carrots	7.3
Butter	1.8 U	Corn (kernel)	1.6 J
Salt	4.8	Cucumber	9.6
Beer	2.7 J	Onions	9.6
Meat, Poultry, Fish, Eggs, Nuts		Peas	4.3
Beef	51.5	Potatoes	2.8 J
Chicken	86.4	Spinach	5.1
Pork	13.5	Tomato	9.9
Eggs	19.9	**Fruit**	
Saltwater finfish	2360	Apple, raw	4.8
Tuna	512	Apple, juice	7.6
Freshwater finfish	160	Banana	2.3 J
Shrimp	1890	Grapes	10.2
Peanut butter	43.6	Grape juice	58.3
Bread		Orange	1.6 J
Corn (meal)	38.6	Orange juice	4.8
Flour	39.1	Peaches	3.4 J
Rice	303	Watermelon	40.2

Adapted from Schoof et al. (1999) with permission from Elsevier.

precipitated or adsorbed on other precipitated compounds, for example in a stratified lake, where can be distinguished two layers; top one, epilimnion and lower hypolimnion. In the top layer water is warm and aerated making oxidative conditions, but in the lower layer the water is colder and poor aerated with reductive conditions. Thus the two layers differ in processes that can occur what is presented in Figure 8.6.

The presence of dissolved oxygen results in higher redox potential and enables oxidation of arsenite to arsenate. The As(V) compounds can remain dissolved in solution or be adsorbed on insoluble matter, for example precipitated iron. Arsenate might also form insoluble compounds, like ferric arsenate, or be transformed back into arsenite, if reductive conditions occur. In case of arsenite, it is not only oxidized to arsenate but also produces arsenic sulphides after reaction with sulphide

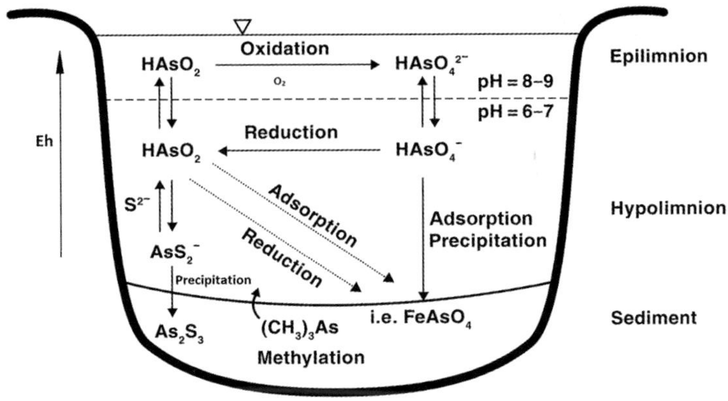

FIGURE 8.6 **Arsenic behaviour in stratified lake.** *Inspired from Dojlido.*

anion. More complicated reactions can involve As(III), as not only production of As(V). In these reaction important role play microorganisms capable to produce in methylation process organoarsenic compounds, like presented in Table 8.3.

8.1.4. Effect on Human Health

The arsenic in a high dose was used as a poison, but it was also found that arsenic even in a small concentration, dosed for a long period, can cause chronic diseases (Box 8.1). Generally it is stated that the As(III) forms are more toxic than the As(V). After an ingestion of the arsenic it can end point in different organs and affect their systems, for example the respiratory, cardiovascular, immune, genitourinary, reproductive, gastrointestinal and nervous systems (WHO, 2001). Therefore some As content was found in different human tissues. It was found that arsenic occurrence is mostly accumulating in hair and nails, and least in brain tissue (Dang et al., 1983). Due to problems with delivering arsenic-free drinking water its content could have been found in hair, nails and urine of the affected people, Table 8.4 (Karim, 2000).

An arsenic effect can be distinguished between a short- and long-term exposure. The short-term exposure reflects to an ingestion of the high dose of arsenic in a short time period and results in a violent vomiting and other gastrointestinal symptoms, which can occur after 30–60 min. Other specific symptoms that occur after the high dose eating are a metallic or garlic-like taste, burning lips and swallowing difficulties. After the initial gastrointestinal problems, multiorgan failures may occur, followed by death. Survivors of an acute arsenic poisoning have been shown to develop a hepatomegaly, melanosis,

TABLE 8.3 Structure of Arsenic Species Identified in Water with Some pKa Values (IARC Monographs, 2004)

Name	Structure	Name	Structure
Arsenous acid (arsenite)	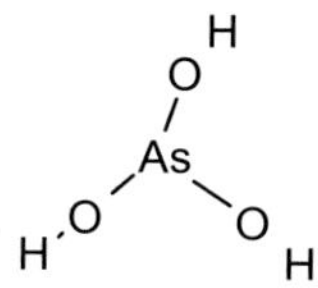 pKa 9.23, 12.13, 13.4	Arsenic acid (arsenate)	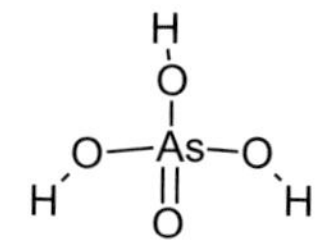 pKa 2.22, 6.98, 11.53
Monomethylarsonous acid – MMA(III)	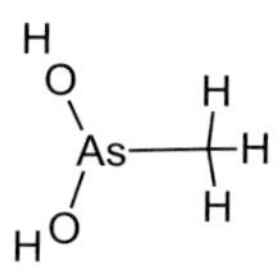	Monomethylarsonic acid MMA(V)	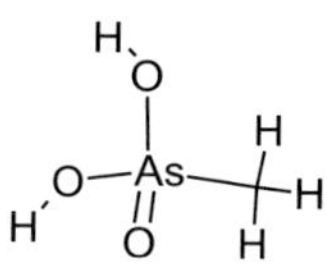 pKa 4.1, 8.7
Dimethylarsinous acid DMA(III)	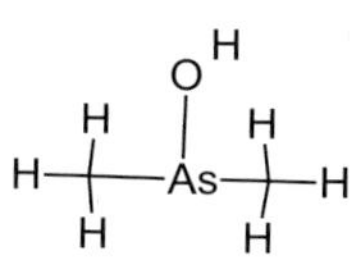	Dimethylarsinic acid DMA(V)	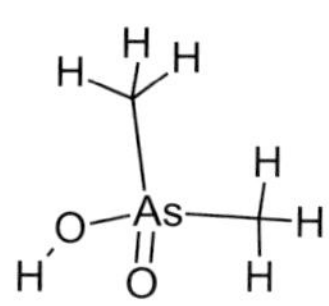 pKa 6.2
Trimethylarsine TMA	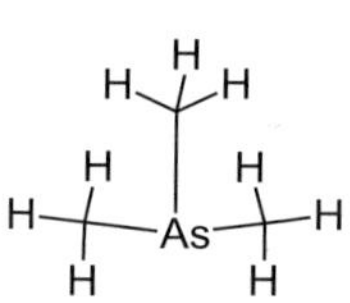	Trimethylarsine oxide TMAO	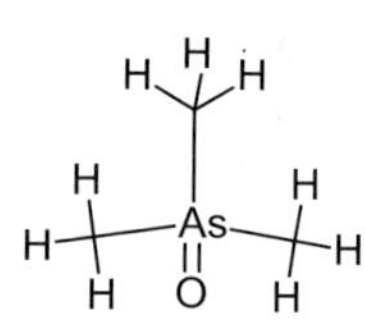

bone marrow suppression, haemolysis, and a polyneuropathy resulting from damage to the peripheral nervous system. The fatal poisoning can take place after intake of 2–8 g of arsenic according to different sources (United Nations, 2003). However there were reported incidents where people were exposed to a smaller than fatal dose. A consumption of water, with arsenic concentration of about 108 mg/l, for 1 week resulted in two of nine person died, four developed encephalopathy and eight showed gastrointestinal symptoms (WHO, 2001; United Nations, 2003).

BOX 8.1

FAMOUS ARSENIC POISONINGS

The arsenic in high doses was used as a poison; thus it was mentioned in some criminal stories as a homicide instrument, although there are some cases less or more scientifically documented, which are an example of the arsenic poisoning on the famous people. These examples with story descriptions are the following:

Napoleon Bonaparte (was the Emperor of the French who conquered almost all Europe). He died in exile on the Island of St Helena. Originally it was suggested that the cause of death was stomach cancer, but in 1961 an elevated arsenic concentration was found in Napoleon's hair. This finding elicited numerous theories of conspiracy, treachery and poisoning. Most recent reports even suggested that an inappropriate medical treatment may have contributed to the exiled Emperor's death (Lugli et al., 2005).

King George III (a king of the Great Britain in the eighteenth century, known also from his madness in the end of his life). Also in this case the arsenic was found in hair, which contained over a 300 times toxic level. Therefore this element was blamed to trigger a king's porphyry, treated as a king's insanity. However the mystery was also that who could poison the king, and an answer was found in a medication, which was given to the king and that contain antimony. The antimony, even when purified, contains significant traces of the arsenic. Thus the arsenic from every medication he was being given to control his 'madness' was triggering more attacks (BBC NEWS Medical Mysteries).

Simon Bolivar (played a key role in the Hispanic America's successful struggle for independence from the Spanish Empire). His cause of death could have been the chronic arsenic poisoning that led to a serious respiratory illness. Considering the many attempts on Bolivar's life throughout his career as a revolutionary, it has been considered a possibility that the death was an assassination. But most of the signs and symptoms point to a slow chronic poisoning, the kind that might be a result from drinking contaminated water. An environmental contact with arsenic would have been entirely possible, because he spent a lot of time in Peru, where discovered Colombian mummies have been positively tested for containing high levels of arsenic, which indicates the possibility that water in Peru may have had unusually high levels of the naturally occurring arsenic species. Moreover he could also ingest the arsenic as a common medical remedy for headaches, wasting and haemorrhoids in those times (University of Maryland Medical Center, 2014).

BOX 8.1 (*cont'd*)

FAMOUS ARSENIC POISONINGS

Charles Francis Hall (was an American Arctic explorer). The autopsy of his body, made by his biographer in 1968, showed that he had received high doses of arsenic in the last two weeks of his life. The person who might have been responsible for Hall's death was Dr Emil Bessels, who was a member of Hall's last expedition and was hated by him. Bessels could have intentionally or unintentionally overdosed him with arsenic, which was commonly used as a medicine in the nineteenth century, but the question of whether or not Hall was murdered remains unanswered (Paddock et al., 1970).

A long-term exposure with the lower levels of arsenic has been also linked to the adverse health effects in a human population. The first reports from the nineteenth century have connected skin effects (including pigmentation changes, hyperkeratosis and skin cancers) with the consumption of arsenic in medicines and drinking water (WHO, 2001, 2011).

Nowadays the chronic health effects have been described and enclose following diseases:

1. Vascular diseases:
   a. Peripheral vascular diseases;
   b. Cardio- and cerebrovascular diseases;
   c. Hypertension;

TABLE 8.4 Concentration of Arsenic in Urine, Hair and Nails of the Affected People in Different Arsenic-Contaminated Water Ingestion Episodes

Location	Concentration in Urine (mg/l)	Concentration in Hair (mg/kg)	Concentration in Nails (mg/kg)
Fairbanks, AL	0.1783	1.0	4.0
Millard County, UT	0.025–0.66	0.10–4.7	–
Antofagasta, Chile	0.025–0.77	4.00–83.4	–
Lassen County, CA	–	0.01–2.00	–
Taiwan	0.0366–0.259	–	–
West Bengal, India	0.03–2.0	1.81–31.05	1.47–52.03
Bangladesh[1]	0.05–9.42	1.1–19.84	1.3–33.98

[1]*Average of the few data available in affected area.*

Reproduced from Karim (2000) with permission ***from Elsevier***

2. Cancer:
 a. Cancers of the lung, bladder and kidney;
 b. Skin cancer and precancerous lesions of the skin.

Moreover the arsenic exposure has been linked to diabetes mellitus according to results of Tseng et al., (1968), who found out a relationship between this disease and a cumulative exposure to arsenic drinking water. Also it is generally accepted that the arsenic long-term exposure is responsible for neurological adverse effects. Moreover the acute arsenic poisoning might cause a reproductive effect in humans, but there is an insufficient evidence to support this conception (WHO, 2001; United Nations, 2003).

The health effects caused by acute arsenic poisoning are called arsenicosis, which has been also responsible for keratosis, skin changes and hyperkeratosis, skin lesions, as shown on the Figure 8.7.

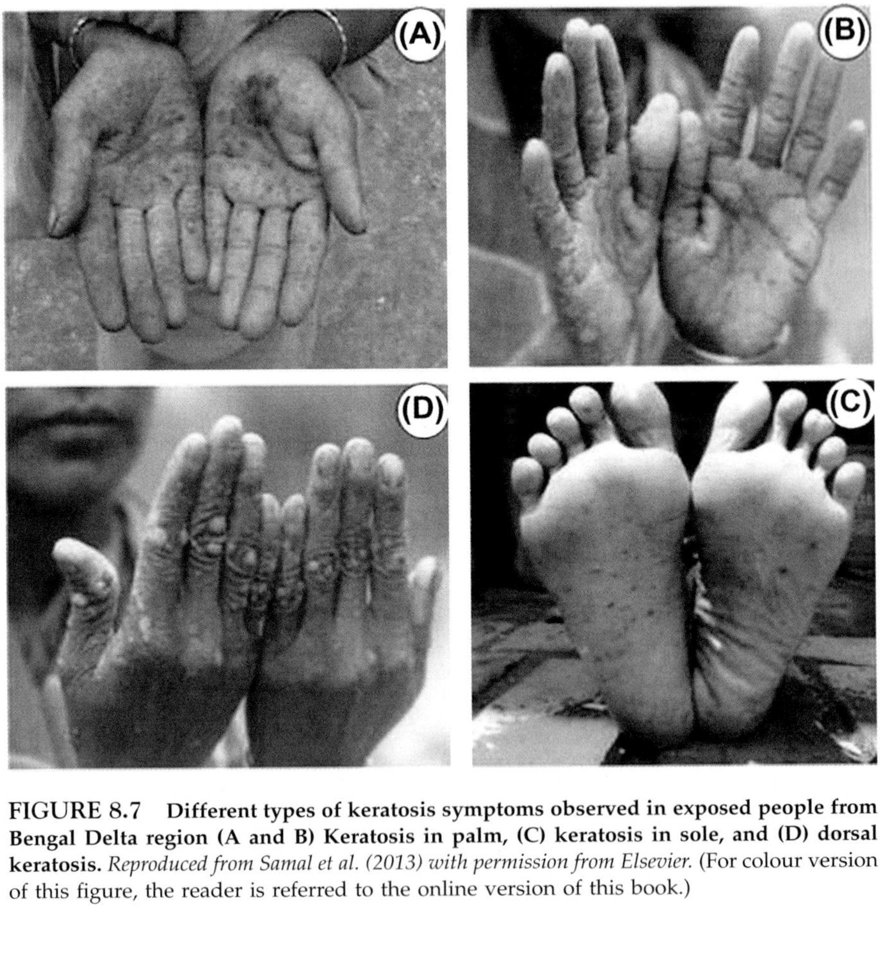

FIGURE 8.7 **Different types of keratosis symptoms observed in exposed people from Bengal Delta region (A and B) Keratosis in palm, (C) keratosis in sole, and (D) dorsal keratosis.** *Reproduced from Samal et al. (2013) with permission from Elsevier.* (For colour version of this figure, the reader is referred to the online version of this book.)

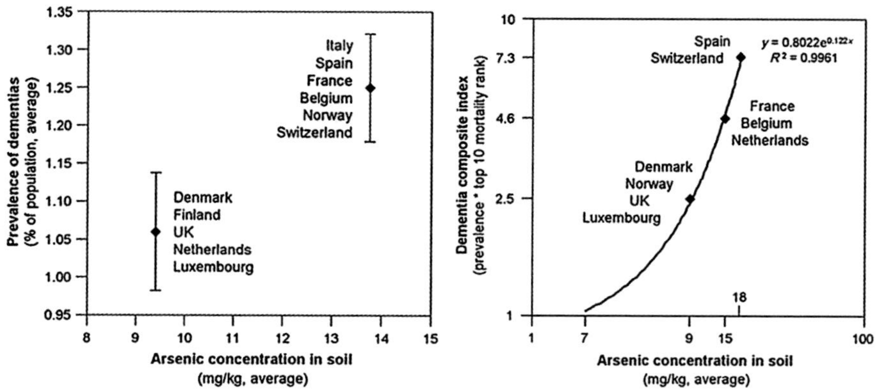

FIGURE 8.8 A combination of geological and epidemiological data which indicates the arsenic influence on the dementia occurrence (on the left) together with obtained model (on the right) which describes Dementia composite index (obtained by multiplying the prevalence data as reported in the Delphi consensus study by the rank order of dementias as a cause of death, i.e. the number of causes of death that are inferior in rank order to Alzheimer's disease and other dementias – AOD), as reported by WHO for European countries in which AOD figures among the 10 leading causes of death *Reproduced from Dani (2010) with permission from Elsevier.*

The sources of As compounds in the human body are limited to not only food and drinks but also surrounding environment. A study showed that there is a relationship between arsenic concentration in topsoil and diseases in the European countries. The study focused on the 'fool's gold', nickname of an arsenopyrite (FeAsS), but other sources of the arsenic, like fossil fuels, are not neglected. The report indicates that the arsenic concentration in topsoil is exponentially related to the prevalence and mortality of the Alzheimer's disease and other dementias. The results show that even a low-level arsenic exposure can affect on the human morbidity and mortality, which shows evidently that this compound is very dangerous even in the small doses (Figure 8.8) (Dani, 2010).

Because arsenic can be dangerous even in small concentrations all over the world are used very rigorous regulations for the maximum concentration of arsenic in the drinking water. However allowable concentrations are varying among different countries and correspond to their development stage, and better developed countries have lower arsenic concentration in drinking water (Table 8.5).

8.2. REMOVAL TECHNOLOGIES

Contradictory to the organic compounds arsenic cannot be just eliminated by an oxidation, which enables to neutralize harmful

TABLE 8.5 The Currently Accepted National Standards for Arsenic in Drinking Water (Danish Ministry of the Environment, 2007a; WHO, 2011; NJDEP, 2011)

Standard	Countries
Countries whose standard is lower than 10 μg/l	Australia (7 μg/l); Denmark (5 μg/l); New Jersey, USA (5 μg/l)
Countries whose standard is 10 μg/l	European Union; Japan; Namibia; Syria; USA
Countries whose standard is 50 μg/l	Bahrain, Bangladesh, Bolivia, China, Egypt, India, Oman and Saudi Arabia, Argentina, Brazil, Chile

compounds, but it has to be physically separated from solution. The most common technologies for arsenic removal are based on the chemical processes of precipitation and coagulation, which involve usage of iron and aluminium salts in precipitation/co-precipitation of arsenic compounds. However they are not used solely, but often have to be supplemented with separation technique to exclude insoluble arsenic-bearing precipitates from solution. General descriptions of technologies, and chemical processes taking part in them, are presented in Table 8.6.

The treatment for arsenic removal, with use of processes from Table 8.9, can be enhanced by oxidation stage, altering arsenic chemical form from arsenite to arsenate. The oxidation stage shall be considered on the beginning of treatment to enable efficient production of arsenate that is easier to remove. Another important improvement of arsenic removal might be done with the help of biological based processes involving bacteria that can play an important role in catalyzing many of the above processes (WHO, 2001).

Nevertheless, each technology has disadvantages and advantages, especially regarding efficiency and costs, which decide on implemented treatment (Table 8.7). Other factors influencing the choice of proper arsenic treatment are the following:

- Local guidelines for arsenic level in drinking water;
- Country development stage, in more developed countries can afford more expensive technologies;
- Authority's requirements and restrictions for water treatment technologies.

Therefore many different technologies can be found around the world. Two leading technologies nowadays are adsorption and coagulation–filtration. Some studies tended to favour coagulation–filtration, regarding costs and hazardous waste handling (Jones et al., in press). However decision about selection of the treatment shall be based on water chemistry and process characteristics. Additionally the general trend in water treatment is to use as less of chemicals and energy as possible, to reduce the costs. Though, these requirements result in more expensive equipment that also involves highly qualified people to maintain it. For better understanding the arsenic removal techniques, involving chemical oxidants and usage of the aluminium and iron, next parts will describe them in detail.

TABLE 8.6 Arsenic Removal Technologies with Processes Involved

Technology	Chemical Processes Involved	Description
Precipitation/ Co-precipitation	Precipitation, coagulation, adsorption	Uses chemicals, iron or aluminium salts to transform dissolved contaminants into an insoluble solid or form another insoluble solid onto which dissolved contaminants are adsorbed. When coagulants are added and form flocs arsenic can become insoluble and form solids, this is known as co-precipitation. This solid can then be removed through sedimentation and filtration.
Filtration	Solid/liquid separation Physical exclusion	Gravity settling (also called sedimentation) can accomplish some of the task of separating arsenic-bearing solids from solution, but filtration is more effective. Most commonly, sand filters are used for this purpose. However some membranes were found to perform similarly, or even better, to media filters by acting as a molecular filter to remove dissolved arsenic, along with many other dissolved and particulate compounds.
Adsorption	Adsorption	Various solid materials, like iron and aluminium oxides or activated carbon, have a strong affinity for dissolved arsenic. Arsenic is strongly attracted to sorption sites on the surfaces of these solids, and is effectively removed from solution. The adsorption media is usually packed into a column. As contaminated water is passed through the column, contaminants are adsorbed.
Ion exchange	Ion exchange, adsorption	It is other form of adsorption that involves weaker bonds resulting in reversible displacement of an ion adsorbed onto a solid surface by a dissolved ion. The ion exchange media is usually packed into a column. As contaminated water is passed through the column, contaminants are removed.

Adapted from (WHO (2001), Environmental Protection Agency of the United States (2002).

TABLE 8.7 A Comparison of Main Arsenic Removal Technologies

Technology	Advantages	Disadvantages	Removal (%)
OXIDATION/PRECIPITATION			
Air oxidation	Relatively simple, low cost but slow process; in situ arsenic removal; Also oxidizes other inorganic and organic constituents in water	Mainly removes arsenic (V) and accelerates the oxidation process	80
Chemical oxidation	Oxidizes other impurities and kills microbes; relatively simple and rapid process; minimum residual mass	Efficient control of the pH and oxidation step is needed	90
COAGULATION/CO-PRECIPITATION			
Alum coagulation	Durable powder chemicals are available; relatively low capital cost and simple in operation; effective over a wider range of pH	Produces toxic sludge; Low removal of arsenic; pre-oxidation may be required	90
Iron coagulation	Common chemicals are available; more efficient than alum coagulation on weight basis	Medium removal of As(III) Sedimentation and filtration needed	94.5
Lime softening	Most common chemicals are available commercially	Readjustment of pH is required	91
SORPTION TECHNIQUES			
Activated alumina	Relatively well known and commercially available	Needs replacement after four to five regeneration	88
Iron-coated sand	Expected to be cheap; No regeneration is required to remove both As(III) and As(V)	Yet to be standardized Produces toxic solid waste	93
Ion exchange resin	Well-defined medium and capacity The process is less dependent on pH of water Exclusive ion-specific resin to remove arsenic	High cost medium; Requires high-tech operation and maintenance; Regeneration creates a sludge disposal problem As(III) is difficult to remove Life of resins	87
MEMBRANE TECHNIQUES			
Nanofiltration	Well-defined and high removal efficiency	Very high-capital cost Pre-conditioning High water rejection	95

Table 8.7 A Comparison of Main Arsenic Removal Technologies—cont'd

Technology	Advantages	Disadvantages	Removal (%)
Reverse osmosis	No toxic solid waste is produced	High-tech operation and maintenance	96
Electrodialysis	Capable of removal of other contaminants	Toxic wastewater produced	95

Reproduced from Parga et al. (2005) with permission from Elsevier.

8.2.1. Oxidation

An oxidation stage in the arsenic treatment is a process of As(III) to As(V) transformation. Such process helps in gaining the arsenic compounds in an ionised form in a wide pH range that increases the affinity for adsorbents. The redox oxidation reaction of arsenic compounds in acidic conditions is described by the following equation (Sharma et al., 2007):

$$As(OH)_3 + H_2O \rightarrow AsO_4^{3-} + 5H^+ + 2e^-$$

$$E^0 = -0.56\ V$$

The arsenite can be oxidized with a help of the common oxidative agents like hydrogen peroxide, ozone or manganese (IV) oxide. Some of the oxidation reactions are presented in Table 8.8.

The oxidation reaction not always results with only As species transformation but might also be combined with other processes. For example the usage of pyrolusite (α-MnO_2) showed a good efficiency of As(III) oxidation followed by adsorption allowing to remove 90% of

TABLE 8.8 Chemical Oxidants and their Reactions with Arsenite (Sharma et al., 2007)

Oxidant	Reaction
Chlorine (Cl_2)	$H_3AsO_3 + HOCl \rightarrow AsO_4^{3-} + Cl^- + 4H^+$
Chlorine dioxide (ClO_2)	$H_3AsO_3 + 2ClO_2 + H_2O \rightarrow AsO_4^{3-} + 2ClO_2^- + 5H^+$
	$5H_3AsO_3 + 2ClO_2 + H_2O \rightarrow 5AsO_4^{3-} + 2Cl^- + 17H^+$
Chloramine (NH_2Cl)	$H_3AsO_3 + NH_2Cl + H_2O \rightarrow AsO_4^{3-} + NH_4^+ + Cl^- + 3H^+$
Hydrogen peroxide (H_2O_2)	$H_3AsO_3 + H_2O_2 \rightarrow AsO_4^{3-} + H_2O + 3H^+$
Ozone (O_3)	$H_3AsO_3 + O_3 \rightarrow AsO_4^{3-} + O_2 + 3H^+$
Permanganate (MnO_4^-)	$3H_3AsO_3 + 2MnO_4^- \rightarrow 3AsO_4^{3-} + 2MnO_2 + 7H^+ + H_2O$
Ferrate (FeO_4^{2-})	$3H_3AsO_3 + 2FeO_4^{2-} + H_2O \rightarrow 3AsO_4^{3-} + 2Fe(OH)_3 + 5H^+$

arsenic at pH 6 (Li et al., 2010). The reaction of oxidation/adsorption of arsenite with the help of MnO_2 can be considered as a complement of oxidation reaction with the help of the permanganate anion (showed in Table 8.8), which occurs according to Li et al., (2010) in the following manner:

$$MnO_2 + H_3AsO_3 \rightarrow MnHAsO_4 + H_2O$$

When it is about the efficient As(III) oxidation the use of advanced oxidative processes has also been proposed, as an involvement of highly reactive radicals seemed to be a good solution. Beside hydrogen peroxide and ozone, the another oxidative species, persulphate anion ($S_2O_8^{2-}$), was proved to be useful for the As(III) oxidation. In general, the oxidation with persulphate can be described by the following reaction:

$$\mathrm{As(III)} + S_2O_8^{2-} \rightarrow \mathrm{As(V)} + 2SO_4^{2-}$$

In particular the persulphate requires activation energy to produce persulphate radicals that are strong oxidizers compared with hydroxyl radicals. For this purpose UV (photochemical) and acoustic (sonochemical) activation can be used (Neppolian et al., 2010) resulting in the oxidation of the As(III) with the help of the produced radicals, and is believed to follow reactions

$$\mathrm{As(III)} + SO_4^{-*} \rightarrow \mathrm{As(IV)} + SO_4^{2-}$$

$$\mathrm{As(IV)} + SO_4^{-*} \rightarrow \mathrm{As(V)} + SO_4^{2-}$$

$$SO_4^{-*} + H_2O \rightarrow HO^* + H^+ + SO_4^{2-}$$

$$SO_4^{-*} + H_2O \rightarrow HO^* + H^+ + SO_4^{2-}$$

$$\mathrm{As(III)} + HO^* \rightarrow \mathrm{As(IV)} + HO^-$$

An advantage of the persulphate activation is creation of hydroxide radicals. The mechanism of the As(III) oxidation with the help of the radicals assumes an occurrence of the highly unstable arsenic form, As(IV), which immediately oxidizes into the As(V) through the reaction with the oxidizing agent. The reaction can be influenced by the presence of dissolved oxygen that was found to increase an oxidation rate, in contrast to the presence of humic acids that will retard arsenite oxidation rate by quenching radical species (Neppolian et al., 2010).

The occurrence of the As(IV) intermediates was also proposed in other arsenite oxidation reactions when advanced oxidation processes (AOP) were applied. For example employment of TiO_2-assisted photocatalysis showed that an adsorbed hydroxyl radical is produced on the titania surface and the As(III) oxidation occurs according to a mechanism that firstly involves adsorption on the TiO_2 surface, where the oxidation takes place (Xu et al., 2005):

$$\{TiO_2\} + As(III) \overset{k_{abs}, k_{des}}{\leftrightarrow} As(III)_{ads}$$

$$\{TiO_2 + {}^{\bullet}OH\} + As(III)_{ads} \rightarrow As(IV)_{ads}$$

$$As(IV)_{ads} \rightarrow As(III)_{ads} + As(V)_{ads}$$

$$As(IV)_{ads} + O_2 \rightarrow As(V)_{ads} + O_2^{-\bullet}/HO_2^{\bullet}$$

The studies indicate that the adsorbed hydroxyl radical on the TiO2 surface, playing a main role in the process, was proposed to oxidize the As(III) to the As(IV) intermediate on the surface. Afterwards, the As(IV) species can disproportionate to As(III) and As(V) at the surface or be oxidized to As(V) by adsorbed oxidative species, like the molecular oxygen or the hydroxyl radical (Xu et al., 2005).

Even though mush work have been done the mechanisms involving As(IV) intermediates are still studied, but generally the AOPs have proved to be applicable for the oxidation of arsenite. However usage of the chemical oxidants or the AOP's in the As(III) oxidation is rather uncommon in a full-scale treatment, even though they stand out with high oxidation rate constants (Figure 8.9).

The most common and easiest used oxidative agent is molecular oxygen that can be delivered to solution in contact with atmospheric air. Although oxygen is not as efficient as other chemical oxidants, but it has

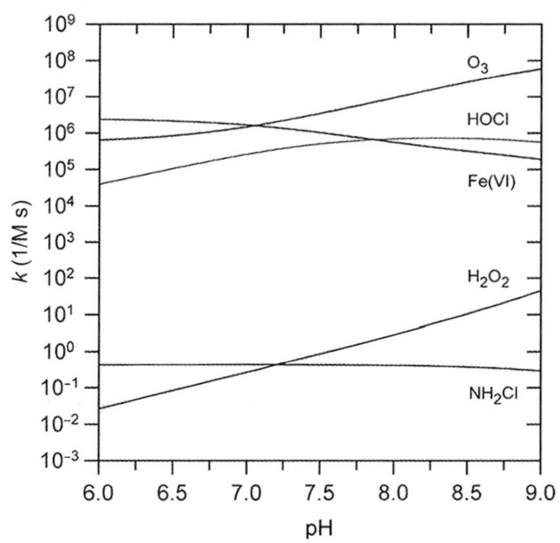

FIGURE 8.9 Observed rate constants (k, 1/M s) as a function of pH for the oxidation of As(III) by different oxidants. *Reproduced from Sharma et al. (2007) with permission from Taylor&Francis.*

TABLE 8.9 Half-lives of the Oxidation Reactions of As(III) in Ground Water (Kim and Nriagu, 2000)

Ground Water	Oxidant	T_c (Half-life)
Higher Fe and Mn concentration	Ozone	4.2 min
	Oxygen	2.2 days
	Air	4.0 days
Lower Fe and Mn concentration	Ozone	4.5 min
	Oxygen	4.8 days
	Air	8.9 days

some advantages – low costs and easy access, oxygen can be delivered in an aeration process. As(III) oxidation half-live ($t½$) in the presence of oxygen is a couple magnitudes higher than for the previously presented chemical oxidants for which it was in range of milliseconds to hours (Sharma et al., 2007). However the oxidation time can be noticeably reduced in the presence of iron and manganese, naturally occurring in some ground waters, as it is presented in Table 8.9.

The results of Kim and Nriagu (2000) are just an example of studies about a role of iron species on arsenite oxidation in the presence of molecular oxygen. It has been proposed that in oxygen-rich water and in the presence of the iron (II) pseudo-Fenton reagent might be produced. The mechanism of iron (II) and dissolved oxygen role in As(III) oxidation is shown in Table 8.10. The attention shall be made into the presence of As(IV) and Fe(VI), which are intermediate species in proposed reactions pathway.

TABLE 8.10 Reactions Proposed for Fe(II)-Enhanced Oxidation of As(III) in Air (Sahai et al., 2007)

Reaction
Fe(II) oxidation generating reactive oxygen species
$Fe(II) + O_2 \rightarrow Fe(III) + O_2^{*-}$
$2Fe(II) + 2O_2^{*-} + 4H^+ \rightarrow 2H_2O_2 + 2Fe(III)$
$Fe(II) + {}^*OH \rightarrow Fe(III) + OH^-$
$Fe(II) + H_2O_2 = Fe(III) + {}^*OH + OH^-$ (Fenton reaction)
$Fe(II) + H_2O_2 \rightarrow 2OH^- + Fe(IV)$ (Additional intermediate oxidant generation)
As (III) oxidation
$Fe(IV) + As(III) \rightarrow Fe(III) + As(IV)$
$As(IV) + O_2 \rightarrow As(V) + O_2^-$
Net redox reactions
$4Fe(II) + O_2 + 10H_2O = 4Fe(III)(OH)_3 + 8H^+$
$2H_3As(III)O_3 + O_2 = 2H_2As(V)O_4 + 2H^+$

Analogous reactions were proposed for the system where zero-valent iron (ZVI, Fe^0) was implemented. The corrosion of ZVI in oxygen-rich environment has revealed some mechanisms, where oxidative species are produced through the following reactions on the iron surface:

$$\equiv Fe^0 + O_2 + xH^+ + 2H_2O \rightarrow \equiv Fe^0\left[Fe^{II}(OH)_{2-x}(H_2O)_xH_2O_2\right]^{x+}$$

$$\equiv Fe^0\left[Fe^{II}(OH)_{2-x}(H_2O)_xH_2O_2\right]^{x+} \rightarrow \equiv Fe^0\left[Fe^{II}(OH)_{2-x}(H_2O)_x\right]^{x+} + H_2O_2$$

$$\equiv Fe^0\left[Fe^{II}(OH)_{2-x}(H_2O)_xH_2O_2\right]^{x+} \rightarrow \equiv Fe^0\left[Fe^{II}(OH)_2Fe^{II}(OH)_{2-x} \times (H_2O)_x\right]^{x+}$$

$$\equiv Fe^0\left[Fe^{II}(OH)_{2-x}(H_2O)_x\right]^{x+} + (2-x)H^+ \rightarrow \equiv Fe^0 + Fe^{2+} + 2H_2O$$

$$Fe^0\left[Fe^{II}(OH)_2Fe^{II}(OH)_{2-x}(H_2O)_x\right]^{x+} + (4-x)H^+ \rightarrow \equiv Fe^0 + 2Fe^{2+} + 4H_2O$$

The produced hydrogen peroxide can react with the main corrosion products – iron (II) ions, and in effect produce hydroxide radicals. The produced oxidative species might be involved in further arsenic oxidation. Based on the proposed mechanism it was claimed that arsenite was oxidized mainly in the solution by Fenton reagent, rather to be removed by a sorption on freshly formed hydrous ferric oxide (Katsoyiannis et al., 2008). However the creation of hydrogen peroxide on the Fe^0 surface has been discussed to be uncommon in these conditions. The new interpretation pointed on an adsorption/surface–oxidation mechanism that combines firstly the adsorption or co-precipitation followed by the arsenite oxidation on the iron surface (Blaney et al., 2007; Pang et al., 2009; Noubactep, 2009a,b).

8.2.2. Co-precipitation and Adsorption of Arsenic

The precipitation process is based on an occurrence of non-soluble arsenic compounds, like arsenic sulphides or arsenate and arsenite salts. Although the process reduces arsenic compounds solubility, it does not guaranty an elimination of arsenic to desirable limit. For example it was possible to gain an arsenic solubility reduction from 6430 to 0.823 mg/l in a process of the arsenic sludge waste neutralization that used high excess of calcium and ferric additives (Palfy et al., 1999). Unfortunately the precipitation of arsenic as metal salts, like calcium or magnesium arsenates, is sensitive to carbon dioxide presence, which is its main drawback. The CO_2 presence causes an increase in the solubility of arsenic compounds by decomposition of solids to more soluble carbonates (Robins, 1981).

Thus the main focus of the arsenic removal has been directed to the co-precipitation and adsorption based mostly on aluminium and iron compounds. Even though it is hard to distinguish between co-precipitation and adsorption, there are some definitions that tried to explain it. The adsorption is a two-dimensional process binding metal on a particle surface, while the co-precipitation is a three-dimensional process incorporating a metal into the bulk particle structure. Moreover reversible processes of the co-precipitation and adsorption, that is dissolution and desorption, differ in mechanisms resulting in a phenomena of less readily release of the co-precipitated metals in a consequence of changes in pH or a ligand concentration (Fulghum, 1988).

Concerning the arsenic compounds adsorption some hydroxides, of aluminium and iron, have been found to be useful and their usage has been widely described.

The aluminium hydroxide can be used as an adsorption media for the arsenate resulting in an arsenate–aluminium hydroxide complex. Depending on the complex charge, determined by the isoelectric point, the adsorption of the arsenate anion is facilitated by a columbic attraction, for positive charge, or competition with columbic repulsion takes place, for negative charge. Therefore the adsorption might be a difficult process depending on not only the amount of adsorbent but also on the conditions, like pH and arsenic oxidation state. Dependence of the adsorption, as a Langmuir constant, on pH is presented in Figure 8.10, where the

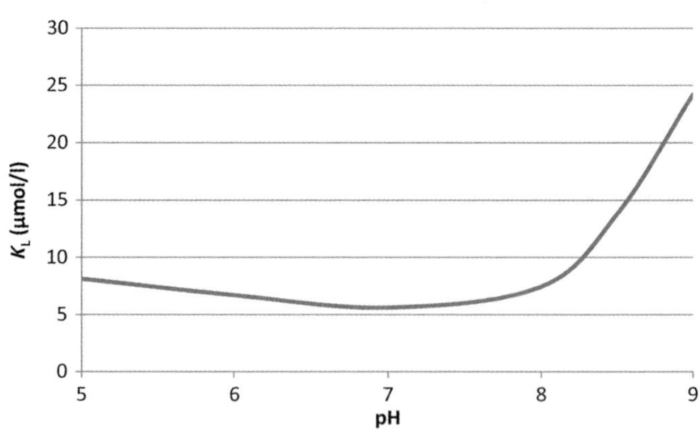

FIGURE 8.10 **Langmuir constant vs pH for aluminium hydroxide.** *Adapted from Anderson et al. (1976) with permission from Elsevier.* (For colour version of this figure, the reader is referred to the online version of this book.)

biggest arsenate adsorption on the aluminium hydroxide takes place in the highest pH, when arsenate anion occurs (Anderson et al., 1976).

The more broadly used, and hence described, are processes based on iron compounds, because of their properties in circum-neutral pH, where iron ions occur as non-dissolved particles of the different iron oxides and hydroxides. For better understating of the iron role in the arsenic removal Figure 8.11 can be studied, where solid phase of iron (III) hydroxide is present in the same conditions like the arsenic (V) compounds – $H_2AsO_4^-$ and $HAsO_4^{2-}$. The phenomena point to the main advantage of iron hydroxides for arsenic co-precipitation purposes in the circum-neutral pH, as their insoluble form has the ability to appear in the same conditions as arsenate anions. The main process that involves iron hydroxides in the arsenic removal is coagulation. However mechanisms of physicochemical reactions between the hydroxides and arsenic compounds are not clear in detail, but four theories are recently taken into consideration:

1. Precipitation, the formation of insoluble compounds such as $FeAsO_4$;
2. Co-precipitation, the incorporation of soluble arsenic species into the growing iron hydroxide phase;

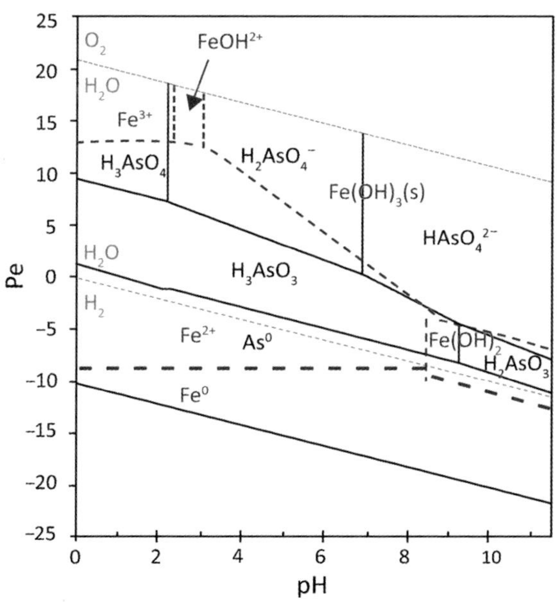

FIGURE 8.11 **Pe–pH diagram for the As–Fe–H_2O system at 25 °C.** Total As = 4 μM; total Fe = 10^{-4} M. *Reproduced from Bang et al. (2005) with permission from Elsevier.* (For colour version of this figure, the reader is referred to the online version of this book.)

3. Occlusion, the entrapment of adsorbed arsenic species in the interior of the growing iron hydroxide phase; adsorption, the binding of soluble arsenic to the external surfaces of the insoluble iron hydroxide (Phenrat et al., 2008).

The arsenic compounds sorption on the iron oxides does not only depends on the pH and redox potential but also is linked to its structure that results from different conditions of formation, like pH, pressure, co-ions and microbial activity. Even though the iron-insoluble compounds might have a similar chemical structure there are big differences in surface complexation kinetics of the arsenite and arsenate sorption on the surface among iron minerals, which can be noticed on the example of iron oxides and hydroxides presented in Table 8.11. Some studies concluded that for *amorphous iron oxide* and goethite, As(V) sorption is more favourable than As(III) at lower pH values, but the opposite is true at higher pH values (above 6.5) (Dixit and Hering, 2003).

Recent years of investigations of iron compounds role in arsenic removal brought some insights about the process improvements and limitations. Freshly precipitated Fe(III) (hydr)oxides, as a result of Fe(II) oxidation, hydrolysis and precipitation were found to have a higher sorption capability than directly dosed Fe(III) compounds (Roberts et al., 2004). Moreover, the presence of Fe(II) ions was found to increase oxidation rate of As(III) in an oxygen-rich environment (Sahai et al., 2007). However, the process has also some limitations that shall be considered in arsenic removal strategy, especially that involve adsorption. It was found that oxyacids (like H_4SiO_4, HCO_3^-, $H_2PO_4^{2-}$ and HPO_4^-), which might

TABLE 8.11 Intrinsic Surface Complexation Constants for Sorption onto Iron Oxide Minerals

Adsorption Reaction	Amorphous Iron Oxide[1]	Geothite[1]	Magnetite[1]	Iron Hydroxide[2]
$\equiv FeOH + H^+ \rightarrow \equiv FeOH_2^+$	7.29	7.47	4.60	
$\equiv FeOH \rightarrow \equiv FeO + H^+$	−8.93	−9.51	−8.20	
ARSENATE ADSORPTION CONSTANT				
$\equiv Fe-OH + AsO_4^{3-} + 3H^+ \rightarrow \equiv Fe-H_2AsO_4 + H_2O$	29.88	31.00		29.31
$\equiv Fe-OH + AsO_4^{3-} + 2H^+ \rightarrow \equiv Fe-HAsO_4^- + H_2O$	24.43	26.81		23.51
$\equiv Fe-OH + AsO_4^{3-} + H^+ \rightarrow \equiv Fe-AsO_4^{3-} + H_2O$	18.10	20.22		10.58
ARSENITE ADSORPTION CONSTANTS				
$\equiv Fe-OH + AsO_3^{3-} + 3H^+ \rightarrow \equiv Fe-H_2AsO_3 + H_2O$	38.76	39.93	38.41	40.20
$\equiv Fe-OH + AsO_3^{3-} + 2H^+ \rightarrow \equiv Fe-HAsO_3^- + H_2O$	31.87	32.40	33.02	

[1]*Dixit and Hering (2003).*
[2]*Kim and Nriagu (2000).*

behave similar to arsenite and arsenate, compete for adsorption sites resulting in lower efficiency of arsenic removal. For example hydroxylapatite ($Ca_5(PO_4)_3OH$; HAP) was found to be the main naturally occurring source of competing ion in Bangladesh ground water. Beside the competition with As compounds the oxyacids ions may inhibit precipitation of ferric oxyhydroxide phase resulting in reduction of co-precipitated arsenic. Effect of different oxyacids on the amount of arsenic uptake in As(III) systems is presented in Figure 8.12 (Ciardelli et al., 2008). Although it was found that oxyacids greatly affected the amount of arsenic uptake in the As(III) systems, only a minimal effect on arsenic uptake in the presence of the HAP was found. Among the sole oxyacids silicate and phosphate were found to reduce arsenic removal, the most has the worst effect, which was 67 and 63%, respectively, which are less by 5 and 9% than the systems without competing oxyacids. The arsenic uptake was even more reduced, to 52%, when all oxyacids were present. However the main oxyacid responsible for decreasing efficacy of As removal is phosphate, which compete for adsorption sites with arsenic compounds and has to be considered for evaluating process efficiency. For this purposes Fe/P ratio is the important factor that can provide a

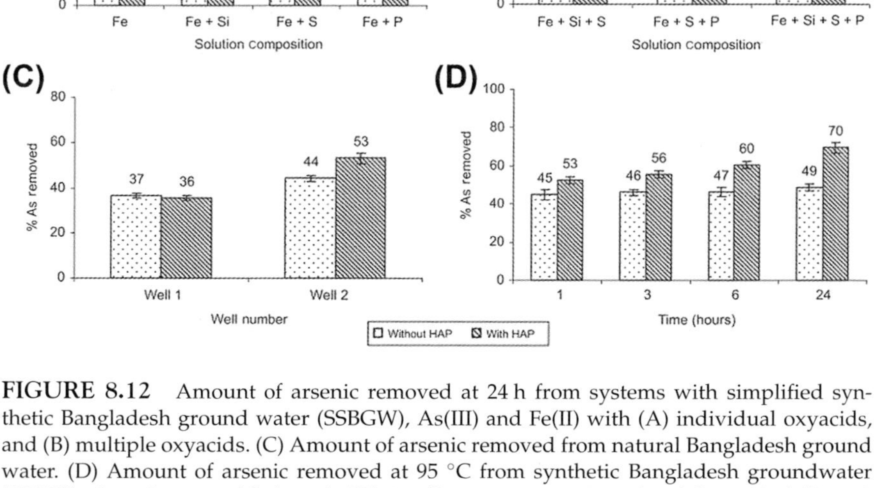

FIGURE 8.12 Amount of arsenic removed at 24 h from systems with simplified synthetic Bangladesh ground water (SSBGW), As(III) and Fe(II) with (A) individual oxyacids, and (B) multiple oxyacids. (C) Amount of arsenic removed from natural Bangladesh ground water. (D) Amount of arsenic removed at 95 °C from synthetic Bangladesh groundwater (SBGW). Arsenic was added as As(III) in all experiments. *Reproduced from Ciardelli et al. (2008) with permission from Elsevier.*

reasonable prediction of arsenic removal in a ground water. When the Fe/P ratio increases (because of either increased Fe or decreased phosphate concentrations), arsenic removal effectiveness is improved dramatically (Figure 8.13). The presence of moderate Fe (>0.1 mg/l) and P (>0.5 mg/l) concentrations in these waters explains the variable and often inefficient removal of arsenic by the filters. The Fe/As ratio typically exceeds 54 for effective treatment of As from silicate- and carbonate-containing ground waters (to <50 μg/l) of varying composition (Chiew et al., 2009).

To perform co-precipitation of arsenic on iron precipitates different iron coagulants, like ferric chloride, ferric sulphate and ferrous sulphate, are applied. It is important, especially from full-scale treatment perspective, to use the most efficient coagulant that will require the lowest dose and changes in pH to minimize the treatment costs. For example a statistical experiment design method was used to investigate the effects of major operating variables such as initial arsenate concentration, coagulant dose and pH. Noticeable is the fact that lower initial arsenic concentration requires higher coagulant dose that has also consequences in costs of the treatment. Moreover, the Fe(III) ions were found to be more effective and economic than Fe(II) ions because of required lower coagulant dose and pH. Among the Fe(III) salts the more efficient was use of ferric sulphate, as it required lower dose, but resulted in worse As removal efficiency than ferric chloride, which could have been a consequence of the lower dose of Fe(III) ions (comparing molar concentration and the fact that 1 M of $Fe_2(SO_4)_3$ produces 2 M of Fe^{3+}). Optimum pH values for maximum

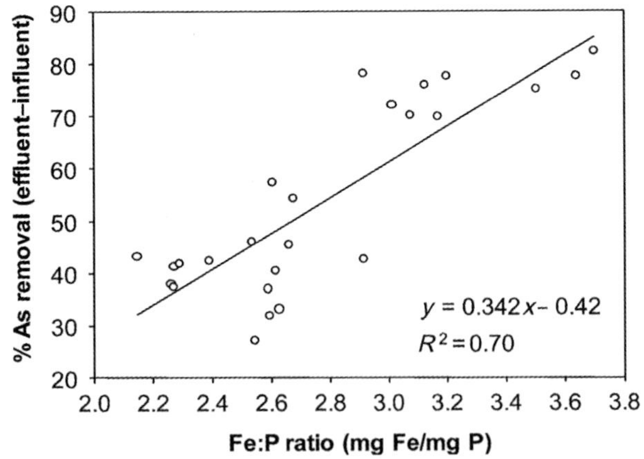

FIGURE 8.13 **Relationship between ratio of Fe/P on arsenic removal efficiency.** If the inlet iron content concentration is high or increases and/or inlet phosphorus concentration is low or reduces, arsenic removal efficiency increases dramatically. *Reprinted (adapted) with permission from Chiew et al. (2009). Copyright 2009 American Chemical Society.*

TABLE 8.12 Obtained Highest Arsenate Removal Efficiencies (Eff.) in the Optimum pH (Opt. pH) and the Optimum Coagulant Dose

		Initial As(V) Concentrations					
		10 μg/l		500 μg/l		1000 μg/l	
Coagulant Type	Opt. pH	Dose (ppm/mM)	Eff. (%)	Dose (ppm/mM)	Eff. (%)	Dose (ppm/mM)	Eff. (%)
$FeCl_3$	7.5	50/0.31	88	31/0.19	100	37/0.23	100
$Fe_2(SO_4)_3$	8	40/0.1	70	28/0.07	100	32/0.08	100
$FeSO_4$	9	60/0.39	63	60/0.39	91	58/0.38	100

Adapted from Baskan and Pala (2009).

arsenate removal for ferric chloride, ferric sulphate and ferrous sulphate were found as 7.5, 8 and 9, respectively (Baskan and Pala, 2009). The performance of using three mentioned coagulants can be compared in Table 8.12, where optimum pH and dose are stated.

The latest investigations focuses on an electrolytic iron generation that is a process where direct current voltage is applied to electrodes releasing iron ions from ZVI anode and water acts as electrolyte (Moreno Casillas et al., 2007). Electrolytic iron generation is a cheap Fe(II) dosage method, especially for coagulation purposes, where it has proved its efficiency in arsenic removal and its application is presented in the last part of the chapter (Majumder and Gupta, 2011; Lacasa et al., 2011; Kobya et al., 2011).

Another simple model was used to describe arsenic removal from the naturally contaminated ground water with usage of the simple household sand filter. The model has proved that the concentration of dissolved iron in ground water was the filtration efficiency decisive factor. Although the model calculations were based on real results of arsenic co-precipitation it neglects the presence of the oxyacids ions, Figure 8.14. Therefore a general positive correlation of As removal with initial Fe concentrations was observed, with distinction to arsenite and arsenate compounds. When comparing arsenite and arsenate removal efficiencies it was found that much more Fe is needed to achieve comparable removal rates for As(III) than for As(V). By fitting the data shown in Figure 8.14(B), it was possible to derive the following empirical equation to estimate the As removal efficiency from concentrations of Fe dissolved in ground water:

As removal (%) $= 13.6 \times \ln(\text{Fe, mg/l}) + 45$ (Berg et al., 2006).

8.2.3. Use of Biotechnology for Arsenic Removal

There are microorganisms that do not only resistant to arsenic toxicity but are able to use it in its own good. In a consequence of biochemical processes involving arsenic compounds by bacteria, like arsenite oxidase,

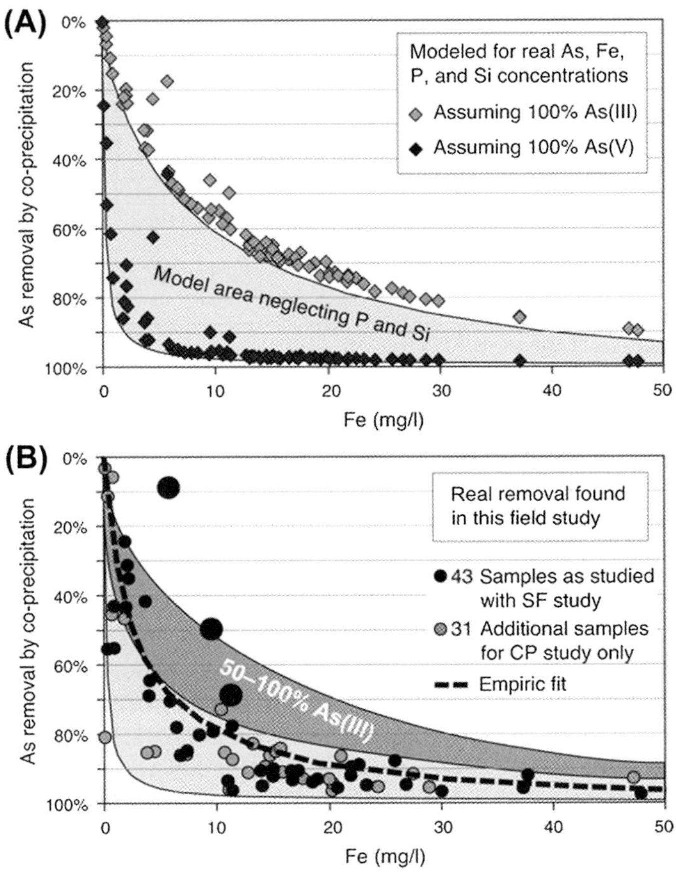

FIGURE 8.14 **Model calculations of As removal and comparison with the results obtained with real co-precipitation experiments.** (A) Theoretical As removal calculated from As(tot), Fe(tot), P and Si concentrations of the 74 samples used for the co-precipitation experiments. Scenarios were computed assuming 100% initial As(III) or As(V), respectively. The shaded area corresponds to the predicted range from the model. (B) Overlay of measured co-precipitation removal with the modelled range derived from Figure A, considering P and Si. The light and dark shaded areas represent the prediction for the abundance of 0–50% As(III) and 50–100% As(III), respectively. The best fit for measured As removal is indicated by the dotted line. Magnified symbols indicate samples with phosphate levels >2.5 mg P/l. (SF corresponds to sand filtration and CP to co-precipitation) *Adapted with permission from Berg et al. (2006). Copyright 2006 American Chemical Society.*

respiratory arsenate reductase or cytoplasmic arsenate reductase, arsenic is being incorporated into microorganisms (Figure 8.15) (Silver and Phung, 2005).

One of the unfortunate consequence of arsenic involvement in bioreactions was suggested to be arsenic release to ground water that takes place in anaerobic conditions, where microbial processes are the main

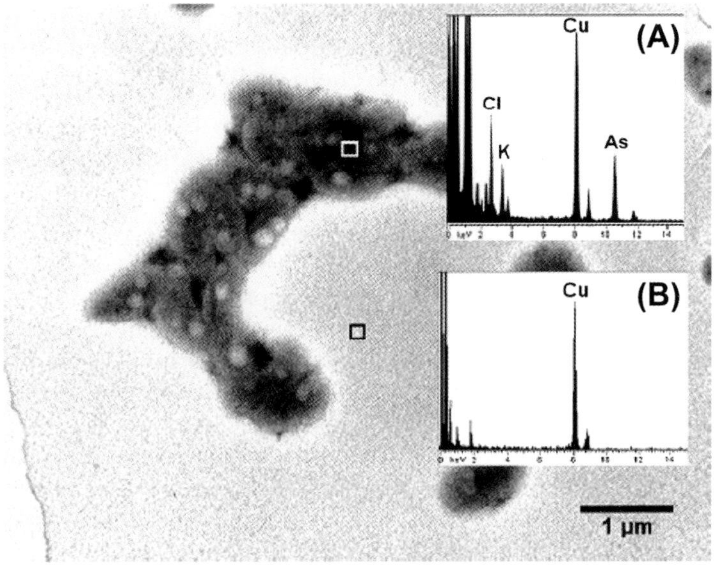

FIGURE 8.15 **Sequestration of arsenic by *Herminiimonas arsenicoxydans*.** Transmission electron microscopy picture of *H. arsenicoxydans* cells grown in As-enriched medium. Energy dispersive X-ray spectra are presented for two spots of analysis: the white square (A) and the black square (B). Cl and K peaks show organic constituents and Cu labels represent peak due to supporting grid. Arsenic content in (A) is 4.5% weight as As_2O_3 including microgrid C-coating quantification. Arsenic is not detected in (B) *Reproduced from Lievremont, D.; Bertin, P.N.; Lett, M. Arsenic in contaminated waters: biogeochemical cycle, microbial metabolism and biotreatment processes. Biochimie 2009; 91 (10): 1229-1237. Copyright © 2009 Elsevier Masson SAS. All rights reserved.*

factors contributing to arsenic mobilization (Lievremont et al., 2009). For example arsenic (V), which is less mobile in ground water than As(III), can be used as energy source, like it was found to take place for the *Bacillus arsenicoselenatis*. In growth of this bacterium As(V) is used as terminal electron acceptor during anaerobic respiration in the following reaction (Maier et al., 2009):

$$\text{lactate} + 2H_2AsO_4^- \rightarrow \text{acetate} + 2H_2AsO_3^- + CO_2 + H_2O$$

$$\Delta G_f^0 = -23.4\ \text{kJ/mole}$$

However other bacteria species were found to help in arsenic removal by their participation in oxidation and sorption processes, which are described in the next paragraphs.

8.2.3.1. Microbiological Oxidation

More desirable than the reaction responsible for releasing As(III) to environment is reaction where As(III) is transformed to As(V) that can be also achieved with the use of bacterium species. The main purpose of the As(III) oxidation, used by some microorganisms, is detoxification and reaction takes place with the help of oxygen (Katsoyiannis and Zouboulis, 2004):

$$H_3AsO_3 + \frac{1}{2}O_2 \xrightarrow{\text{bacteria}} H_3AsO_4$$

The variety of bacteria species performing above reaction (arsenite oxidizers) were reported in literature. Most of them were found in environment where naturally arsenic compounds occur, soil and mines, enabling to evolve bacteria organisms that can survive in such location, Table 8.13. The arsenite oxidizers were investigated to tolerance for arsenic and arsenite oxidation rate, which are presented in Table 8.13.

Although a number of As-oxidizing bacteria were found in the environment, only few of them can be used for drinking water purpose. The main reason is their influence on human health; so the species may not be pathogenic. Moreover the bacteria shall also have very low nutritional requirements, preferable would be those which use As(III) as an energy source. One of species fulfilling the requirements is, for example, a CASO1 population, selected from a mining site. The population was found to be a mix of different organisms, among which two main are related to *Thiomonas* and *Ralstonia pickettii*. The main features of the bacteria mixture, which make them beneficial for the water treatment processes are the following: (1) it is autotrophic and needs only a low nutrient supply; (2) it preserves its ability to oxidize As(III) over a large interval of pH, temperature and As(III) concentration and (3) high As(III)-oxidizing rates were obtained in a fixed-bed reactor (Figure 8.16) (Battaglia-Brunet et al., 2002).

Another promising biotechnology of arsenic oxidation is implementation of iron-oxidizing bacteria, like *Gallionella ferruginea* and *Leptothrix ochracea*. The bacteria were found to effectively oxidize As(III) during biotic oxidation of iron, that is the basis of simple and efficient method for iron and arsenic treatment occurring in sand filtration. Moreover, deposited iron oxides in the filter medium together with microorganisms offer favourable environment for arsenic to be adsorbed and removed from the aqueous stream (Katsoyiannis and Zouboulis, 2004). However more detailed research of Fe-rich biofilm developed during biological Fe removal indicates that other genus, than those responsible for iron oxidation – *Leptothrix* and *Gallionella*, was found to promote As(III) (Casiot et al., 2006). Nowadays investigations focus on identification of bacteria strains found in arsenic treatment systems, mechanisms involving arsenic compounds and possibilities in their implementation in kind of treatment (Box 8.2).

TABLE 8.13 Summary of Studies of Arsenite Oxidizers (Salmassi et al., 2002)

Microorganism	Isolation Environment	Ambient As	Experimental As (mM)	V_{max} – Kinetic Constant
Bacillus arsenoxydans	Cattle-dipping fluids	NR	20–100	NR
Pseudomonas arsenoxydans	Cattle-dipping fluids	NR	20–100	100 μM/mg N/h
Xanthomonas arsenoxydans				31 μM/mg dry wt/h
Achromobacter arsenoxydans				
Soil consortium	Soil	NR	2.5–40	NR
Alcaligenes faecalis	Raw sewage	NR	10–20	0.05 μmol/mg protein/min
Alcaligenes strain	Soil	NR	0.24–2.4	6.7 μl O_2/min (c. 10^{10} cells)
Pseudomonas arsenitoxidans	Mine water	NR	13.3–26.7	NR
Seawater consortium	Coastal water	<70 nM	0.0013	3–9.2 μM/h
As1–As6	Mine water	27–173 μM	0.67–6.7	NR
Sulfolobus acidocaldarius	–[1]	Na	1	NR
Alcaligenes faecalis Purified enzyme	–[1]	NA	0.20	0.023 μmol/mg protein/min
ULPAs1	Contaminated water	0.47 mmol/kg dry wt	1.33	NR
NT-25, NT-26	Gold mine arsenopyrite rock	NR	5–10	NR
Agrobacterium albertimagni	Macrophyte surfaces	2.7 μM	0.002–5	$(1.81 \pm 0.58) \times 10^{-12}$ μmol/cell/min
Strain AOL15				0.043 ± 0.017 μmol/mg protein/min

[1]*Previously isolated bacterium; NA – not applicable; NR – not reported.*

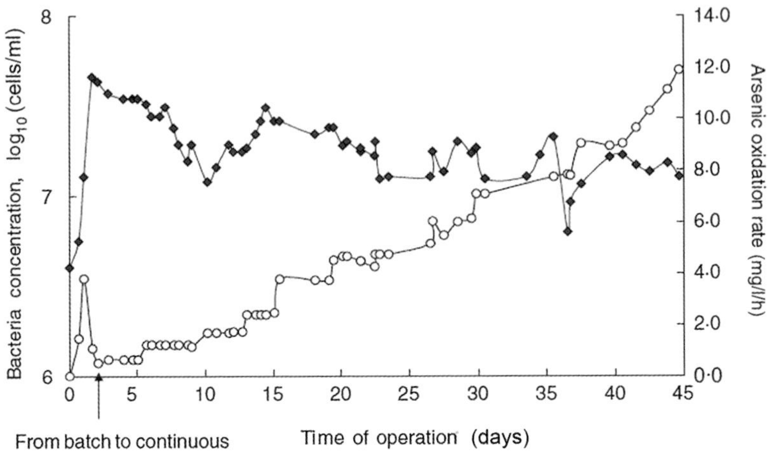

FIGURE 8.16 Results from continuous experiment in a glass reactor with development of the CASO1 bacteria biofilm (◆ – total bacteria counted under an optical microscope; ○ – As(III) oxidation rate). *Reproduced from Battaglia-Brunet et al. (2002) with permission from Wiley.*

BOX 8.2

THE ARSENIC-EATING BACTERIA

The discovery of the first microorganism on the Earth able to thrive and reproduce using the toxic arsenic compounds changed a fundamental knowledge about life. It has expanded the definition of the life, what can be used in the field of astrobiology and gives researchers more broadly an area in seeking signs of the new life in the solar system. It was found that microorganisms in the Mono Lake, USA, can substitute the arsenic for phosphorus in their cell components. The phosphorus is a part of the chemical backbone of the DNA and RNA, the structures that carry genetic instructions for the life, and is considered as an essential element for all living cells. The arsenic is chemically similar to phosphorus, and thanks to that the arsenate behaves similarly to phosphate. Thus the microorganisms, like the strain GFAJ-1, use the arsenic to build their cell parts. The key issue that the researchers investigated was the arsenic role in a microbe growing. It was found that the arsenic actually became incorporated into organisms' vital biochemical machinery, such as the DNA, proteins and the cell membranes. The results of this study bring new point of view into ongoing research in many areas, including the study of Earth's evolution, organic chemistry, biogeochemical cycles, disease mitigation and Earth system research. The finding also will open new frontiers in microbiology and other areas of research (Brown and Weselby). However the critics of the discovery suggest that the author's

BOX 8.2 (*cont'd*)

THE ARSENIC-EATING BACTERIA

own data hint at an organism that is simply absorbing and isolating the arsenate while making use of the trace phosphates in its environment. Opponents point that the arsenate-grown cells were analyzed in their resting phase, which require less phosphate for survival than does an active growth and one calculation in the paper showed that the DNA in this cells actually contained 26 times more phosphorus than arsenic (Katsnelson, 2010). Of course discussion about this new discovery has just begun, but its development and a new research will for sure bring more light in this subject.

8.2.3.2. Biological Sorption

Beside the indirect enhancement of adsorption and co-precipitation processes by arsenite oxidation, microorganisms can also play solely an important part in sorption process of arsenic compounds. Their role can be twofold, by direct sorption on the microorganisms or sorption on compounds produced by them, bioproducts.

Direct sorption of arsenic was found to take place on the different microorganisms and medias, like anaerobic or fungal biomass (Pokhrel and Viraraghavan, 2008; Chowdhury and Mulligan, 2011). For example *Lactobacillus acidophilus* (1 mg dry wt/ml) enabled to remove 30, 60, 300, 420 and 600 ppb from 50, 100, 500, 1000 and 2000 ppb of As(III)-containing water solution, respectively, within 3 h at pH 7. The increase of biomass content (to 2 mg dry wt/ml) enhanced removal of As(III) by 1.66, 1.33, 1.16, 1.42 and 1.33 times, respectively (Singh and Sarma, 2010). The biosorption, with the help of Fe(III)-treated *Staphylococcus xylosus*, was found also to be useful for the treatment of Cr(VI) and As(V) mixture. The difference in effectiveness of arsenic and chromium removal from single, with only one of the contaminant, and multi-compound, with both contaminants, mixtures was found. In binary mixtures, the competitions between the inorganic ions were found, as addition of one compound decreases uptake capacity of the other. Therefore biosorption with the use of biomass, on example of effective simultaneous removal of Cr(VI) and As(V), has shown that the process is spontaneous and driven by randomness sorption of the species at the solid/solution interface, which might result in difficulties of optimizing the process for the full-scale treatment (Aryal et al., 2011). Nevertheless the possibility of arsenic biosorption has shown to be implacable for different microbiological strains.

Another source of biomass found to have a potential for remediation of (inorganic) arsenic-containing water was found to be an anaerobic sludge from an anaerobic wastewater treatment plant. The studies showed that the sludge had proved to be a cost-effective and ecofriendly biosorbent because of its availability, particulate shape, sufficient mechanical strength, fast sorption rate and ease of disposal. However the mechanisms responsible for efficient arsenic removal are not known, but it is believed that biosorption may occur thanks to the binding mechanism based on the protein/amino acid–arsenic interaction (Chowdhury and Mulligan, 2011).

In case of the bioproducts formed by microorganisms the obvious example is biogenic schwertmannite (form of the iron precipitate – $Fe_8O_8(OH)_{4.42}(SO_4)_{1.79}$) produced by *Acidithiobacillus ferrooxidans*. The biogenic compound was found to play important role in the adsorption of As(III) from water. The process can occur thanks to formation of schwertmannite particles as a result of ferrous sulphate oxidation by *Acidithiobacillus ferrooxidans* cells making possible effective arsenic removal from ground water with a high adsorption capacity of As(III) in the pH range of 7–10. The investigations focus on the mechanisms of As(III) incorporation into the schwertmannite structure, which indicates involvement of ligand exchanges between As(III) species and surface hydroxyl group and sulphate, to get more insight before applying the method for water production (Liao et al., 2011).

8.3. ARSENIC IN STABILE SOLID STATE

When finding the proper technology for arsenic removal in full scale the faith of arsenic-bearing by-products of the treatment shall also play a significant role. Depending on the applied treatment arsenic compounds might be trapped on the surface of used adsorbent or be up-concentrated, like it takes place in case of membrane filtration. In case of applying iron co-precipitation the treatment produces sludge-containing arsenic and iron separated from treated water during a separation processes. It should be taken into account that the sludge shall not be a source of recontamination, which means that arsenic should be in the least soluble form. To evaluate arsenic mobility in such materials a toxicity characteristic leaching procedure (TCLP) is used. The TCLP is a standardized method enabling to characterize arsenic dissolution from the sludge in range of pH, by applying leachants containing different amounts of acetic acid or sodium hydroxide. The leaching behaviour of arsenic–iron hydroxide sludge, studied with the help of more sophisticated techniques, helped in gaining better understanding about formation and properties of the arsenic–iron hydroxide sludge. Therefore it was suggested that arsenic–iron hydroxide sludge surface is inner or outer spherically sorbed by arsenic, rather than the precipitate of insoluble iron–arsenic compounds such as $Fe(AsO_4)$. Desorption and resorption of arsenic onto the leached sludge are the main phenomena controlling arsenic leachability and the relationship between arsenic in TCLP leachate and sorbed on the iron hydroxide surface can be modelled by the Langmuir isotherm (Phenrat et al., 2008). The findings therefore brought up a hypothetical interfacial structure of inner–sphere surface complex (Figure 8.17). Under the inner–sphere complex model, arsenic is presumed to bind directly to either surface oxide ion or an underlying Fe ion by effectively replacing one of the surface OH_x groups.

Another study of arsenic sludge containing high concentration of Fe and As showed that arsenic was associated with either the amorphous iron oxyhydroxides or crystalline iron oxides (goethite or hematite). Generally arsenic concentration in the TCLP extract was lower than the TCLP limits indicating that iron oxyhydroxides are able to efficiently solubilise arsenic compounds, as it is strongly correlated with that of iron. However pH batch leaching studies showed a strong dependence of pH and Eh on As leaching from the sludge. Therefore at low pH As release

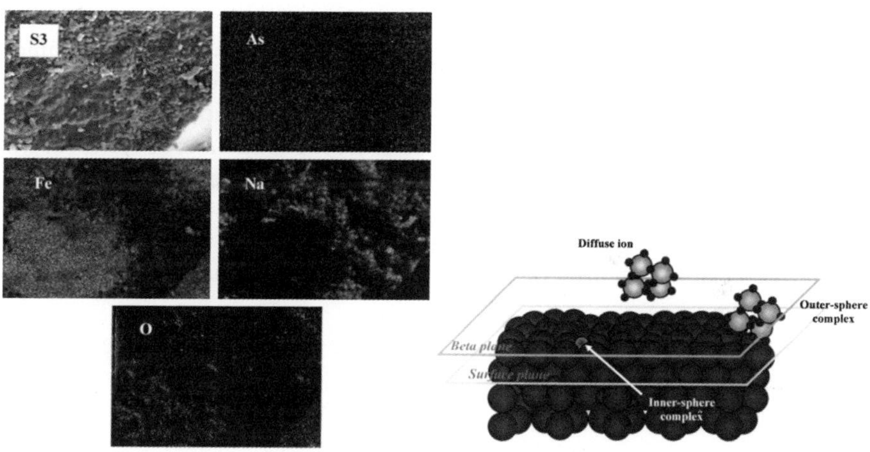

FIGURE 8.17 **Hypothetical interfacial structure (right) analogous to the SEM photograph of the sludge (left).** Left: X-ray dot maps illustrate the distribution of Fe, O, As and Na atoms on the surface of the sludge. The white dots indicate the presence of the element of interest while the black background indicates the absence of the element of interest. Right: This model consists of ferric hydroxide (the group of dark grey spheres) as an adsorbent; arsenic oxyanion as an adsorbate that has lost the water molecules from its primary hydration sphere and binds directly to the surface, forming an inner-sphere complex (a single light grey sphere); Na^+ and Cl^- as adsorbates that are bonded to the surface but retain their water of hydration, forming an outer-sphere complex (a group of spheres attaching the surface); and other ions that are attracted to the surface via electrostatic interactions, but are not bonded directly to the surface (diffused ions). *Reused from Phenrat et al. (2008) with permission from ASCE (originally adapted from Benjamin, M. (2002). Water chemistry, McGraw-Hill, Singapore).*

occurred in either dissolution of the amorphous Fe oxyhydroxide bearing phase or the less resistant carbonate phase in the acidic region and in high pH a desorption of As occurred. Therefore it is important to treat the produced sludge properly, that is to keep pH below neutral values and maintain high redox conditions for minimum As solubility for the disposal in the environment (Shafiquzzaman et al., 2010).

8.4. EXAMPLE OF THE SIMPLE ARSENIC REMOVAL TECHNOLOGY

The most demanding requirements regarding arsenic presence in drinking water can be met in Denmark, which is 5 μg/l. Moreover the Danish authorities have limit possibilities for arsenic removal treatment from contaminated ground water by prohibiting range of techniques used for this purpose (Danish Ministry of the Environment, 2007b). Therefore most of the mentioned technologies, from Tables 8.6 and 8.7, available around the world cannot be implemented. As a result of such requirements only simple techniques, based on iron co-precipitation or adsorption, may be used for arsenic removal. Although there is still place for improvement in the conventional treatment, the process can be enhanced to bring higher removal efficiency or reduced costs, especially operational expenses. Especially in case of iron-based arsenic removal the improvements can be performed in each part of the treatment, that is iron dosage, aeration and separation techniques (Figure 8.18).

An example of the method that implements some of the improvements is a novel Danish treatment system, MicroDrop Aqua (MDA) (Lebech

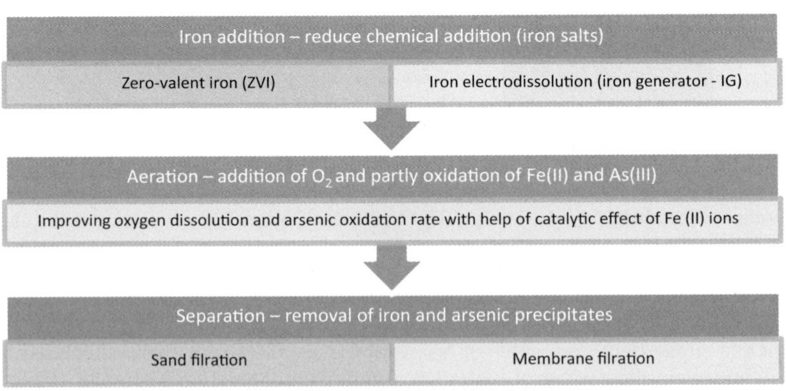

FIGURE 8.18 **Conventional arsenic removal and the ways for its improvement implemented by the novel approach.** (For colour version of this figure, the reader is referred to the online version of this book.)

November 16, 2006). The system was invented for removing arsenic trace level without chemical addition and is based on the innovative iron addition followed by specially designed aeration unit, as presented in Figure 8.19. The process is based on the iron dosage, but with the help of ZVI or electrolytic iron dissolution (Fe-ED). A next step of the method is an aeration device where water is stripped on the polymer grid pipes, resulting in increasing of contact surface between water and air. Moreover, the iron dosage before aeration unit causes building up of precipitated iron surface on the polymer material. Thanks to the MDA system it is possible to efficiently co-precipitate arsenic on the iron oxyhydroxides that can be separated from the treated stream with one of the filtration techniques.

8.4.1. System Performance

The performance of two different sites, where the MDA systems were implemented, is presented in Figure 8.20. The systems differed in iron dosage methods, that is one had the iron capsule with ZVI and other the iron generator, but both were using the MDA aeration unit and sand filter. Nevertheless the treatments did not differ in general mechanism of arsenic removal. Ground water is firstly supplemented with iron in anoxic conditions, in the IC or IG, from where it is mixed with air in aeration unit resulting in partly ferrous ion and arsenite oxidation, but foremost saturating water with dissolved oxygen that was enabling oxidation reactions.

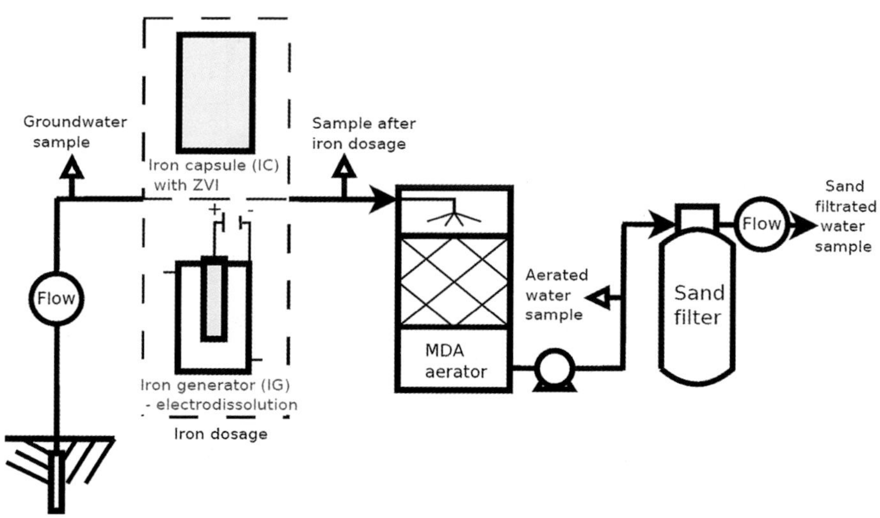

FIGURE 8.19 **Scheme of MDA arsenic removal treatment; iron capsule with ZVI or iron generator with Fe-ED used for iron dosage is followed by the MDA aeration unit and sand filter.** (For colour version of this figure, the reader is referred to the online version of this book.)

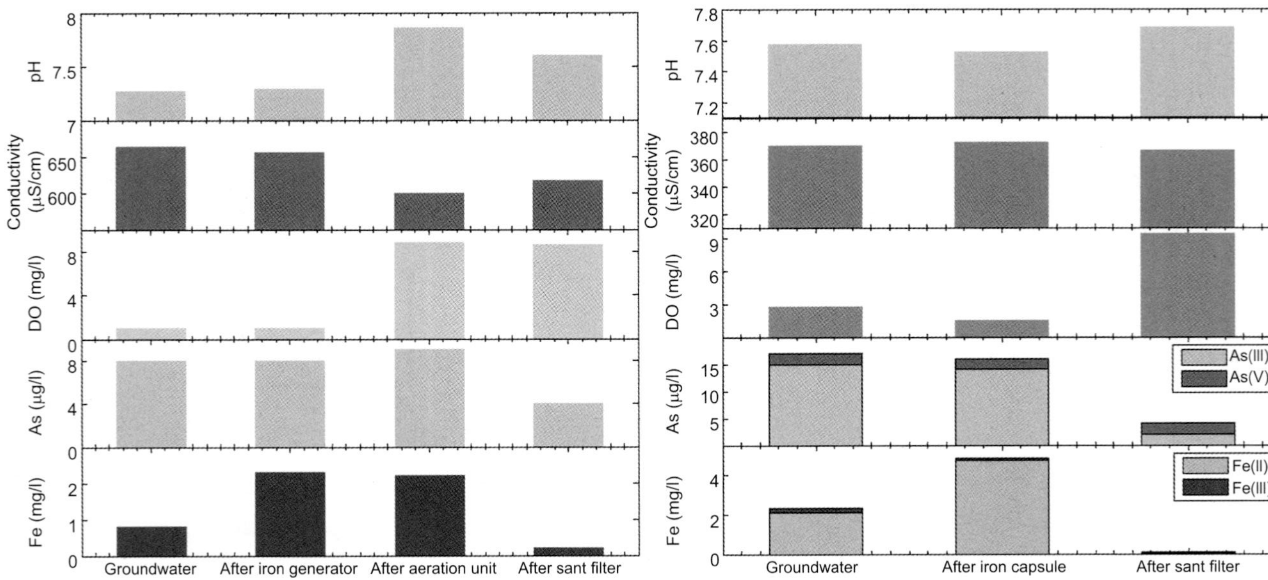

FIGURE 8.20 Water parameters at each stage of treatment with use of Fe-ED (IG current of 2.1A and flow 0.28 l/s) – on the left – and ZVI – on the right; from top: pH; conductivity; dissolved oxygen; As concentration; Fe concentration. (For colour version of this figure, the reader is referred to the online version of this book.)

TABLE 8.14 Comparison of the Sand Filter Backwash Sludge Collected from Different Waterworks

Iron Dosage Method	As (mg/g)	Fe (mg/g)	Mn (mg/g)	Mg (mg/g)	Ca (mg/g)	Fe/As Ratio (mg/mg)
Iron generator	3.3	225	0.29	3.0	65	68
Iron capsule	1.4	226	6.6	0.5	22	166

Both inorganic species, iron and arsenic, are removed in sand filter. By running the system for a long period it was possible to revealed an iron dose necessary for efficient arsenic co-precipitation in a sand filtration. The dose, as an Fe/As ratio, could have been obtained from backwash sludge that is produced on regularly basis during sand filter maintenance. As it can be noticed in Table 8.14 the systems differed in the Fe/As ratio found in the backwash sludge and indicate that iron electrodissolution might have required lower iron dosage. However the sand filter performance and especially the role of other ions cannot be neglected, as it was discussed in previous section 8.2.2.

8.4.2. Iron Addition

One of the conventional methods for iron addition is use of ferrous salts, examples where introduced in Table 8.12. The described system however makes use of other iron ions supplementation, that is ZVI or Fe-ED. The devices and their mounting employing the novel iron dosing are presented in Figure 8.21. The principles of both processes are the same and are based on ferrous ion release, a result of the following reaction:

$$Fe^0 \rightarrow Fe^{2+} + 2e^-$$

The main difference between the implementation of the ZVI and Fe-ED is control of produced ions. The ZVI, when solely used, is driven by spontaneous redox reactions, where no control of formed Fe^{2+} takes place. Moreover the ZVI has irregular distribution of anodic (iron-releasing) and cathodic (electron acceptor) sites that take part in iron release. Therefore the implementation of the ZVI might not give the desirable effect at once and long period of process optimization might be required to find proper amount of ZVI source and residence time in the iron capsule for achieving specific iron dosage, as it is shown in Figure 8.22.

More promising is the usage of Fe-ED, that is driven by applied current to iron sacrifice anode; therefore amount of electrons taking part in the process can be controlled. The control is based on the Faraday law that enable to calculate the iron dose from applied current. The Fe dose, as a

FIGURE 8.21 **Danish waterworks, where the system with the ZVI (pictures on the left) and the Fe-ED (on the right) were mounted; iron generator on the left, iron generator connected with aeration unit on the right picture.** *Courtesy of MicroDrop Aqua ApS, Denmark.* (For colour version of this figure, the reader is referred to the online version of this book.)

mass of iron released from anode in time (N), can be found with the help of the following equation:

$$N = \frac{m_{Fe}}{t} = C_{Fe,produced} \times Q = \frac{I \times M}{F \times z}$$

where Q – water flow rate (l/s), $C_{Fe,produced}$ – concentration of produced Fe (mg/l); m_{Fe} – mass of released iron (g); I – current of energy source applied (A), t – time (s); F – Faraday constant (96,485 C/mol); M – molar mass of iron (56 g/mol); z – the valence number of ions of the substance – in case of Fe^{2+} release it is 2.

By using the above equation it is possible to optimize the arsenic removal, as there is clear correlation between applied current into the iron anode and quality of produced water (Figure 8.23).

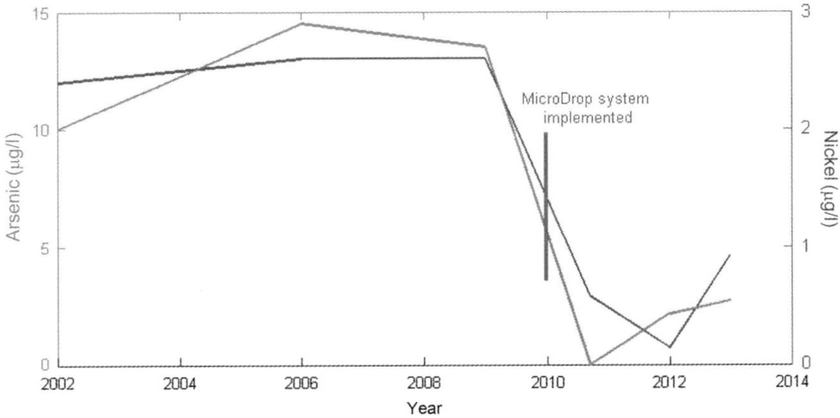

FIGURE 8.22 Arsenic and nickel removal efficiency before and after the MicroDrop system with the ZVI, as presented in Figure 8.19, was implemented in the Danish waterworks (ground water contains about 18 μg/l). (For colour version of this figure, the reader is referred to the online version of this book.)

The long usage of the iron generator also showed occurrence of carbonate scale deposits on the cathode surface, beside iron hydroxides, like hematite, schwertmannite and goethite, found on the anode surface (Figure 8.24).

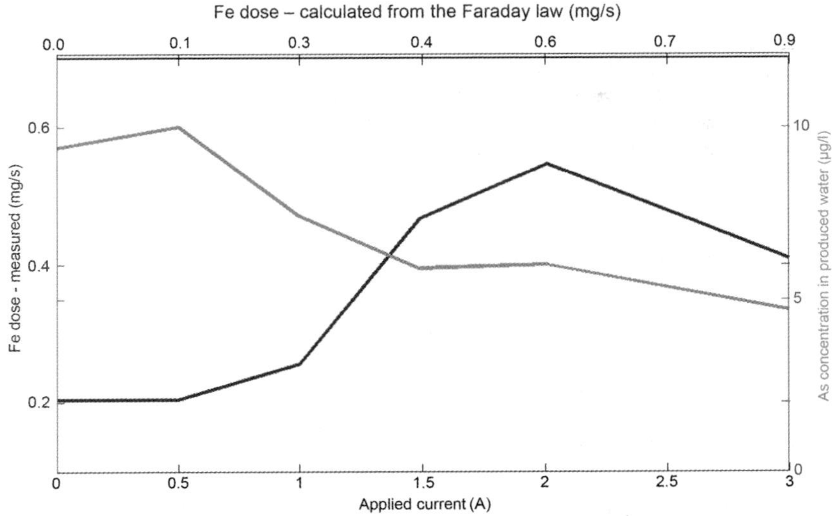

FIGURE 8.23 Correlation between the applied current (theoretical Fe dose) and produced water quality in waterworks where the Fe-ED process was used as presented in Figure 8.19. (For colour version of this figure, the reader is referred to the online version of this book.)

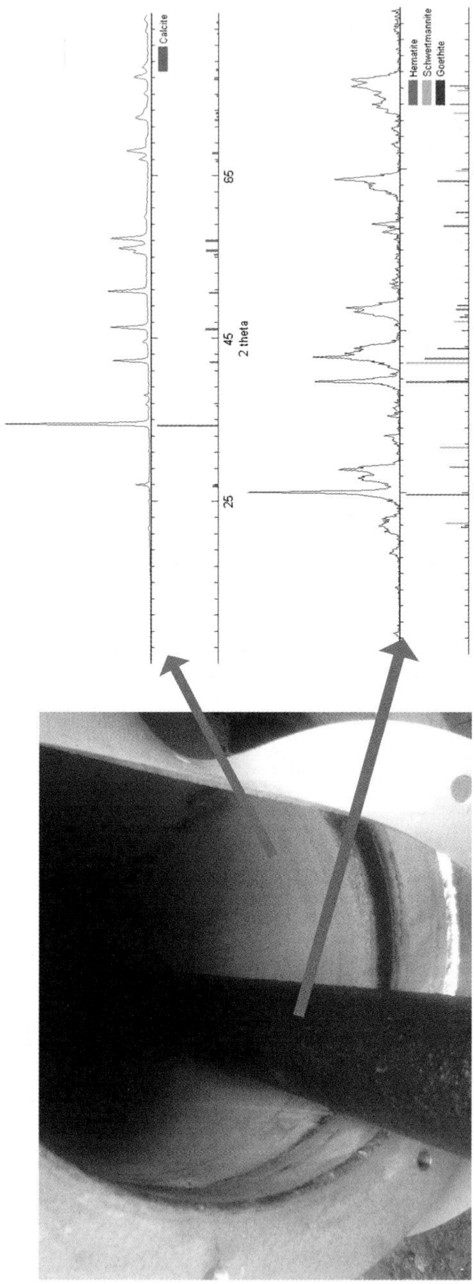

FIGURE 8.24 XRD analysis of cathode (stainless steel wall) and iron rod anode residuals after long implementation of the Fe-ED process. (For colour version of this figure, the reader is referred to the online version of this book.)

The scale surface building up, as a consequence of calcium and magnesium bicarbonates presence in treated water, has been found to be responsible for decreasing efficiency of the iron release process (Kowalski et al., in press). The mechanisms responsible for scaling are based on reduction reaction, where electrons are utilized, that results in increase of the interfacial pH sufficiently to precipitate the carbonates, which is presented by the following reactions (Yanagisawa and Matsumura, 2011):

$$3H_2O + 3e^- \rightarrow (3/2)H_2(g) + 3OH^-$$

$$Ca^{2+} + HCO_3^- + OH \rightarrow CaCO_3 + H_2O$$

The presence of scale-forming compounds has to be considered not only in the Fe-ED process but also when the ZVI is used. Even though the carbonate scale developed on the iron surface has been known for a long time as Tillman layer and was desirable in iron piping systems as anticorrosion protective layer it is unwanted when the aim is to have the release of iron (Tillman and Heublein, 1912). In a consequence of carbonate scale the iron release from the ZVI might be limited or even totally blocked, when the scale surface will cover most of the iron surface. However in case of the iron generator, where anode and cathode are separated, scale does not occur on the anode surface; therefore iron release cannot be stopped totally, which is the main advantage over the ZVI. On the other hand the calcite surface on the cathode was found to be responsible for higher energy demand required for applying current. The solution for those and other issues regarding implementation of the iron electrodissolution requires development of an iron generator design and maintenance procedure.

8.4.3. Aeration Unit

The aeration device is build from three main parts; distribution pipes on the top and polymer grid and collection tank in the bottom (Figure 8.25). The first generation of the aerator was combined with iron dosage device that took place in an iron tray containing ZVI, but in the nowadays systems the tray is replaced with separate unit as it is shown in Figure 8.19. When the aeration unit is put at place it contains clean and black polymer grid fillings in time and, due to the preceded iron dosage, is covered by precipitated iron, specifically two-line ferrihydrite, that is amorphous Fe(III) oxyhydroxide, as shown in Figure 8.26.

The surface of ferrihydrite is believed to be responsible for enhancing the arsenic removal process, which is an extra advantage beside efficient supplementation of oxygen into water. To understand the role of ferrihydrite surface results from recirculation of As-bearing solution in

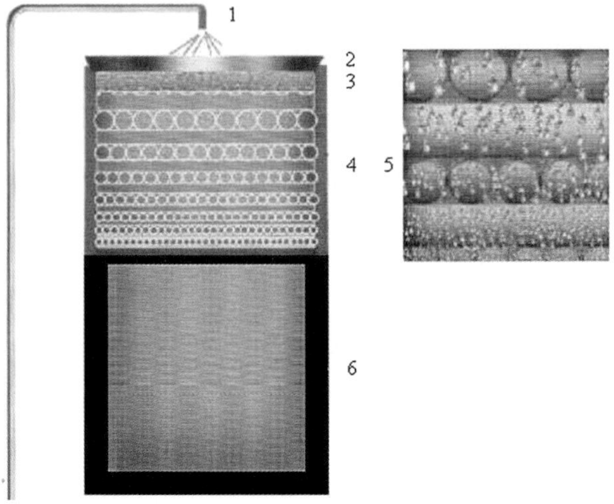

FIGURE 8.25 The MDA aeration unit (on the left): (1) inlet of ground water; (2) distribution tray; (3) iron coil tray; (4) polymer grid pipe aeration unit; (5) polymer grid pipe oxidation unit; (6) buffer tank (Søgaard and Sønderby, 2007; Lebech, 2006). (For colour version of this figure, the reader is referred to the online version of this book.)

aeration device shall be analyzed (Figure 8.27). During the unusual handling of an arsenic-contaminated solution changes in iron and arsenic concentration can be noticed, whereas simultaneously oxygen level and pH remain constant. It was found that the level of As(III) was decreasing in time more rapidly than it occurred in control experiments that were performed with the absence of ferrihydrite-coated surfaces. The decrease of arsenite in aeration tower is rather because of adsorption as in the pH

FIGURE 8.26 **Pictures of polymer grid pipe coated with iron hydroxides, and clean.** (For colour version of this figure, the reader is referred to the online version of this book.)

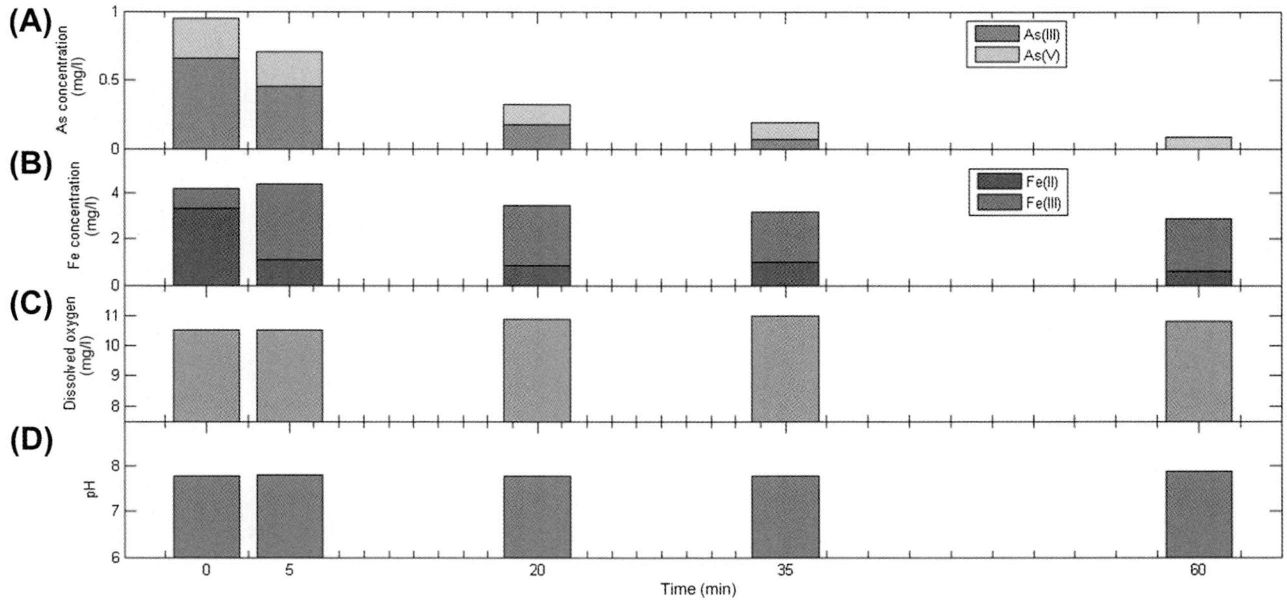

FIGURE 8.27 **Results of recirculation experiment, where As-bearing ground water was used.** The values corresponds to quality of solutions sampled from the bottom tank of the aeration unit for a given time; from top: (A) arsenic speciation, (B) iron speciation, (C) dissolved oxygen level and (D) pH. (For colour version of this figure, the reader is referred to the online version of this book.)

range 6–9 As(III) is sorbed to a similar or greater extent than As(V) on iron oxides (Dixit and Hering, 2003). However the oxidation of arsenite on the ferrihydrite surface cannot be neglected, especially that conditions, high oxygen level and the presence of ferrous ion, might promote the transformation from arsenite to arsenate.

The As(III) to As(V) transformation was observed in laboratory experiments in the presence of the FeOOH-coated polymer bodies and without them and after addition of Fe dose (10 ppm of Fe(II)) are shown in Figure 8.28. Generally in experiments with ferrihydrate surfaces a decrease of total arsenic concentration, especially in the experiment where no iron dosage, was used could be noticed. The results clearly show that adsorption of arsenic species on ferrihydrite took place. However in the case of iron addition in the presence of the iron oxyhydroxide the adsorption was a reversible process, unlike it was observed in case of solely used iron oxide surface. Moreover the combination of the ferrihydrite and iron dosage resulted in the fastest arsenite oxidation and full desorption of arsenic species, which were found to be arsenate. It shall be noted that after the end of presented batch tests not only a decrease of

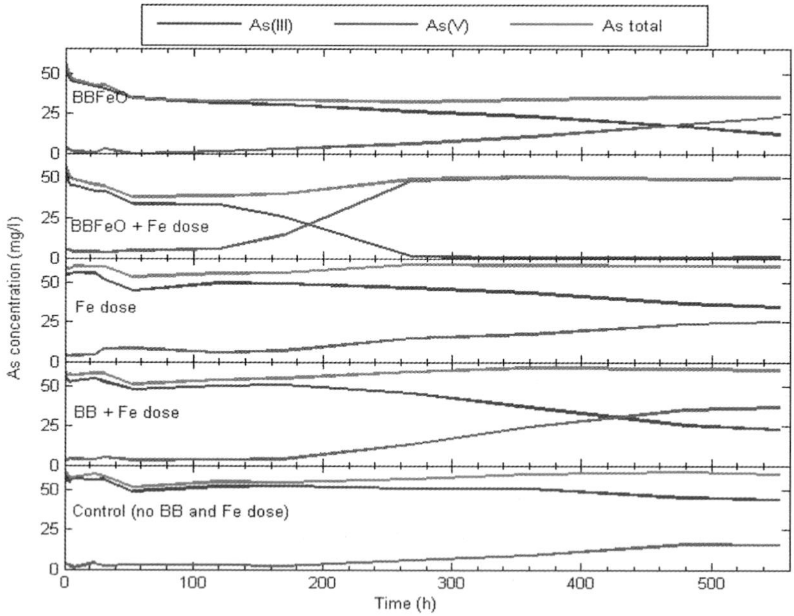

FIGURE 8.28 Arsenic speciation in time in aerated solution (DO = 7.5 mg/l, initial pH = 7 ± 0.2) with and without Fe dose (10 ppm) and the plastic bodies, coated with ferrihydrite (BBFeO) and clean (BB), compared with control experiments, where no iron dosage and surfaces were used. (For colour version of this figure, the reader is referred to the online version of this book.)

dissolved oxygen was observed but also pH dropped from initial 7 to about 4, in samples where no BBFeO was used. The reason of such phenomena might be explained by the role of ferrihydrite in influencing pH. Firstly, the adsorption reaction for As(V) and formation of surface complexes employing H^+ according to reactions presented in Table 8.11 could have been involved. Another interpretation of the pH change is the fact that As compounds adsorption on ferrihydrite can result in H^+/OH^- release. It was demonstrated that arsenate adsorption resulted in a netto release of OH^- at pH 4.6 and 9.2 that agrees with adsorption reactions. However, arsenite adsorption resulted in netto H^+ release at pH 4.6, which can be explained in the following reactions (Jain et al., 1999):

$$\equiv FeOH + H_3AsO_3 \rightarrow \equiv FeH_2AsO_3 + H^+$$

Therefore, making use of H^+ in surface complexation reactions on the surface of plastic bodies (BBFeO) might neutralize the produced H^+ that comes from oxidation reactions from Table 8.10 (Sahai et al., 2007).

8.4.4. Summary – Arsenic Removal

As it was shown in previous paragraphs introduction of the innovative systems is an example of simple water processing. However the mechanisms involved in removal of arsenic with addition of iron compounds shall consider the presence of other ions that can influence the treatment. Therefore it is not only important to know the role of each species but also to manage their behaviour by introducing solutions that prevents undesired phenomena. The noticeable example is interference of scale-forming compounds (Ca^{2+} and HCO_3^-) on the iron dosage technique. There are differences in scaling mechanism between implementation of the ZVI and Fe-ED that favour the second technique, due to control of scale occurring location. Knowledge, in detail of iron and oxygen demand is crucial for proper arsenic treatment optimization and depends on the raw water quality. These concerns are presented in Figure 8.29, where simplified behaviour of mentioned ions at each step of the novel treatment is shown. However these are only some of the issues that shall be taken into account when introducing the full-scale treatment that shall deliver arsenic-free water to people.

Combination of presented processes is only one example of many technologies that are a response to the arsenic contamination of drinking water sources affecting millions of people around the world. Ion exchange, activated alumina, reverse osmosis, membrane filtration, modified coagulation/filtration and enhanced lime softening are water treatment technologies for As removal recommended by the U.S. Environmental Protection Agency (see Table 8.6). However, none of these technologies are currently applied on a broad scale in developing

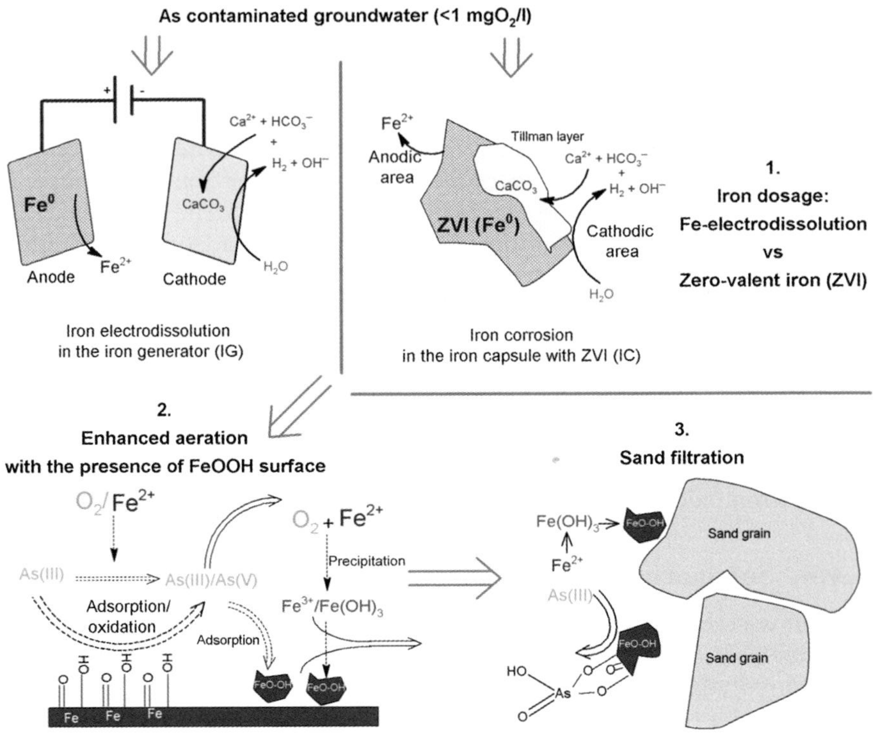

FIGURE 8.29 Behaviour of As, Ca and Fe compounds at each step of the MDA system for arsenic removal. (For colour version of this figure, the reader is referred to the online version of this book.)

TABLE 8.15 Comparison of Implementation of the Iron Capsule (ZVI) and Iron Generator (Electrodissolution) as Iron Dosage Methods

Parameter	Iron Capsule (ZVI)	Iron Generator (Fe-ED)
Costs consideration Capital expenditure – CAPEX; Operating expense – OPEX	Low CAPEX and OPEX: can be done from the scrap	High CAPEX – design and manufacturing of iron generator; Medium OPEX – energy for applying current; changing anode
Optimization	Not available	Adjusting current depending on the required As removal
Maintenance	Changing the ZVI source	Cleaning cathode from scale Changing the anode material
Recommended end user	Small waterworks Suitable for developing countries	Large-scale waterworks and industrial water treatment

BOX 8.3

ARSENIC REMOVAL IN PRACTICE

Based on the As and Fe content in produced sludge and raw ground water find:

1. An Fe dose required for As removal to below 5 μg/l (assume 4 μg As/l)
2. For water supply producing 5 m^3/h (daily production: 75 m^3) find consumption of $FeCl_2$ (150 g/l)
3. Current for Fe electrodissolution in iron generator to gain specific Fe dose (from point 1), according to Faraday law (assuming that only Fe^{2+} is produced)

	Raw Water	**Generated Sludge**
Iron	0.24 mg/l	260 mg/g dry mass
Arsenic	20 μg/l	4 mg/g dry mass

countries because they require sophisticated technical systems and are therefore unpractical in low-income regions. Thus recent trend is to provide technology that will be able to be constructed, operated and maintained by small not qualified communities. The requirements might be fulfilled by the presented method, as it is based on the simple processes. Nevertheless the experience from running the novel treatment in the Danish waterworks brings some insights into the system advantages and limitations regarding iron dosage method (Table 8.15) (Box 8.3).

References

Amini, M., Abbaspour, K.C., Berg, M., Winkel, L., Hug, S.J., Hoehn, E., Yang, H., Johnson, C.A., 2008. Statistical modeling of global geogenic arsenic contamination in groundwater. Environ. Sci. Technol. 42 (10), 3669–3675.

Andersen, E.S., Jespersgard, P., Østergard, O.G., 2007. Databog fysik, kemi. F & K, Kobenhavn.

Anderson, M.A., Ferguson, J.F., Gavis, J., 1976. Arsenate adsorption on amorphous aluminum hydroxide. J. Colloid Interface Sci. 54 (3), 391–399.

Aryal, M., Ziagova, M., Liakopoulou-Kyriakides, M., 2011. Comparison of Cr(VI) and As(V) removal in single and binary mixtures with Fe(III)-treated *Staphylococcus xylosus* biomass: thermodynamic studies. Chem. Eng. J. 169 (1–3), 100–106.

Bang, S., Johnson, M.D., Korfiatis, G.P., Meng, X., 2005. Chemical reactions between arsenic and zero-valent iron in water. Water Res. 39 (5), 763–770.

Baskan, M.B., Pala, A., 2009. Determination of arsenic removal efficiency by ferric ions using response surface methodology. J. Hazard. Mater. 166 (2–3), 796–801.

Battaglia-Brunet, F., Dictor, M., Garrido, F., Crouzet, C., Morin, D., Dekeyser, K., Clarens, M., Baranger, P., 2002. An arsenic(III)-oxidizing bacterial population: selection, characterization, and performance in reactors. J. Appl. Microbiol. 93 (4), 656–667.

BBC NEWS Medical Mysteries King George III: Mad or Misunderstood? 2011. http://news.bbc.co.uk/2/hi/3889903.stm (accessed 06.06.11.).

Berg, M., Luzi, S., Trang, P.T.K., Viet, P.H., Giger, W., Stüben, D., 2006. Arsenic removal from groundwater by household sand filters: comparative field study, model calculations, and health benefits. Environ. Sci. Technol. 40 (17), 5567–5573.

Blacksmith Institute 2010 Report Blacksmith Worlds Worst Polluted Top Six Toxic Threats, 2011. http://www.worstpolluted.org/2010-report.html (accessed 06.06.11.).

Blaney, L., Sarkar, S., SenGupta, A.K., 2007. Comment on "arsenic removal from groundwater by household sand filters: comparative field study, model calculations, and health benefits". Environ. Sci. Technol. 41 (3), 1051–1052.

Brown, D., Weselby, C., 2011. NASA-Funded Research Discovers Life Built with Toxic Chemical. http://www.nasa.gov/topics/universe/features/astrobiology_toxic_chemical.html (accessed 06.06.11.).

Casiot, C., Pedron, V., Bruneel, O., Duran, R., Personné, J.C., Grapin, G., Drakidès, C., Elbaz-Poulichet, F., 2006. A new bacterial strain mediating as oxidation in the Fe-rich biofilm naturally growing in a groundwater Fe treatment pilot unit. Chemosphere 64 (3), 492–496.

Chakraborti, D., Rahman, M.M., Das, B., Murrill, M., Dey, S., Chandra Mukherjee, S., Dhar, R.K., Biswas, B.K., Chowdhury, U.K., Roy, S., Sorif, S., Selim, M., Rahman, M., Quamruzzaman, Q., 2010. Status of groundwater arsenic contamination in Bangladesh: a 14-year study report. Water Res. 44 (19), 5789–5802.

Chiew, H., Sampson, M.L., Huch, S., Ken, S., Bostick, B.C., 2009. Effect of groundwater iron and phosphate on the efficacy of arsenic removal by iron-amended bios and filters. Environ. Sci. Technol. 43 (16), 6295–6300.

Chon, H., Ohandja, D., Voulvoulis, N., 2010. Implementation of E.U. Water Framework Directive: source assessment of metallic substances at catchment levels. J. Environ. Monit. 12 (1), 36–47.

Chowdhury, M.R.I., Mulligan, C.N., 2011. Biosorption of arsenic from contaminated water by anaerobic biomass. J. Hazard. Mater. 190 (1–3), 486–492.

Ciardelli, M.C., Xu, H., Sahai, N., 2008. Role of Fe(II), phosphate, silicate, sulfate, and carbonate in arsenic uptake by coprecipitation in synthetic and natural groundwater. Water Res. 42 (3), 615–624.

Dang, H.S., Jaiswal, D.D., Somasundaram, S., 1983. Distribution of arsenic in human tissues and milk. Sci. Total Environ. 29 (1–2), 171–175.

Dani, S.U., 2010. Arsenic for the fool: an exponential connection. Sci. Total Environ. 408 (8), 1842–1846.

Danish Ministry of the Environment, 2007a. BEK 1449 from 11, App. 1b.

Danish Ministry of the Environment, 2007b. Executive order no. 1451, Executive order for water extraction and waterworks.

Dixit, S., Hering, J.G., 2003. Comparison of arsenic(V) and arsenic(III) sorption onto iron oxide minerals: implications for arsenic mobility. Environ. Sci. Technol. 37 (18), 4182–4189.

Dojlido, J.R. 1995. Chemia wód powierzchniowych. Wydaw. Ekonomia i Środowisko, Białystok.

Environmental Protection Agency of the United States, 2002. Arsenic Treatment Technologies for Soil, Waste, and Water.

Fulghum, J.E., 1988. Discrimination between adsorption and coprecipitation in aquatic particle standards by surface analysis techniques: lead distributions in calcium carbonates. Environ. Sci. Technol. 22 (4), 463–467.

Hector, A., Casillas, M., Cocke, D.L., Gomes, J.A., Morkovsky, P., Parga, J.R., Peterson, E., Garci, C., 2007. Electrochemistry behind Electrocoagulation using Iron Electrodes. ECS Trans. 6 (9), 1–15. http://dx.doi.org/10.1149/1.2790397.

IARC Monographs on the Evaluation of Carcinogenic Risks to Humans Some Drinking-Water Disinfectants and Contaminants, Including Arsenic, 2004. World Health Organization International Agency for Research on Cancer, Lyon, France.

Jain, A., Raven, K.P., Loeppert, R.H., 1999. Arsenite and arsenate adsorption on ferrihydrite: Surface charge reduction and net OH-release stoichiometry. Environ. Sci. Technol. 33, 1179–1184.

Jones, C.J., Laky, D., Galambos, I., Avendano, C., Colvin, V.L. Life cycle analysis of two hungarian drinking water arsenic removal technologies. Water Sci. Technol. Water Supply, in press. http://dx.doi.org/10.2166/ws.2013.165

Karim, M.M., 2000. Arsenic in groundwater and health problems in Bangladesh. Water Res. 34 (1), 304–310.

Katsnelson, A., 2010. Microbe gets toxic response. Nature 468, 741.

Katsoyiannis, I.A., Ruettimann, T., Hug, S.J., 2008. pH dependence of fenton reagent generation and As(III) oxidation and removal by corrosion of zero valent iron in aerated water. Environ. Sci. Technol. 42 (19), 7424–7430.

Katsoyiannis, I.A., Zouboulis, A.I., 2004. Application of biological processes for the removal of arsenic from groundwaters. Water Res. 38 (1), 17–26.

Kim, M., Nriagu, J., 2000. Oxidation of arsenite in groundwater using ozone and oxygen. Sci. Total Environ. 247 (1), 71–79.

Kobya, M., Gebologlu, U., Ulu, F., Oncel, S., Demirbas, E., 2011. "Removal of arsenic from drinking water by the electrocoagulation using Fe and Al electrodes". Electrochimica Acta. 56, 5060–5070. http://dx.doi.org/10.1016/j.electacta.2011.03.086.

Kowalski, K.P., Arturi, K., Søgaard, E.G. Removal of arsenic from contaminated groundwater with application of iron electrodissolution, aeration and sand filtration. Water Sci. Technol., in press. doi:10.2166/ws.2013.198

Lacasa, E., Cãnizares, P., Sáez, C., Fernández, F.J., Rodrigo, M.A., 2011. Removal of arsenic by iron and aluminium electrochemically assisted coagulation. Sep. Purif. Technol. 79, 15–19. http://dx.doi.org/10.1016/j.seppur.2011.03.005.

Lebech, F., November 16, 2006. Method for Removing Contaminant Trace Species, Especially Arsenic, from Water. U.S. Patent. 11/920,146(7,947,183).

Li, X., Liu, C., Li, F., Li, Y., Zhang, L., Liu, C., Zhou, Y., 2010. The oxidative transformation of sodium arsenite at the interface of α-MnO_2 and water. J. Hazard. Mater. 173 (1–3), 675–681.

Liao, Y., Liang, J., Zhou, L., 2011. Adsorptive removal of As(III) by biogenic schwertmannite from simulated As-contaminated groundwater. Chemosphere 83 (3), 295–301.

Lievremont, D., Bertin, P.N., Lett, M., 2009. Arsenic in contaminated waters: biogeochemical cycle, microbial metabolism and biotreatment processes. Biochimie 91 (10), 1229–1237.

Lugli, A., Lugli, A.K., Horcic, M., 2005. Napoleon's autopsy: new perspectives. Hum. Pathol. 36 (4), 320–324.

Maier, R.M., Pepper, I.L., Gerba, C.P., 2009. Environmental Microbiology. Academic Press, London.

Majumder, C., Gupta, A., 2011. Prediction of Arsenic Removal by Electrocoagulation: Model Development by Factorial Design. J. Hazard. Toxic Radioact. Waste 15 (1), 48–54. http://dx.doi.org/10.1061/(ASCE)HZ.1944-8376.0000065.

Neppolian, B., Doronila, A., Ashokkumar, M., 2010. Sonochemical oxidation of arsenic(III) to arsenic(V) using potassium peroxydisulfate as an oxidizing agent. Water Res. 44 (12), 3687–3695.

NJDEP, 2011. Safe Drinking Water Act Rules, N.J.A.C. 7:10.

Noubactep, C., 2009a. Comment on "effect of groundwater iron and phosphate on the efficacy of arsenic removal by iron-amended biosand filters". Environ. Sci. Technol. 43 (22), 8698.

Noubactep, C., 2009b. Comment on pH dependence of fenton reagent generation and As(III) oxidation and removal by corrosion of zero valent iron in aerated water. Environ. Sci. Technol. 43 (1), 233.

Paddock, F.K., Loomis Jr., C.C., Perkons, A.K., 1970. An inquest on the death of Charles Francis Hall. N. Engl. J. Med. 282 (14), 784–786.

Palfy, P., Vircikova, E., Molnar, L., 1999. Processing of arsenic waste by precipitation and solidification. Waste Manage. 19 (1), 55–59.

Pang, S., Jiang, J., Ma, J., 2009. New insight into the oxidation of arsenite by the reaction of zerovalent iron and oxygen. Comment on "pH dependence of Fenton reagent generation and As(III) oxidation and removal by corrosion of zero valent iron in aerated water". Environ. Sci. Technol. 43 (10), 3978–3979.

Parga, J.R., Cocke, D.L., Valenzuela, J.L., Gomes, J.A., Kesmez, M., Irwin, G., Moreno, H., Weir, M., 2005. Arsenic removal via electrocoagulation from heavy metal contaminated groundwater in la Comarca Lagunera Mexico. J. Hazard. Mater. 124 (1–3), 247–254.

Phenrat, T., Marhaba, T.F., Rachakornkij, M., 2008. Leaching behaviors of arsenic from arsenic-iron hydroxide sludge during TCLP. J. Environ. Eng. 134 (8), 671–682.

Pokhrel, D., Viraraghavan, T., 2008. Arsenic removal in an iron oxide-coated fungal biomass column: analysis of breakthrough curves. Bioresour. Technol. 99 (6), 2067–2071.

Roberts, L.C., Hug, S.J., Ruettimann, T., Billah, M., Khan, A.W., Rahman, M.T., 2004. Arsenic removal with iron(II) and iron(III) in waters with high silicate and phosphate concentrations. Environ. Sci. Technol. 38 (1), 307–315.

Robins, R.G., 1981. The solubility of metal arsenates. Metall. Trans. B 12 (1), 103–109.

Sahai, N., Lee, Y.J., Xu, H., Ciardelli, M., Gaillard, J., 2007. Role of Fe(II) and phosphate in arsenic uptake by coprecipitation. Geochim. Cosmochim. Acta 71 (13), 3193–3210.

Salmassi, T.M., Venkateswaren, K., Satomi, M., Nealson, K.H., Newman, D.K., Hering, J.G., 2002. Oxidation of arsenite by *Agrobacterium albertimagni*, AOL15, sp. nov., isolated from Hot Creek, California. Geomicrobiol. J. 19 (1), 53–66.

Samal, A.C., Kar, S., Maity, J.P., Santra, S.C., 2013. Arsenicosis and its relationship with nutritional status in two arsenic affected areas of West Bengal, India. J. Asian Earth Sci. (77), 303–310.

Schoof, R.A., Yost, L.J., Eickhoff, J., Crecelius, E.A., Cragin, D.W., Meacher, D.M., Menzel, D.B., 1999. A Market Basket Survey of inorganic arsenic in food. Food Chem. Toxicol. 37 (8), 839–846.

Shafiquzzaman, M., Azam, M.S., Nakajima, J., Bari, Q.H., 2010. Arsenic leaching characteristics of the sludges from iron based removal process. Desalination 261 (1–2), 41–45.

Sharma, V.K., Dutta, P.K., Ray, A.K., 2007. Review of kinetics of chemical and photocatalytical oxidation of Arsenic(III) as influenced by pH. J. Environ. Sci. Health A Tox. Hazard. Subst. Environ. Eng. 42 (7), 997–1004.

Silver, S., Phung, L.T., 2005. Genes and enzymes involved in bacterial oxidation and reduction of inorganic arsenic. Appl. Environ. Microbiol. 71 (2), 599–608.

Singh, A.L., Sarma, P.N., 2010. Removal of arsenic(III) from waste water using *Lactobacillus acidophilus*. Bioremediation J. 14 (2), 92–97.

Smedley, P.L., Kinniburgh, D.G., 2002. A review of the source, behaviour and distribution of arsenic in natural waters. Appl. Geochem. 17 (5), 517–568.

Søgaard, E.G., Sønderby, C., 2007. Elimination of arsenic substances from groundwater for production of drinkable water using microdrop reactor system. In: Chemical Water and Wastewater Treatment: The International Gothenburg Symposium on Chemical Treatment of Water and Wastewater, IX, p. 389.

Steenhuysen, J., 2011. Arsenic Could Kill Millions in Bangladesh | Reuters. http://www.reuters.com/article/2010/06/18/us-arsenic-water-idUSTRE65H5SB20100618 (accessed 06.06.11.).

Tillmans, J., Heublein, O., 1912. Investigation of carbon dioxide which attacks calcium carbonate in natural waters (in German). Gesund Ing. 35 (34), 669.

Tseng, W.P., Chu, H.M., How, S.W., Fong, J.M., Lin, C.S., Yeh, S., 1968. Prevalence of skin cancer in an endemic area of chronic arsenicism in Taiwan. J. Natl. Cancer Inst. 40 (3), 453–463.

United Nations, 2003. United Nations Synthesis Report on Arsenic in Drinking Water.

University of Maryland Medical Center. "Doctors reconsider health and death of 'El Libertador,' general who freed South America." ScienceDaily. www.sciencedaily.com/releases/2010/04/100428110816.htm (accessed February 4, 2014)

Van Halem, D., Bakker, S.A., Amy, G.L., Van Dijk, J.C., 2009. Arsenic in drinking water: a worldwide water quality concern for water supply companies. Drink. Water Eng. Sci. 2 (1), 29–34.

Wang, S., Mulligan, C.N., 2006. Occurrence of arsenic contamination in Canada: sources, behavior and distribution. Sci. Total Environ. 366 (2–3), 701–721.

WHO, 2001. World Health Organization of the United Nations Synthesis Report on Arsenic in Drinking Water, 2011.

WHO, 2011. Guidelines for Drinking-Water Quality. World Health Organization, Geneva.

Xu, T., Kamat, P.V., O'Shea, K.E., 2005. Mechanistic evaluation of arsenite oxidation in TiO_2 assisted photocatalysis. J. Phys. Chem. A 109 (40), 9070–9075.

Yanagisawa, N., Matsumura, T., 2011. Scale removing and preventing by using high-frequency electrolysis system. Trans. Geothermal Resources Council. 35 (2), 1269–1272.

Index

Note: Page numbers with "*f*" denote figures; "*t*" tables; "*b*" boxes.